"In a world besieged by pollution, climate change, and declining biodiversity, it is vital to assess whether human rights can help address these challenges. This comprehensive new book by leading experts Sumudu Atapattu and Andrea Schapper offers a clear and timely assessment."

David Boyd, UN Special Rapporteur on Human Rights and Environment & Associate Professor, Law, Policy, and Sustainability, University of British Columbia, Canada

"Environmental degradation and human rights violations are intrinsically intertwined. In this textbook, leading experts Sumudu Atapattu and Andrea Schapper discuss their relationship comprehensively."

Judge Navanethem Pillay, Former UN High Commissioner for Human Rights

"This clear, accessible, student-friendly and comprehensive treatment of human rights and the environment is both timely and welcome. The book addresses, with expertise and clarity, the foundational aspects of the field, offering students a solid understanding of its key institutional and conceptual characteristics. The book also introduces students to important critical themes and provides helpful questions for reflection and discussion. This textbook is likely to prove a most valuable resource for anyone teaching or studying the increasingly important nexus between human rights and the environment."

Anna Grear, Professor of Law and Theory, School of Law and Politics, Cardiff University, UK

"Climate change and environmental destruction are the biggest threats to all species living on the globe. Combined they constitute the worst global and intergenerational injustice. This textbook contributes an exceptional insight to what law brings to addressing the injustices as well as the limitations of the legal regime. In a pedagogical and highly sophisticated manner, the authors combine human rights and environmental justice framed by the UN Sustainable Development Goals. This textbook is not only important to stimulate curiosity and critical reflection for students, it will also be highly inspiring reading for all to whom the survival of the earth is a major concern."

Morten Kjaerum, Director, Raoul Wallenberg Institute of Human Rights and Humanitarian Law, Sweden

"In 2012, the current United Nations Special Rapporteur on Human Rights and the Environment, David Boyd, suggested we are witnessing an 'environmental rights revolution'. In their new book, Professors Sumudu Atapattu and Andrea Schapper, both prominent scholars in the areas of environmental justice, sustainable development and environmental law, make a timely and important contribution to this 'revolution'. They provide: a comprehensive survey that frames the issues surrounding human rights and environmental protection; a historical perspective on the emergence of environmental human rights; the advantages and disadvantages associated with the rights-based approach; and a description of regional developments. The book also usefully focuses on specific human rights that are critical for environmental protection, while it discusses climate change and human rights and emerging, but critically important, issues that remain at the periphery of State concern, such as rights of nature and the extraterritorial application of environmental rights."

Louis J. Kotzé, Research Professor,
North-West University, South Africa

"Atapattu and Schapper rigorously dissect the recognition of environmental rights across legal systems and jurisdictions. Focusing on the positive force for good of environmental rights as well as the inherent limitations of the environmental rights agenda, this book is an indispensable resource for students and scholars of this emerging legal phenomenon."

Ole W Pedersen, Reader in Environmental Law,
Newcastle Law School, UK

"As the thematic boundaries of the environmental rights landscape continues to expand, the need for a comprehensive exploration of the topic is ever more needed. The authors masterfully situate cutting-edge normative developments within a solid conceptual framework that illuminates the interaction of human rights and the environment."

Marcos Orellana, Director, Environment and Human Rights Division
Human Rights Watch and Adjunct Associate Professor,
George Washington University School of Law, USA

"In this book, Atapattu and Schapper provide an insightful, interdisciplinary introduction to this new and growing field. The text provides an excellent blend of theory and practice—engaging students in the conceptual, doctrinal, socio-political and practical challenges of using human rights law for environmental protection and sustainable development. The case study on climate change is an effective approach that focusses the student on the details of the most complex and critical environmental challenge. The relationship between climate change and human rights is important enough to warrant this attention in its own right. The case study also allows students to gain a deeper appreciation of the complexities of a rights-based approach to environmental protection than would a text that surveys all environmental issues."

David Hunter, Professor of Law, American University
Washington College of Law, USA

Human Rights and the Environment

The field of human rights and the environment has grown phenomenally during the last few years and this textbook will be one of the first to encourage students to think critically about how many environmental issues lead to a violation of existing rights.

Taking a socio-legal approach, this book will provide a good understanding of both human rights and environmental issues, as well as the limitations of each regime, and will explore the ways in which human rights law and institutions can be used to obtain relief for the victims of environmental degradation or of adverse effects of environmental policies. In addition, it will place an emphasis on climate change and climate policies to highlight the pros and cons of using a human rights framework and to underscore its importance in the context of climate change. As well as identifying emerging issues and areas for further research, each chapter will be rich in pedagogical features, including web links to further research and discussion questions for beyond the classroom.

Combining their specialisms in law and politics, Atapattu and Schapper have developed a truly inter-disciplinary resource that will be essential for students of human rights, environmental studies, international law, international relations, politics, and philosophy.

Sumudu Atapattu is the Director of Research Centers and International Programs at the University of Wisconsin Law School, USA. She teaches seminar classes on "International Environmental Law" and "Climate Change, Human Rights and the Environment." She is affiliated with UW-Madison's Nelson Institute for Environmental Studies and the Center for South Asia and is the Executive Director of the Human Rights Program. She serves as the Lead Counsel for Human Rights at the Center for International Sustainable Development Law, Montreal, and is affiliated faculty at the Raoul Wallenberg Institute for Human Rights, Sweden. Her publications include *Emerging Principles of International Environmental Law* (2006), *Human Rights Approaches to Climate Change: Challenges and Opportunities* (2016), *International Environmental Law and the Global South* (2015), and *The Cambridge Handbook on Environmental Justice and Sustainable Development* (forthcoming).

Andrea Schapper is a Lecturer in International Politics and Programme Director of the M.Sc. in International Conflict and Cooperation at the University of Stirling, UK. She is the Co-Director of the Centre for Policy, Conflict and Cooperation and member of the Human Security, Conflict and Cooperation interdisciplinary research group at Stirling. She coordinates and teaches modules on "Human Rights in International Politics," "The United Nations in a Globalized World," "International Organizations" and "Political Concepts and Ideas." Previously, when she worked at the University of Darmstadt in Germany, Andrea also taught a seminar on "Climate Change and Human Rights." Her publications include, among others, a monograph titled *From the Global to the Local: How International Rights Reach Bangladesh's Children* (2014) and the special journal issue *Human Rights and Climate Change: Mapping Institutional Interlinkages* (2014). Her journal articles have been published in, among others, *International Relations*, *Human Rights Quarterly*, *International Environmental Agreements: Politics, Law and Economics*, *Cambridge Review of International Affairs*, and the *Journal of International Relations and Development*.

Key Issues in Environment and Sustainability

This series provides comprehensive, original and accessible texts on the core topics in environment and sustainability. The texts take an interdisciplinary and international approach to the key issues in this field.

Sustainability
Key Issues
Helen Kopnina and Eleanor Shoreman-Ouimet

Ecomedia
Key Issues
Edited by Stephen Rust, Salma Monani and Sean Cubitt

Ecosystem Services
Key Issues
Mark Everard

Sustainability Science
Key Issues
Edited by Ariane König and Jerome Ravetz

Sustainable Business
Key Issues (2nd Ed)
Helen Kopnina and John Blewitt

Sustainable Consumption
Key Issues
Lucie Middlemiss

Human Rights and the Environment
Key Issues
Sumudu Atapattu and Andrea Schapper

Human Rights and the Environment

Key Issues

Sumudu Atapattu and Andrea Schapper

Routledge
Taylor & Francis Group
LONDON AND NEW YORK

earthscan
from Routledge

First published 2019
by Routledge
2 Park Square, Milton Park, Abingdon, Oxon OX14 4RN

and by Routledge
52 Vanderbilt Avenue, New York, NY 10017

Routledge is an imprint of the Taylor & Francis Group, an informa business

British Library Cataloguing-in-Publication Data
A catalogue record for this book is available from the British Library

Library of Congress Cataloging-in-Publication Data
Names: Atapattu, Sumudu A., author. | Schapper, Andrea, author.
Title: Human rights and the environment: key issues/Sumudu Atapattu and Andrea Schapper.
Description: Abingdon, Oxon; New York, NY: Routledge, 2019. | Series: Key issues in environment and sustainability | Includes bibliographical references and index.
Identifiers: LCCN 2018052185 (print) | LCCN 2018055535 (ebook) | ISBN 9781315193397 (eBook) | ISBN 9781138722743 (hardback) | ISBN 9781138722750 (pbk)
Subjects: LCSH: Environmental law, International. | Climatic changes—Law and legislation. | International law and human rights.
Classification: LCC K3585 (ebook) | LCC K3585 .A863 2019 (print) | DDC 344.04/6—dc23
LC record available at https://lccn.loc.gov/2018052185

ISBN: 978-1-138-72274-3 (hbk)
ISBN: 978-1-138-72275-0 (pbk)
ISBN: 978-1-315-19339-7 (ebk)

Typeset in Goudy
by Deanta Global Publishing Services, Chennai, India

For my beloved grandson
Charlie
who makes the fight for a healthy environment worth
every minute – S.A.

For my beloved father
Helmut Schapper
who was the first person in the world to teach me about
human rights – A.S.

Contents

Acknowledgments xiii
List of abbreviations xv
List of tables xix
List of cases xx

PART I
Introduction and evolution 1

1 Human rights and environmental protection: framing the issues 3
2 Emergence of a human right to a healthy environment 35
3 Pros and cons of a human rights-based approach to environmental protection 63
4 Regional systems of human rights 85

PART II
Human rights of special relevance to environmental protection 107

5 Substantive rights 109
6 Procedural rights 129
7 Constitutional developments 154
8 Selected national cases 175

PART III
Climate change and human rights 203

9 From UNFCCC to Paris Agreement: a human rights assessment 205

10 Social movements and civil society 231

11 Vulnerability and climate change 249

12 Mitigation, adaptation, and loss and damage 267

PART IV
Emerging issues related to environmental rights 285

13 Extraterritorial application of environmental rights 287

14 Business, human rights, and the environment 308

15 Inter-generational rights, animal rights, and rights of nature and ecosystems 325

16 Human rights and the environment: square pegs in round holes? 340

Index 349

Acknowledgments

An undertaking of this nature is the culmination of months of hard work, with the assistance of many people along the way. Sumudu would like to thank her research assistants Alisha Esselstein and Swastika Goel who spent many months researching material for the volume. She would like to thank the University of Wisconsin Law School, the United States for its financial support and encouragement, and giving her the opportunity to teach a seminar course on climate change and human rights that fits very well within its "law in action" approach. UW Law School must be one of the few law schools in the United States that offer such an innovative course. She would also like to thank the incredible staff at the UW Law Library, especially Kris Turner and Sunil Rao. Without the support of the UW Law School this project would probably not have seen the light of day. Sumudu would also like to thank her family for their constant love and support.

Andrea would like to thank the Human Rights and Climate Change Working Group for providing her with invaluable insights into the activities of social movements and civil society networks advocating for recognition of human rights at the international climate negotiations. Many of the group members have taken time out of their very busy schedules to share their experiences in interviews, and part of the research results are included in this textbook. Andrea would also like to thank her husband Tony for his never-ending patience, his encouragement, and his love.

We would also like to Annabelle Harris and Matthew Shobbrook of Routledge for their support and encouragement and for agreeing to our many (seemingly never-ending) requests to extend the submission deadline. We are also grateful to the anonymous reviewers whose comments helped us improve the textbook.

The field of human rights and the environment has grown phenomenally over the last few years and we tried our best to condense the material in an accessible manner. We seem to be at a historic moment when the international community is at the verge of recognizing a human right to a healthy environment. We hope this book contributes in a small way towards that effort.

Needless to say, we take responsibility for any errors.

Sumudu Atapattu
Madison, Wisconsin, United States

Andrea Schapper
Stirling, Scotland

Abbreviations

ACHPR	African Commission on Human and People's Rights
AfDB	African Development Bank
AENOR	Asociación Española de Normalización y Certificación (Spanish Association for Standardization)
AI	Amnesty International
AOSIS	Alliance of Small Island States
ASEAN	Association of Southeast Asian Nations
AWG-LCA	Ad hoc Working Group on Long-term Cooperative Action
BINGO	Business and Industry CSO
CABEI	Central American Bank for Economic Integration
CAN	Climate Action Now
CAT	Committee Against Torture
CBDR	Common but Differentiated Responsibility
CC	Climate Change
CDM	Clean Development Mechanism
CED	Convention on Enforced Disappearances
CEDAW	Convention on the Elimination of All Forms of Discrimination against Women
CEDIA	Centro Para el Desarrollo del Indígena Amazónico
CELDF	Community Environmental Legal Defend Fund
CER	Certified Emission Reductions
CERD	Committee on Elimination of Racial Discrimination
CESCR	Committee on Economic, Social and Cultural Rights
CIEL	Center for International Environmental Law
CHR	Commission on Human Rights
CMW	Committee on Migrant Workers
CNDH	Comisión Nacional de los Derechos Humanos (National human Rights Commission of Mexico)
COP	Conference of the Parties
CO2	Carbon Dioxide
CPJ	Committee to Protect Journalists
CRC	Convention on the Rights of Child

CRDP	Concurrent Retirement and Disability Payments
CRGE	Climate Resilient Green Economy
CRPD	Convention in the Rights of Persons with Disabilities
CSO	Civil Society Organization
CSR	Corporate Social Responsibility
DEQ	Department of Environmental Quality
DFID	Department for International Development
DNA	Deoxyribonucleic Acid
ECE	Economic Commission for Europe
ECLAC	Economic Commission for Latin America and Caribbean
ECOSOC	Economic and Social Council of the United Nations.
EEPCo	Ethiopian Electric Power Corporation
EHP	Environmental Health Perspectives
EHRD	Environmental Human Rights Defenders
EIA	Environmental Impact Assessment
EIB	European Investment Bank
EITI	Extractive Industries Transparency Initiative
ENGO	Environmental NGO
EPA	Environmental Protection Agency
ESC	Economic, Social and Cultural Rights
ESD	Education for Sustainable Development
ET	Extraterritorial
ETO	Extraterritorial Obligations
EU	European Union
EQA	Environmental Quality Act
EXIM	Export Import Bank
FHI	Family Health International (formerly)
FPIC	Free, Prior, and Informed Consent
GA	General Assembly
GAP	Great Ape Project
GCF	Green Climate Fund
GDP	Gross Domestic Product
GE	Green Economy
GEF	Global Environment Facility
GENISA	Generadora del Istmo S.A.
GHG	Greenhouse Gases
GI-ESCR	Global Initiative for Economic Social and Cultural Rights
GLAN	Global Legal Action Network
GoE	Government of Ethiopia
GPDD	Global Program for Disability and Development
GPE	Global Pact for the Environment
HR	Human Rights
HRC	Human Rights Council
HRCCWG	Human Rights and Climate Change Working Group
HRW	Human Rights Watch

IACHR	Inter-American Commission on Human Rights
IACtHR	Inter-American Court of Human Rights
ICBC	Industrial and Commercial Bank of China
ICC	International Chamber of Commerce
ICCPR	International Convenant on Civil and Political Rights
ICESCR	International Convenant on Economic Social and Cultural Rights
ICHRP	International Council on Human Rights Policy
ICJ	International Court of Justice
IDMC	Internal Displacement Monitoring Centre
IE	Independent Expert
IEL	International Environmental Law
IET	International Emissions Trading
IFAD	International Fund for Agricultural Development
IFC	International Finance Corporation
IHRL	International Human Rights Law
IIPFCC	International Indigenous Peoples Forum on Climate Change
ILC	International Law Commission
ILO	International Labour Organization
INDCs	Intended Nationally Determined Contributions
IO	International Organizations
IOE	International Organization of Employers
IOM	International Organization for Migration
IPCC	Intergovernmental Panel on Climate Change
IPO	Indigenous People's Organizations
ISHR	International Service for Human Rights
ITLOS	International Tribunal for the Law of the Sea
JI	Joint Implementation
KPCS	Kimberley Process Certification Scheme
LDCF	Least Developed Country Fund
MCA	Montana Codes Annotated
MDG	Millennium Development Goals
MFHR	Marangopoulos Foundation for Human Rights
MLJ	Malayan Law Journal
MNCs	Multinational Corporations
MOSOP	Movement for the Survival of the Ogoni People
NAPs	National Adaptation Plans
NCP	National Contact Point
NDC	Nationally Determined Contributions
NEA	National Education Association
NGO	Nongovernmental Organization
NHRI	National Human Right Institutions
OAS	Organization of American States
OECD	Organization for Economic Co-operation
OHCHR	Office of the United Nations High Commissioner for Human Rights

PCB	Polychlorinated Biphenyl
PETA	People for Ethical Treatment of Animals
PIL	Public Interest Litigation
REDD	Reducing Emissions from Deforestation and Forest Degradation
RINGO	Research and Independent NGO
SDGs	Sustainable Development Goals
SDM	Sustainable Development Mechanism
SIDS	Small Island Developing States
SLR	Sri Lanka Reporter
SPE	Special Purpose Entities
SPJV	Seven-Up Pete Joint Venture
SR	Special Rapporteurs
TANs	Transnational Advocacy Networks
TEK	Traditional Ecological Knowledge
TUNGO	Trade Union NGO
UCC	United Church of Christ, Council for Climate Justice
UDHR	Universal Declaration of Human Rights
UN	United Nations
UNCTAD	United Nations Conference on Trade and Development
UNDP	United Nations Development Programme
UNDRIP	United Nations Declaration on the Rights of Indigenous Peoples
UNDESA	United Nations Department of Economic and Social Affairs
UNEP	United Nations Environment Programme
UNFCCC	United Nations Framework Convention on Climate Change
UNGA	United Nations General Assembly
UNGP	United Nations Guiding Principles
UNHCR	United Nations High Commissioner for Refugees
UNHRC	United Nations Human Rights Council
UNICEF	United Nations Children's Fund
UNITA	União Nacional para a Independência Total de Angola
UNTC	United Nations Treaty Collection
UOI	Union of India
UPR	Universal Periodic Review
USCB	United States Council for Business
USCIB	United States Council for International Business
WAPDA	Water and Power Development Authority
WB	World Bank
WCED	World Commission on Environment and Development
WEDO	Women's Environment and Development Organization
WHO	World Health Organization
WMO	World Meteorological Organization
YOUNGO	Youth NGO (International Youth Climate Movement)

Tables

1.1	Sustainable Development Goals, 2015	22
7.1	Number of constitutions with environmental rights	156
14.1	IFC Performance Standards	315
14.2	UN Guiding Principles on Business and Human Rights	319

Cases

International cases and complaints to human rights treaty bodies

- Apirana Mahuika et al. v. New Zealand Communication No. 547/1993, U.N. Doc. CCPR/C/70/D/547/1993 (2000)
- Case Concerning the Gabcikovo Nagymaros Project (Hungary v. Slovakia) International Court of Justice, 1996
- Claude Reyes et al. v. Chile (2006) [Online] Available at: www.corteidh.or.cr/docs/casos/articulos/seriec_151_ing.pdf (Accessed: 1 October 2018)
- Delia Saldias de Lopez v. Uruguay CCPR/C/13/D/52/1979, UN Human Rights Committee (HRC), 29 July 1981
- E.H.P et al. v. Canada No. 67/1980, U.N. Doc. CCPR/C/OP/1 at 20 (1982)
- EP et al. v. Columbia Communication No. 318/1988, U.N. Doc. CCPR/C/39/D/318/1988 (1990).
- Ecuador v. Columbia (Ariel Herbicide Spraying) No. 2011/31, 21 October 2011
- Georgia v. Russian Federation 2008 I.C.J. 140
- Ilmari Länsman et al. v. Finland 11 June 1992 Communication No. 511/1992, U.N. Doc. CCPR/C/52/D/511/1992 (1994) [Online] Available at: http://hrlibrary.umn.edu/undocs/html/vws511.htm) (Accessed: 1 October 2018)
- Jouni E. Länsman et al. v. Finland Communication No. 671/1995, U.N. Doc. CCPR/C/58/D/671/1995 (1996)
- Lubicon Lake Band v. Canada CCPR Communication No. 167/1984
- Pulp Mills on the River Uruguay (Argentina v. Uruguay): ICJ Rep (2010)
- Reparations for Injuries Suffered in the Service of the U.N. Order 1948, I.C.J 121
- Susila Malani Dahanayake and 41 other Sri Lankan citizens v. Sri Lanka Communication No. 1331/2004, UN Doc. CCPR/C/87/D (2006)
- United States v. Canada (Trail Smelter Case) Arbitral Trib., 3 U.N. Rep. Int'l Arb. Awards 1905 (1941)
- Vaihere Bordes and John Temeharo v. France, Communication No. 645/1995, UN Doc. CCPR/C/57/D (1996)

Regional cases

- Alejandre v. Cuba 42 F. Supp. 2d 1317 (S.D. Fla. 1999)
- Buckley v. United Kingdom ECHR 25 Sep 2006
- Bankovic and Others v. Belgium and 16 Other Contracting States Application no. 52207/99, 12 December 2001
- Community of La Oroya v. Peru Petition 1473-06, Inter-American Commission on Human Rights H.R., Report No. 76/09, OEA/Ser.L/V/II., doc. 51 (2009), petition 1473-06, August 5, 2009
- Florea v. Romania (2010) (Application no. 37186/03), Judgment of 14 September 2010 (2010) (in French). Summary available in English at: http://sim.law.uu.nl/sim/caselaw/Hof.nsf/1d4d0dd240bfee7ec12568490035df05/83ff6f5c75014d78c1257797002d1151?OpenDocument
- Guerra and others v. Italy Application No. 14967/89; (1998) 26 EHRR 357; [1998] ECHR 7
- Hatton and Others v. United Kingdom ECHR 8 July 2003
- Ilascu and Others v. Moldova and Russia 48787/99, Council of Europe: European Court of Human Rights, 8 July 2004
- Kawas Fernandez v. Honduras Judgment of 3 April 2009 (Merits, Reparations and Costs)
- Kichwa People of Sarayaku v. Ecuador (2012) Kichwa v. Ecuador, Merits, Reparations and Costs, Judgment, Inter-Am. Ct. H.R. (ser. C) No. 245, 58 (June 27, 2012).
- LCB v. United Kingdom Merits, App No 23413/94, Case No 14/1997/798/1001, [1998] ECHR 49, ECHR 1998-III, (1999) 27 EHRR 212, IHRL 3274 (ECHR 1998), 9th June 1998, European Court of Human Rights
- Leander v. Sweden App no 9248/81, A/116, (1987) 9 EHRR 433, IHRL 69 (ECHR 1987), 26th March 1987, European Court of Human Rights
- Lopez Ostra v. Spain Application No. 16798/90; (1995) 20 EHRR 277; [1994] ECHR 46; (1994) 303-C Eur. Ct. H.R. (ser A)
- Mayagna (Sumo) Awas Tingni Community v. Nicaragua (2001) [Online] Available at: www.corteidh.or.cr/index.php/en/jurisprudencia (Accessed: 1 October 2018)
- Maya Communities of the Toledo District v. Belize Report No. 78/00; Case 12.053 (2000)
- Mary and Carrie Dann v. US Case 11.140, Report No. 75/02, Inter-Am. C.H.R., Doc. 5 rev. 1 at 860 (2002)
- Mossville Environmental Action Now v. United States Report No. 43/10, Petition 242-05, 17 March 2010
- Observer and Guardian v. United Kingdom ECHR 26 November 1991
- Oneryildiz v. Turkey Application No. 48939/99; (2005) 41 EHRR 20; [2004] ECHR 657; 18 BHRC 145, (2005); [2004] Inquest LR 108
- Paul and Audrey Edwards v. United Kingdom Application no. 46477/99 (2002)
- Powell and Rayner v. United Kingdom ECHR 21 February 1990
- Saldano v. Argentina Petition IACHR Report 38/99, 11 March 1999, Ann. Rep. IACHR 1998, 289

- Saramaka People v. Suriname (2007) [Online] Available at: www.corteidh.or.cr/index.php/en/jurisprudencia (Accessed: 1 October 2018)
- Social and Economic Rights Action Center and the Center for Economic and Social Rights v. Nigeria Comm. No. 155/96 (2001)
- Stoine Hristov v. Bulgaria (2009) Available in French only. Referred to in Manual on Human Rights and Environment (2012)
- Velasquez Rodriguez v. Honduras (1998) [Online] Available at: www.corteidh.or.cr/index.php/en/jurisprudencia (Accessed: 1 October 2018)
- Yanomami v. Brazil Comunidad Yanomami Caso N° 7615. Resolución N° 12/85 (1985)
- Yakye Axa v. Paraguay (17 June 2005) [Online] Available at: www.corteidh.or.cr/index.php/en/jurisprudencia (Accessed: 1 October 2018)

Domestic cases

- Adong bin Kuwau and others v. Kerajaan Negeri Johor and Anor (1997) [1997] 1 MLJ 418
- Anjum Irfan v. Lahore Development Authority (2002) (Lahore Air Pollution Case) (WP 25084/1997, dd 2002)
- Aparicio Benito v. Spain (2006) Application no. 36150/03, Judgment of 13 November 2006
- AP Pollution Control Board II v. MV Nayudu (2000) [Online] Available at: https://indiankanoon.org/doc/1543623/ (Accessed: 1 October 2018)
- Bakun Hydro-Electric Dam case (1996) referred to in Raman (1997)
- Bokhari v. Federation of Pakistan (2003) [Online] Available at: www.elaw.org/content/pakistan-dr-amjad-h-bokhari-vs-federation-pakistan-constitutional-petition-452003-amicus-cur (amicus curiae by Dr Pavez Hassan) (Accessed: 1 October 2018)
- Bulankulama and others v. Ministry of Industrial Development and Others (1999) S.C. Application No. 884/99 (F/R) [Online] Available at: www.elaw.org/content/sri-lanka-bulankulama-v-min-industrial-development-eppawala-case-sc-application-no-88499-fr (Accessed: 1 October 2018)
- Centre for Environmental Justice v. Ministry of Agriculture, Environment, Irrigation and Mahaweli Development et al. (Bundala National Park Case) (2006) [Online] Available at: www.elaw.org/content/sri-lanka-centre-environmental-justice-v-ministry-agriculture-environment-irrigation-and-mah (petition) (Accessed: 1 October 2018)
- Clean Air Council, SB through his Guardian and BB through his Guardians v. USA, Donald Trump as President and others (2017) Case 2:17-cv-04977-PD [Online] Available at: https://cleanair.org/wp-content/uploads/Complaint-17-4977.pdf (Accessed: 1 October 2018)
- Coard et al. v. United States (1999) Report No. 109/99 - Case 10.951, Inter-American Commission on Human Rights (IACHR), 29 September 1999
- Comunidad de Chanaral v. Codeco Division el Saldor (2018) [Online] Available at: www.camara.cl/pdf.aspx?prmTIPO=OFICIOFISCALIZACIONRESPUESTA&prmID=69883&prmNUMERO=212&prmRTE=1152 (Accessed: 1 October 2018)

- Communities for a Better Environment v. City of Richmond (2010) Court of Appeal of California 184 Cal. App. 4th 70 (2010)
- Delia Saldias de Lopez v. Uruguay (1981) CCPR/C/13/D/52/1979, UN Human Rights Committee (HRC), 29 July 1981
- Dr Mohiuddin Farooque and another v. Bangladesh, represented by the Secretary, Ministry of Irrigation, Water Resources and Flood Control, Government of the People's Republic of Bangladesh, and others (1997) 50 DLR (HCD) (1998) 84
- Earthlife Africa Johannesburg v. Minister of Environmental Affairs et al. Case no. 65662/16 (2017)
- Environmental Foundation Ltd. v. Urban Development Authority et al. [Online] Available at: www.right2info.org/resources/publications/case-pdfs/sri-lanka_environmental-foundation-ltd.-v.-urban-development-authority-of-sri-lanka (Accessed: 1 October 2018)
- Gbemre v. Shell Petroleum Development Company and others Suit No. FHC/B/CS/53/05; (2005) AHRLR 151 (NgHC 2005)
- Greenpeace Nordic Association v. Norway Ministry of Petroleum and Energy (2017) [Online] Available at: http://climatecasechart.com/non-us-case/greenpeace-nordic-assn-and-nature-youth-v-norway-ministry-of-petroleum-and-energy/ (including an unofficial translation in English) (Accessed: 1 October 2018)
- Gunaratne v. Homagama Pradeshiya Sabha (1998) 2 Sri L.R. 11 [Online] Available at: www.lawnet.gov.lk/wp-content/uploads/2016/11/002-SLLR-SLLR-1998-V-2-GUNARATNE-v.-THE-HOMAGAMA-PRADESIYA-SABHA-AND-OTHERS.pdf (Accessed: 1 October 2018)
- In re Greenpeace Southeast Asia and others v. "Carbon Majors" (2015) Case No. CHR-NI-2016-0001 (2015)
- Ioane Teitiota v. The Chief Executive of the Ministry of Business, Innovation and Employment (2015) NZSC 107, New Zealand: Supreme Court, 20 July 2015
- Irfan v. Lahore Development Authority No. 25084 (Lahore HC 2003) [Online] Available at: https://elaw.org/content/pakistan-anjun-irfan-v-lahore-development-authority-wp-250841997-dd-20020614-lahore-air-poll (Accessed: 1 October 2018)
- Juliana et al. v. USA et al. (2015) Juliana et al. v. USA et al. (April 2016) United States District Court, For the District of Oregon, 6:15-cv-1517-TC, Magistrate Judge Coffin, Order and Findings and Recommendations [Online] Available at: https://static1.squarespace.com/static/571d109b04426270152febe0/t/576195342fe1316f09d2eb8d/1466012983313/16.04.08.OrderDenyingMTD.pdf (Accessed: 25 September 2018)
- Juliana et al. v. USA et al. (November 2016) United States District Court, For the District of Oregon, 6:15-cv-1517-TC, Judge Aiken, Opinion and Order [Online] Available at: https://static1.squarespace.com/static/571d109b04426270152febe0/t/5824e85e6a49638292ddd1c9/1478813795912/Order+MTD.Aiken.pdf (Accessed: 25 September 2018)
- Karnataka Industrial Areas Development Board v. Sri C. Kenchappa and Others (2006) Case No: Appeal (civil) 7405 of 2000 [Online] Available at: https://indiankanoon.org/doc/992326/ (Accessed: 1 October 2018)

- Kottabadu Durage Sriyani Silva v. Chanaka Iddamalgoda (2003) [Online] Available at: www.hrcsl.lk/PFF/Freedom_From_Torture/Sriyani%20Silva%20V.%20Iddamalgod...pdf (Accessed: 1 September 2018)
- Lliuya v. RWE (2015) AG Case No. 2 O 285/15 Essen Regional Court
- Marangopoulos Foundation for Human Rights v. Greece Complaint (2007) No. 30/2005
- Massachusetts v. Environmental Protection Agency (2007) 549 U.S. 497 (2007)
- McArthur River Mine Case (2006) [Online] Available at: www.edonsw.org.au/mcarthur_river_diversion_northern_territory_mining_impacts (Accessed: 1 October 2018)
- M.C. Mehta v. Kamal Nath and Others (1996) Available at: https://indiankanoon.org/doc/1514672/
- M.C. Mehta v. Union of India and Others (1991) WP 860/1991 (1991.11.22) [Online] Available at: www.elaw.org/content/india-mc-mehta-v-union-india-wp-8601991-19911122-environmental-education-case (Accessed: 1 October 2018)
- Mendoza Beatriz Silva and others v. National Government of Argentina and others (2008) File M. 1569. XL
- Metropolitan Manila Development Authority v. Concerned Residents of Manila Bay G.R. Nos. 171947-48(2008)
- Minors Oposa v. Secretary of the Department of Environment and Natural Resources 33 ILM 173 (1994)
- Montana Environmental Information Center v. Department of Environmental Quality (USA) No. 97-455 (1999)
- Mundy v. Central Environmental Authority and others SC Appeal 58/2003 (2003)
- Narmada Bachao Andolan v. Union of India 10 S.C.C. 664 (2000) Writ Petition (civil) 328 of 2002 [Online] Available at: https://indiankanoon.org/doc/1642722/ (Accessed: 1 October 2018)
- Native Village of Kivalina v. ExxonMobil Corp. 696 F.3d 849 (9th Cir. 2012)
- Netherlands v. Urgenda Foundation (Urgenda appeal) (2018) [Online] Available at: https://uitspraken.rechtspraak.nl/inziendocument?id=ECLI:NL:GHDHA:2018:2610 (Accessed: 1 October 2018)
- People's Union for Civil Liberties and Another v. UOI and Others (2005) [Online] Available at: www.advocatekhoj.com/library/judgments/index.php?go=2009/february/192.php (Accessed: 1 October 2018)
- Petro Flores v. Codelco, Division Salvador, (1988) Rol 2.052 (Sup. Ct. Chile, 23 June 1988) Translated in Georgetown International Environmental Law Review 2, 251–253 (1989)
- Prakash Mani Sharma v. His Majesty's Government Cabinet Secretariat and Others (2003) WP 2237/1990 (2003.3.11) [Online] Available at: www.elaw.org/content/nepal-prakash-mani-sharma-v-his-majesty%E2%80%99s-government-cabinet-secretariat-and-others-wp-22371 (Accessed: 1 September 2018)

- Prakash Mani Sharma and Others v. Nepal Drinking Water Corporation and Others (2001) WP 2237/1990 (10 July 2001) [Online] Available at: www.elaw.org/content/nepal-advocate-prakash-mani-sharma-and-others-vs-nepal-drinking-water-corporation-and-others (Accessed: 1 October 2018)
- Salt Miners Labor Union v. Industries and Mineral Development (1993) Punjab Lahore Human Rights Case No. 120 of 1993 Supreme Court of Pakistan 1994 SCMR 2061 [Online] Available at: www.globalhealthrights.org/wp-content/uploads/2013/02/SC-1994-Salt-Miners-v.-Director-Industries-and-Mineral-Development.pdf (Accessed: 1 September 2018)
- Senih Ozay v. Ministry of the Environment and Eurogold Madencilik (1996) Ref. No 1996/5477; Ruling No.1997/2312
- Shehla Zia and others v. WAPDA (1994) PLD 1994 SC 693
- Subhash Kumar v. State of Bihar and Others (1991) AIR 420, 1991 SCR (1) 5 (1991) [Online] Available at: https://indiankanoon.org/doc/1646284/ (Accessed: 1 October 2018)
- Sugathapala Mendis and Another v. Chandrika Kumaratunga and Others (Water's Edge Case) (2008) [Online] Available at: www.lawnet.gov.lk/2008/12/31/sugathapala-mendis-and-another-v-chandrika-kumaratunga-and-others-waters/
- Suray Prasad Sharma Dhungel v. Godavari Marble Industries and Others (1992) WP 35/1992
- Tan Tek Seng v. Suruhanjaya Perkhidmatan Pendidikan and Another (1996) 1 MLJ 261
- Tarsasag a Szabadsagjogokert v. Hungary ECHR, 14 April 2009
- Taskin et al. v. Turkey Application No. 46117/99; (2006) 42 EHRR 50
- Thomson v. Minister for Climate Change Issues (2017) NZHC 733
- Thorgeir Thorgeirson v. Iceland (1992) (Application no. 13778/88) [Online] Available at: www.humanrights.is/static/files/Itarefni/torgeir_torgeirson_gegn_islandi.pdf (Accessed: 1 October 2018)
- Urgenda Foundation v. The State of The Netherlands (2015) District Court of Hague, C/09/456689 / HA ZA 13-1396 (2015)
- Vellore Citizens Welfare Forum v. Union of India (1996) [Online] Available at: https://indiankanoon.org/doc/1934103/ (Accessed: 1 September 2018)
- Victor Ramon Castrillon Vega v. Federación Nacional de Algodoneros y Corporation Autonoma Regional del Cesar (1997) Accion de Tutela Case No 4577 (Colom.), discussed in Kravchenko and Bonine, 2008, 70
- VZW Klimatzaak v. Kingdom of Belgium (2016) (summary) [Online] Available at: climatecasechart.com/non-us-case/vzw-klimaatzaak-v-kingdom-of-belgium-et-al/ (Accessed: 1 October 2018)
- Weerasekera et al. v. Keangnam Enterprises Ltd. (2009) CA (PCH). Apn No. 40/2004(dd.2009)
- Withanage v. Geological Survey and Mines Bureau and Others (2004) CA Application of 21 October 2004
- Wolfson v. Brammer (2011) 822 F. Supp. 2d 925 (D. Ariz. 2011)

Part I

Introduction and evolution

1 Human rights and environmental protection

Framing the issues

Introduction

The past two decades have witnessed an upsurge in the recognition of environmental rights around the globe and the use of human rights law and institutions to seek redress for environmental injustices. How did this happen? International human rights law does not recognize a separate right to a healthy environment, yet almost all human rights institutions now recognize that environmental degradation and pollution can lead to a violation of the rights protected under various human rights treaties. It is now generally recognized that human rights can be infringed by environmental degradation.

An example will make this clearer. Suppose there is an industry that produces toys in a particular neighborhood. There are several houses in close proximity to the industrial plant. The operators do not follow any safety standards and dump untreated wastes into a nearby waterway and emit fumes into the atmosphere. People in the area regularly fall sick with respiratory problems, and those who use the water from the river have started getting rashes and other skin diseases and diarrheal problems. It is quite clear that the source of these problems is the industry and its unsanitary practices. Due to this situation, several rights of the people in the area are affected: their right to health, right to privacy and family life, right to water, right to an adequate standard of living and, in very serious cases, even the right to life could be affected.

However, whether this situation amounts to a violation of rights depends on several things: (a) is the industry state owned or privately owned?; (b) is it violating any standards recognized by law in its operations?; (c) does it have a license to operate?; (d) are there any laws and standards in place?; (e) have local authorities taken any action?; (f) are there any extenuating circumstances?; (g) have people complained?; (h) what action, if any, did the authorities take to remedy the situation?; and (i) if it is privately owned, have the government authorities taken any action to control its activities or provide a remedy? These are some of the questions that human rights institutions have examined in similar situations. While environmental issues can impact the enjoyment of rights, they do not necessarily lead to a violation of those rights in every situation. As we will discuss in later chapters, sometimes it is hard to establish the link between environmental issues and protected rights, particularly in the absence of a distinct right to a healthy or safe environment.

This chapter discusses three frameworks that are useful (and have been used) to articulate, interpret, and give context to environmental rights: (a) human rights; (b) environmental justice; and (c) sustainable development, including the Sustainable Development Goals (SDGs) adopted by the international community in 2015. As we shall see, they overlap to some extent, but have unique features that are useful and have been used in relation to environmental rights. Together, they provide a good overarching framework to discuss environmental rights which each one alone cannot do. Before we turn to the three frameworks, we will elaborate on what we mean by "environmental rights."

What are environmental rights?

Scholars adopt various definitions of environmental rights. For some, they mean just procedural rights of access to information, participation in decision-making, and access to remedies. For others, they mean a substantive (stand-alone) right to a healthy environment. Yet others adopt a hybrid approach to mean all the rights that are used in relation to environmental issues. For example, human rights that are used this way include the right to life in extreme situations, but more often include the right to health, right to food and water, and the right to privacy and family life.

Human rights law and environmental protection inter-relate in four different ways (Anton and Shelton, 2011: 130); the first approach is to include human rights guarantees in environmental instruments. These are mainly procedural in nature. However, some have questioned the efficacy of this approach because most environmental treaties have weak compliance mechanisms. The second approach is to invoke human rights law and institutions when their enjoyment is threatened by environmental harm. Here, of course, the focus is on human beings and protecting their rights, rather than on protecting the environment. The third approach is to adopt a new human right to an ecologically balanced environment, and not purely in relation to human beings, but a sustainable environment in the long run. Many environmentalists oppose the focus on human beings and some human rights advocates oppose it because they feel that it lacks content and its recognition could dilute existing core rights (Alston, 1984). The final approach is to frame this in terms of duties, rather than rights (Shelton, 2001). Whatever approach is adopted, the link between human rights and environmental degradation is undeniable. Adopting a human rights-based approach to environmental issues has both advantages and disadvantages, which we will discuss in Chapter 2.

For the purposes of this book, we will adopt a broad definition of environmental rights that includes both substantive rights and procedural rights. We will discuss these in detail in Chapters 5 and 6, but environmental rights include the right to live in a healthy (or satisfactory, clean, favorable) environment (this could take various forms), and the other human rights that could be infringed by living in a polluted environment – the right to food, water, health, privacy, property, adequate standard of living, and culture. The procedural rights include the right to information, right to participate in the decision-making process, and the right to remedies. To give effect to these rights, states may have to promulgate legislation, put institutions in place, monitor the activities of private individuals and punish them in the event of wrongdoing, ensure the preparation of environmental impact assessment reports, publicize those reports and facilitate public participation, and provide a remedy in the event these obligations are violated (see *Ogoniland* case, discussed in Chapter 4).

Even the meaning of the word "rights" is contested (Anton and Shelton, 2011). As Roscoe Pound observed: "There is no more ambiguous word in legal and juristic literature than the word right." Rights can mean legal rights, moral rights, or rights that originate from historic or customary practices, but are not incorporated in a legal document. For example, the eldest child in a family might have the "right" to inherit the family heirlooms in a particular culture or may have the "right" to get the first bite of the first harvest. While these rights are important for that culture or family, they are more like traditions or practices, rather than legal rights.

In this volume, we will not engage in the debate on what rights mean, even though that debate is important. We will simply adopt the human rights framework (as described in the next section) as incorporated in various human rights instruments and as interpreted by various human rights bodies and courts, while appreciating that "rights" are not confined to the legal documents.

Rights imply duties – just as individuals enjoy rights, states have the duty to protect them. There could be situations where rights conflict. These conflicts must be resolved by reference to other principles. Similarly, when rights are infringed, remedies must be available, whether they are judicial or administrative in nature, and some redress must be available to the victim. If no redress mechanism is available, they are not rights in the legal sense.

The tale of three frameworks

Human rights law

Historically, individual rights can be traced to the Magna Carta of 1215. Other documents such as the English Bill of Rights (1689), the French Declaration on the Rights of Man and Citizen (1789), and the U.S. Constitution and Bill of Rights (1791) have all influenced the development of human rights. In addition, some of the rights can be traced to major world religions, which frame them more as duties owed towards others.

The premise of human rights is based on the dignity of the human person and the rights that are recognized accrue to them by virtue of being born human. Some rights such as the right to vote and the freedom of movement are confined to citizens, while other rights are enjoyed by all individuals irrespective of nationality. However, rights are not confined to human beings alone. Animals and corporations have rights, and a more recent development is where nature, including rivers and forests, is accorded legal rights. Some of these developments will be discussed in Chapter 15. Even the notion of "a human being" is now undergoing transformation, in light of technological advancements taking place that have ramifications for human rights.

> Human rights are often said to resemble fundamental or natural rights in the natural law tradition – that is, they are rights that do not owe their existence to a legal source and do not depend on legal sanctions. Instead, human rights are claimed to transcend legal systems not in accord with those rights and to obtain in all places and at all times regardless of what the positive law provides. Legal systems should embody rights. If one does not, it is deficient.
>
> (Anton and Shelton, 2011: 121)

The origins of modern human rights law lie in the UN Charter which was adopted in 1945 in the aftermath of the atrocities committed during World War II when the international community vowed "never again." The Preamble specially refers to the determination "to save succeeding generations from the scourge of war, which twice in our lifetime has brought untold sorrow to mankind," and "to reaffirm faith in fundamental human rights, in the dignity and worth of the human person, in the equal rights of men and women and of nations large and small" (UN Charter, 1945: Preamble). One of the purposes of the UN is "to achieve international co-operation in solving international problems of an economic, social, cultural, or humanitarian character, and in promoting and encouraging respect for human rights and for fundamental freedoms for all without distinction as to race, sex, language, or religion" (UN Charter, 1945: Article 1.3). The Charter contains a separate chapter on human rights.

The UN Charter was closely followed by the Universal Declaration of Human Rights in 1948 (UDHR, 1948). Although not binding on its signatories, it had a huge influence on the development of human rights law. The UDHR affirmed both civil and political rights, and economic, social, and cultural rights. Eleanor Roosevelt, the U.S. first lady, played a key role in drafting the Declaration. This founding document, adopted in less than two years, was to be followed by a binding document (de Schutter, 2012). However, the Cold War intervened, and it took the international community almost two decades to adopt binding obligations embodied in a treaty. In the end, the obligations in the UDHR were bifurcated along civil and political rights, and economic, social, and cultural rights, and two Covenants were opened for signature in 1966 embodying each set of rights, even though the official UN position is that all rights are universal, indivisible, inter-related, and interdependent (Vienna Declaration on Human Rights, 1993); the two treaties are: the International Covenant on Civil and Political Rights (ICCPR, 1966) and the International Covenant on Economic, Social and Cultural Rights (ICESCR, 1966).

Since then, human rights instruments have skyrocketed, with a plethora of human rights treaties and soft law instruments and institutions adopted, not just at the global level, but at the regional level as well. In addition to the two

Covenants mentioned above – ICCPR and ICESCR – there are instruments governing specialized groups: the Convention on the Rights of the Child (CRC), the Convention on the Elimination of All Forms of Discrimination against Women (CEDAW), the Convention for the Protection of Migrant Workers and Members of their Families, the UN Declaration on the Rights of Indigenous Peoples, and the Convention on the Protection of Persons with Disabilities. Moreover, there are treaties that govern specialized issues: the Genocide Convention, the Convention on the Elimination of Racial Discrimination, the International Convention for the Protection of All Persons from Enforced Disappearance, and the Convention Against Torture. Human rights law has developed into a sophisticated body of treaties and institutions both at the international level and the regional level. It has also influenced the development of national law. However, none of the international human rights treaties embodies a right to a healthy environment as a distinct right. Three regional human rights treaties do. These will be discussed in Chapter 4.

One of the most significant developments in relation to human rights is the right given to individuals under certain treaties to bring a claim against his/her own state for the violation of his/her rights before the committee established to monitor compliance with the treaty. Under general international law, only states have standing to bring claims before international dispute settlement mechanisms such as the International Court of Justice (Article 34 of the ICJ Statute specifically embodies this principle). Giving individuals the right to bring a claim against their state before an international forum is a significant development. Before such a claim can be brought, however, local remedies have to be exhausted. In other words, any remedies that are available at the national level (including administrative remedies) must be exhausted before an international forum can be approached to seek a remedy.

The nature of human rights

Human rights law operates vertically between states as duty-bearers and its people as rights-holders. It is usually the link of nationality that binds people to their state. States by ratifying human rights treaties undertake to ensure that the rights embodied in those treaties are afforded to their citizens. Thus, for example, Article 2 of the International Covenant on Civil and Political Rights (ICCPR) provides that:

1 Each State Party to the present Covenant undertakes to respect and to ensure to all individuals within its territory and subject to its jurisdiction the rights recognized in the present Covenant, without distinction of any kind, such as race, color, sex, language, religion, political or other opinion, national or social origin, property, birth or other status.

2 Where not already provided for by existing legislative or other measures, each State Party to the present Covenant undertakes to take the necessary steps, in accordance with its constitutional processes and with the provisions

> of the present Covenant, to adopt such laws or other measures as may be necessary to give effect to the rights recognized in the present Covenant.

Article 2 makes it clear that the rights embodied there should be accorded to all individuals within the territory of the state and subject to its jurisdiction without distinction of any kind. Moreover, states are required to adopt legislation and other measures to give effect to the rights recognized in the Covenant. Thus, as a general principle, these rights do not operate extraterritorially. We will discuss this issue in more detail later.

We mentioned earlier that human rights law operates vertically between states and people within the state – usually citizens, although Article 2 above makes it clear that the rights are not limited to citizens. Thus, people living in a particular country temporarily for work or as permanent residents should be accorded these rights, although some rights, such as the right to vote, are often restricted to citizens. While, as a general principle, states are not responsible for the damage caused to people by private individuals or corporations, states become responsible if they failed to control these activities or if states fail to punish the wrongdoers. Here states become responsible for their own failure. This principle becomes important in the environmental context as most polluting activities are carried out by private companies, often multinational companies based in the Global North.

Typology of human rights

International human rights law adopts a typology of obligations with regard to human rights – respect, protect, and fulfill. While this typology is usually applied in relation to economic, social, and cultural rights, it can equally be applied in relation to civil and political rights (de Schutter, 2012).

The obligation to *respect* requires States to refrain from interfering directly or indirectly with the enjoyment of the right. The obligation to *protect* requires States to take measures that prevent third parties from interfering with the enjoyment of the right. The obligation to *fulfill* contains obligations to facilitate, provide, and promote and requires states to adopt appropriate legislative, administrative, budgetary, judicial, promotional, and other measures towards the full realization of the right (General Comment No 14 on the Right to Health; Anton and Shelton, 2011).

Even though reference is often made to various generations of rights – the first generation refers to civil and political rights, the second generation refers to economic, social, and cultural rights, while the third generation refers to collective rights – the official UN position is that "*all human rights are universal, indivisible and interdependent and interrelated*" (Vienna Declaration on Human Rights, 1993, Article 5, emphasis added).

Environmental justice

In light of increasing environmental challenges, such as pollution, climate change, and loss of biodiversity, and their disproportionate impact on certain communities, many scholars have started to think about these problems from a justice perspective. Political philosophers and theorists were among the first to discuss how just relations between state actors, societies, and generations in the context of environmental problems can be shaped. Who bears the responsibility for, and who are the "recipients" (Page, 2006: 50) of environmental justice? What obligations does environmental justice entail?

The all-encompassing and multi-level nature of environmental justice is described succinctly by Kuehn who points out that environmental justice disputes arise at the international, national, and local levels:

> Disputes at the international level include allegations that governments and multinational corporations are exploiting indigenous peoples and the impoverished conditions of developing nations. At the national level, although an overwhelming number of studies show differences by race and income in exposures to environmental hazards, debate continues about the strength of that evidence and the appropriate political and legal responses to such disparities. At the local level, many people of color and lower-income communities believe they have not been treated fairly regarding the distribution of the environmental benefits and burdens.
>
> (2000: 10681)

Although there are many varying understandings of, and debates about, environmental justice, most scholars agree on three common denominators when defining it. First, they depart from a relational understanding of justice. This means unjust situations arise between different actors, and normative claims are raised prescribing how justice can be achieved between them. Against this background, it is relevant to establish the duty-bearers and recipients of justice, including which actors, i.e., individuals, groups, or states, can raise claims against others (Beckman and Page, 2008). Second, environmental problems often exacerbate already existing inequalities (Humphreys, 2014) and intensify economic and social disparities that might have their root causes elsewhere. Third, environmental injustice often relates to situations in which those who are the least responsible for a problem are the ones that are most affected and most vulnerable to its consequences, but at the same time have the fewest resources to adapt to it (Morrow, 2010).

What is difficult when thinking about environmental justice are considerations about space and time. Many injustices have their roots in the colonial era or started to evolve during times of industrialization; they have been reinforced with globalization processes and are reflected in current institutions often echoing past political situations. And although economic activities many decades ago in industrializing countries might have caused environmental problems, severe effects resulting from these activities are felt today in developing countries, or will be, threatening future generations all over the world. Therefore, it is very difficult to establish cause and effect, and thus identify who is responsible for creating environmental injustice.

The environmental justice framework essentially seeks to identify the relationship between environmental degradation and the disproportionate impact on various groups, especially minorities, people of color, women, indigenous groups, etc. These groups have historically been marginalized, and environmental problems have a disproportionate impact on them. There are various definitions of environmental justice. Dinah Shelton equates it to "Aesop's elephant," where in the fable of Aesop, several blind men who touch an elephant describe it in different ways depending on where they touch (Shelton, 2009). This aptly describes the various interpretations of environmental justice adopted by scholars, activists, and victims:

> Over the past decade during which communities, academics, regulated firms, and government officials have struggled with issues of the relationship of environmental quality to race and class, the quest to explain the essence of the problems underlying environmental justice disputes has been manifested in the varying terminology and definitions used to refer to such disputes.
>
> (Kuehn, 2000: 10681)

Rather than navigate through the many definitions of environmental justice, we will use the four-part characterization of environmental justice proposed by Kuehn: (a) distributive justice; (b) procedural justice; (c) corrective justice; and (d) social justice (Kuehn, 2009). Distributive justice requires equal treatment and equal access to resources and lowering of environmental risk, while procedural justice requires the participation of all stakeholders in decisions that affect them. Corrective justice requires punishing wrongdoers and remedying harm inflicted on individuals and communities. Social justice is the least developed and most tenuous aspect (Gonzalez, 2015), but it overlaps with the social pillar of sustainable development, discussed later in this chapter. It coincides with the goal of achieving a more just society, and some have described environmental justice as a "marriage of the movement for social justice with environmentalism," integrating environmental concerns into a broader agenda that emphasizes social, racial, and economic justice (Kuehn, 2000). This pillar shows that environmental justice cannot be separated from struggles for other forms of justice that often underlie the reasons for environmental problems. As Sheila Foster points out, focusing on distributive justice alone neglects the social structures and agents that are causing environmental problems (Foster, 1998; Atapattu, 2016).

Although there have been many justice struggles over the years all over the world, the modern environmental justice movement originated in the United States as a direct response to the practice of locating polluting industries and hazardous activities in areas of low-income and minority communities (Kuehn, 2000; Foster, 1998). These polluting activities had a huge impact on the lives of people and impinged on their rights. This situation ultimately resulted in an Executive Order on Environmental Justice issued by President Clinton in 1994. The order called on each federal agency to make "achieving environmental justice part of its mission by identifying and addressing, as appropriate, disproportionately high and adverse human health or environmental effects of its programs, policies and activities on minority populations and low-income populations in the US" (Executive Order No. 12898, 1994). It established an Interagency Working Group on Environmental Justice to, *inter alia*, provide guidance to federal agencies on criteria for identifying disproportionately high adverse effects on minority and low-income populations, examine data and studies on environmental justice, develop interagency model projects, and hold public hearings (Executive Order No. 12898, 1994). This order is in existence even today.

Environmental injustice and racism were the basis of *Mossville Environmental Action Now v. United States*, currently pending before the Inter-American Commission of Human Rights. The petitioners alleged that the Mossville residents who are predominately African-American, are subject to a disproportionate pollution burden, what they refer to as environmental racism, in breach of the American Declaration of the Rights and Duties of Man (1948). They further alleged that the State is responsible for the violation of Mossville residents' rights to life, health, and private life guaranteed under the American Declaration. This is the first case that has been brought against the United States on the basis of environmental justice, and in particular, environmental racism, before an international forum.

Structuring environmental justice along the justice relationships of responsible and affected actors, we can differentiate between international, intra-societal, and inter-generational injustice.

International injustice

International injustice occurs when actions of one state have a transboundary effect and harm people in another country. In the area of climate change, the main concern is that developed countries have extensively utilized carbon-intensive industries to foster growth, and developing countries (as well as emerging economies) shall not be able to do the same in the future because we have arrived at a time when we need to carefully watch our greenhouse gas emissions. Moreover, many developing countries are confronted with the consequences of climate change most severely by facing changes in precipitation, extreme weather events, increasing floods, and intensified droughts (OHCHR, 2009). Hence, there is an imbalance between the contribution to climate change and responsibility (of developed states) and emerging harm resulting from that and lack of resources to adapt (of developing states). Henry Shue calls these multiple layers of injustice that are often historically grown "compound injustice" (2014: 36). Normative claims to diminish international climate injustice are that greenhouse gas emissions have to be tremendously reduced, adaptation and mitigation costs have to be more equally distributed and should relate to historic emission responsibilities (e.g., the "polluter pays" principle), energy and other consumption patterns need to alter, and fair institutions need to be created.

We can see these normative claims reflected in climate agreements. The 1992 United Nations Framework Convention on Climate Change (UNFCCC) acknowledges in its preamble that

> the largest share of historical and current global emissions of greenhouse gases has originated in developed countries, that per capita emissions in developing countries are still relatively low and that the share of global emissions originating in developing countries will grow to meet their social and development needs.

Against the background of this historical responsibility of developed countries, the UNFCCC stipulates that the climate system should be protected "on the basis of equity" and according to countries' "common but differentiated responsibilities and respective capabilities" (UNFCCC, 1992: Article 3). This basically means that while all countries have the responsibility to preserve a functioning climate system, the major share of mitigation and adaptation costs have to be borne by developed countries that have used carbon-intensive industries to become wealthy. The attempt to reduce greenhouse gas emissions through international agreements and to distribute costs according to capabilities as stipulated in the UNFCCC can be understood as concrete steps to diminish international climate injustice.

Climate change is probably the most prominent example of a global injustice. However, there are other environmental challenges leading to international environmental injustice. Cross-border pollution is becoming an increasingly acute problem worldwide. As wind patterns and water flows do not respect national borders, pollution in one country can lead to disastrous environmental effects in another one. In Asia, cross-border pollution has become a major concern as China, India, Indonesia, and other countries have become the world's hub for factories. Downwind air pollution and decreasing air quality are serious problems and lead to tensions between countries in East Asia and Southeast Asia. China's economic growth and energy production relies heavily on fossil fuels, such as coal, and together with emissions from the manufacturing sectors, has led to severe environmental damage in Japan and South Korea caused by toxic chemical smog. Haze pollution originating from forest fires in Indonesia is another example, sometimes having adverse effects not just in neighboring states, but also as far away as Australia. Many of the emerging economies argue that they have the right to foster economic development and growth just like developed countries have in the past. These examples demonstrate how important and challenging international co-operation is, particularly with regard to achieving environmental justice.

Intra-societal injustice

Many scholars maintain that an international justice approach is not enough to comprehensively understand how environmental challenges affect people differently. They argue that we need to take a closer look inside the state, within society, to realize that justice concerns often relate to more or less affluent groups. Such intra-societal injustice concerns refer to the relationship between different groups within societies, or the relationship between the state and society respectively. Particular social groups and local communities are unequally exposed to the impacts of environmental problems to which they have contributed very little. Moreover, those who have continuously been neglected and excluded by their governments often suffer the most from intensified inequalities (Humphreys, 2014: 138) that are deepened in the face of certain environmental challenges and a changing climate (Newell, 2012). Thus, environmental or climate inequality "reinforces and, at the same time reflects, other forms of hierarchy and exploitation along lines of class, race and gender" and may lead to situations of "double-discrimination" (Newell, 2005: 70 and 87).

This means some population groups can be even more affected by situations of environmental injustice if they belong to a certain class, race, or gender. In this context, the term "environmental racism" has been coined. It means that there are situations in which racial minority groups within society are even more exposed to toxic waste, flooding, pollution, or natural resource extraction. Robert Bullard defines environmental racism in the following way: "Environmental racism refers to any policy, practice or directive that differentially affects or disadvantages (whether intended or unintended) individuals, groups or communities based on race or color" (Bullard, 2001: 160).

Several studies have clearly demonstrated a link between hazardous waste disposal and race, as well as the socio-economic and educational level of communities. In the United States and other countries, hazardous waste materials are more likely to be disposed in communities in which less resistance and political protest is expected. The *Cerrell Report* (Cerrell Associates, 1984) is often cited in this regard. It is a document produced by the consulting firm Cerrell Associates several decades ago in 1984 in order to advise the California Waste Management Board. The report exemplifies how public decisions on waste management exacerbate societal injustice. One important statement in the report is that the most significant obstacle to waste management is public opposition. Since middle- and high-income communities are more likely to organize protests, Cerrell Associates advised the California Waste Management Board to situate trash-to-energy-incinerators in poorer communities:

> Members of middle or higher-socioeconomic strata (a composite index of level of education, occupational prestige, and income) are more likely to organize into effective groups to express their political interests and views. All socioeconomic groupings tend to resent the nearby siting of major facilities, but the middle and upper-socioeconomic strata possess better resources to effectuate their opposition. Middle and higher-socioeconomic strata neighborhoods should not fall at least within the one-mile and five-mile radii of the proposed site.
>
> (Cerrell Associates, 1984: 26)

In this case, according to the Energy Justice Network, 3 waste incinerators (of the 43 incinerators planned) were built and all of them were situated in low-income communities of color (Energy Justice Network, 2007: 1). Decisions like these exacerbate already existing situations of social injustice, as it means that certain societal groups are even more exposed to pollution, hazardous waste, or toxic fumes. Thus, these communities can potentially become even more disadvantaged due to increased health risks or decreasing land and house prices. Discriminatory siting, however, is not the only factor contributing to environmental injustice. Inequality in environmental law enforcement and compliance through governmental agencies can also contribute to intensified social injustice between different communities.

In 1991, the First National People of Color Environmental Leadership Summit was held in Washington, D.C. This marked an important point in the history of the environmental justice movement. First, it led to a broadening of its agenda to include issues like resource allocation, public health, and community empowerment. Second, it strengthened the transnational movement as delegates from grassroots organizations and national institutions around the world participated in the event. And third, the delegates at the Summit adopted 17 Principles of Environmental Justice that were circulated at the Earth Summit in Rio de Janeiro and that are still relevant for the work of governmental and non-governmental organizations (NGOs) today. In the Principles, the movement states that it considers all governmental acts of environmental injustice a violation of international law, in particular of the Universal Declaration of Human Rights (Principle 10) and that it opposes all destructive operations of multinational companies (Principle 14). Key to the Principles of Environmental Justice is also the right to participate of those affected by environmental decisions (Principle 7) (Environmental Justice Network, 1991).

Climate change and climate policies also affect various groups within society differently. A report by the UN Human Rights Council identified women, children, and indigenous peoples – and also the elderly and persons with disabilities – in developing countries to be particularly vulnerable to such challenges (OHCHR, 2009: 15–18). As these challenges are becoming more and more urgent to address, part of the environmental justice movement started to particularly focus on climate change. A group of environmental and indigenous rights organizations and networks came together in 2002 and adopted the Bali Principles of Climate Justice on the basis of the 17 Principles of Environmental Justice. The Bali Principles link questions of climate-related challenges with local communities' ability for sustaining their life (Schlosberg, 2009: 6). Concrete demands in this respect do not only refer to equity, but also to political participation in decision-making concerning political responses to climate change on the basis of comprehensive information, access to judicial remedies, and compensation (Schlosberg, 2009: 16). This means, similarly to the Principles of Environmental Justice, that in the Bali Principles of Climate Justice participation of affected communities is also seen as significant for diminishing societal injustice.

Increasingly, such claims for procedural justice and rights are also brought forward in the context of climate policy implementation. In 2010, for instance, under pressure from certain civil society networks of the climate justice movement, procedural rights were institutionalized as a requirement for the implementation of forest management programs for Reducing Emissions from Deforestation and Forest Degradation (REDD+) at the Conference of the Parties (COP) of the United Nations Framework Convention on Climate Change (UNFCCC) in Cancun (Cancun Agreements, 2010: Appendix I, paragraph 2). Currently, an introduction of procedural rights is also discussed for the Clean Development Mechanism and the Sustainable Development Mechanism at the international climate conferences. Both are mechanisms to mitigate greenhouse gas emissions. Concrete policy projects under the framework of the CDM, for example, have led to rights infringements among local community groups in the past. Therefore, the idea is to strengthen their voices in the creation of these policies to prevent further social injustice.

Inter-generational injustice

The relationship between past, present, and future generations is another dimension of environmental justice that is increasingly considered, especially in the context of climate change. Our own past and contemporary lifestyles, characterized by the consumption of fossil fuels and high greenhouse gas emissions, have led and will further lead to injustice toward future generations who might possibly not be able to enjoy a clean and healthy environment anymore (Hiskes, 2005; Caney, 2006; Page, 2006; Hiskes, 2009; Shue, 2011). These generations are seen as the future citizens of contemporary state units (Hiskes, 2009). If you think about justice between generations, you need to start making different decisions. This means the current generation of decision-makers is responsible for not imposing risks and dangers on future generations who cannot be held accountable for greenhouse gas emissions from the past. A historic responsibility relating to previous emissions, e.g., during the time of industrialization, cannot be assumed anymore (Page, 2008), but we need to be aware that our CO2-intensive lifestyle is currently leading to droughts, floods, extreme weather events, and changes in precipitation. Most severely affected are poor populations in developing countries. How will their future look? Will their children and grandchildren still be able to enjoy the right to water, the right to food, and the right to health?

Demands to diminish inter-generational injustice – mostly brought forward by political theorists and philosophers – comprise the establishment of environmental rights. Whereas some scholars refer to a human right to a healthy environment (Hiskes, 2009), others put an emphasis on climate rights (Brandstedt and Bergman, 2013), energy rights (Caney, 2011), emission rights (Moellendorf, 2011; Caney, 2011), the right to ecological space (Hayward, 2007), or rights-protecting institutions (Shue, 2014). All of these approaches differ in one way or the other. Hayward (2007), for instance, argues that there cannot be any emission rights. There is no human right to emit, but rather a human right to be protected from emissions and pollution. The concept of ecological space, therefore, entails that the use of natural resources is limited, in particular with a view to protecting basic human rights, whereas emission rights conflict with rights protection of future generations. This means that poor countries should be supported in a low-carbon development path through international co-operation and technology transfer by developed countries (Hayward, 2007: 447–9).

The overall idea behind all of these considerations, mainly originating in political theory and philosophy, is to introduce constitutional environmental human rights that oblige today's representatives to adopt policies that take the interests of future generations into account. Although these considerations are about fostering ways of indirect and inter-generational democratic representation (Beckman and Page, 2008), several concerns remain. Who exactly can speak for future generations, how can they be represented, and in which way can we assess their interests? (Dudai, 2009: 302). Protecting the rights of future generations in a changing environment means reshaping current political systems and institutions (Beckman and Page, 2008: 529). This is quite an ambitious endeavor, and often too far away from current political realities and agendas.

All of the above-mentioned dimensions of environmental injustice – international, intra-societal, and inter-generational injustice – can, of course, also overlap. This would mean that future generations of certain societal groups in developing countries would be predominantly exposed to these forms of injustice. This is particularly relevant with respect to certain examples we have discussed in this chapter, including pollution, waste disposal, and the consequences of climate change, like floods, droughts, or extreme weather events.

Inter-species injustice

Whereas international, intra-societal, and intergenerational justice concerns focus on relationships between human beings in states, societies, or as part of generations, considerations on inter-species justice go beyond that. They take into account how we can strengthen environmental justice in relation to the non-human world, including animals, species, and eco-systems. Scholars argue that considering the scarcity of natural resources and the urgent need for sustainable solutions to environmental challenges and climate change, we can only understand fragile human inter-relations as part of the ecosystems that maintain them biophysically as well as socially. In her book, *Frontiers of Justice*, Martha Nussbaum also highlights the inter-relatedness of the human and the non-human world. She emphasizes how human decisions often have an adverse effect on non-human species leading to situations of inter-species injustice:

> Our choices affect the lives of non-human species every day, and often cause them enormous suffering. Animals are not simply part of the furniture of the

> world; they are active beings trying to live their lives; and we often stand in their way. That looks like a problem of justice.
>
> (2006: 22)

In *Frontiers of Justice*, Nussbaum employs a capability approach to the concept of animal dignity. Dignified existence, according to Nussbaum, includes freedom from pain and cruelty, adequate nutrition, freedom from fear, or freedom to behave and act in a way that is characteristic for this species. This means she links dignity with "a rich plurality of life activities" (2006: 346) and argues that animals' right to be treated with dignity is the foundation of interspecies justice.

We find environmental justice considerations relating to the non-human world not only in political theory and philosophy, but also in practice. In several countries, such as New Zealand, India, Colombia, and Bolivia, governments and constitutional courts have granted rights to ecosystems and natural resources, such as rivers. In Bolivia, the Law of the Mother Earth was adopted by the Plurinational Legislative Assembly in 2010. The law establishes legal personhood to Mother Earth and defines her as a subject of collective public interest:

> For the purpose of protecting and enforcing its rights, Mother Earth takes on the character of collective public interest. Mother Earth and all its components, including human communities, are entitled to all the inherent rights recognized in this Law.
>
> (La Asamblea Legislativa Plurinacional, 2010)

The legal rights to nature that were established include the right to life, to regeneration, to clean air and water, and to biodiversity, balance, and restoration.

Viewing Mother Earth as inter-related with all of her components as part of ecosystems that need to be preserved highlights the departure **from an anthropocentric to an eco-centric approach**. Traditionally, human rights approaches focus on the dignity and protection of human beings and this means they are **anthropocentric**, i.e., human-centered. In contrast to that, **eco-centric** approaches focus on the entire ecosystem and consider humans as only one component of it, next to millions of other species. Followers of an eco-centric approach often criticize capitalism and suggest that we need to radically rethink our political economy. They argue that after we have extensively exploited our natural resources, we now need to fundamentally change our production and consumption patterns in order to become more sustainable. Economic growth might not be possible anymore, and we should rather think about concepts of zero-growth or de-growth (Meadows, Randers, and Meadows, 2004). However, this argument does not take the North–South disparity into consideration and the need for Southern countries to achieve a decent standard of living for their people and for the Global North to cut down their consumption.

A hybrid approach seeks to combine anthropocentric and eco-centric thinking and would make more modest reform suggestions to protect the environment, such as recycling, reducing greenhouse gas emissions, and conserving wildlife and biodiversity. It would not necessarily grant legal personhood to ecosystems but rather encourage human beings to represent non-human components in decision-making processes.

Sustainable development

The third framework we will use in this textbook is sustainable development. The modern understanding of the concept of sustainable development is based on the report of the World Commission on Environment and Development (WCED, 1987). The report entitled *Our Common Future* is also known as the Brundtland Report, named after the Commission's Chairperson, Gro Harlem Brundtland, a former Prime Minister of Norway. The history behind the appointment of the WCED, which has origins in the Stockholm Conference on the Human Environment held in 1972 (the first ever global conference on the environment), bears testimony to the historic divisions in the global community, mostly along North–South lines.

Despite a flurry of activity following the Stockholm Conference, including the establishment of the United Nations Environment Programme (UNEP), developing countries felt that environmental protection was another form of colonialism and oppression by developed countries to stall their quest for economic development. Developed countries, on the other hand, were feeling the negative impacts of economic development in the form of environmental degradation and needed the co-operation of developing countries to tackle the emerging transboundary environmental problems. Developing countries, however, were not convinced.

The increasing North–South divide in relation to environmental protection led to a polarization in the UN General Assembly, almost paralyzing any action on environmental protection. No consensus could be reached and ultimately the UN General Assembly appointed the World Commission on Environment and Development (WCED) to find ways to reconcile environmental protection with economic development. The mandate given to the Commission was threefold: (a) to re-examine the critical environment and development issues and to formulate proposals to address them; (b) to propose new forms of international co-operation on these issues that will influence policies and events; and (c) to raise the levels of understanding and commitment to action of individuals, voluntary organizations, businesses, institutes, and governments.

The central message of the WCED report revolves around the concept of sustainable development, defined as "Development that meets the needs of the present generation without compromising the ability of the future generations to meet their needs" (WCED, 1987). By explicitly making reference to future generations as well as the needs of the present generation, we can see a strong link to considerations of both intra- and inter-generational justice, as discussed in the previous section.

The Brundtland formulation of sustainable development refers to balancing two pillars – economic development and environmental protection – even though the report discussed many social aspects, such as the need to address poverty and improving the living standards of people. A third pillar – social development – was added by the international community at the Copenhagen Summit for Social Development in 1995. Some add "culture" as a fourth pillar, even though it is part of the social pillar. Thus, sustainable development now comprises three pillars: environmental protection, economic development, and social development. To meet the needs of the current generation and those of future generations, we need to balance economic, social, and environmental objectives (Khor, 2012: 70). However, this depiction of sustainable development – as three pillars or the three legs of a stool – has been critiqued as once again perpetuating the myth that the environment is just another pillar and not the floor on which both the society and economy rest (Dawe and Ryan, 2003). While appreciating this critique and agreeing with it, for the purpose of this book, we will adopt the definition of sustainable development as comprising three dimensions as identified here.

The history of sustainable development dates back to the Stockholm Conference in 1972. One key outcome was the recognition that poverty needs to be addressed in order to protect the environment. Twenty years later, at the 1992 United Nations Conference on Environment and Development, also known as the Earth Summit, world leaders adopted the Rio Declaration on Environment and Development. Principle 3 of the Declaration refers to the right to development that needs to be implemented with a view to equally meeting developmental and environmental needs of present and future generations (Rio Declaration: Principle 3). Another product of the Earth Summit is Agenda 21, a UN action plan for sustainable development in the 21st century. Among others, it makes suggestions on how to conserve and manage resources but also how to strengthen the role of certain groups in development processes, including children and women, NGOs, indigenous peoples, and local communities. The 1992 Earth Summit is of particular importance, because not only did it result in the formulation of Agenda 21 and the Rio Declaration, but it also led to the adoption of a number of highly relevant conventions, including the United Nations Framework Convention on Climate Change (UNFCCC), the United Nations Convention on Biological Diversity, and the Statement of Forest Principles.

After another decade, the World Summit on Sustainable Development (also called Rio+10) was held in Johannesburg in 2002 and affirmed commitment to implementing Agenda 21 and the Millennium Development Goals (Johannesburg Declaration on Sustainable Development, 2002). The Copenhagen Declaration on Social Development had a considerable influence on the Johannesburg Declaration adopted at Rio+10 which added the third pillar of social development to sustainable development:

> We recognize that poverty eradication, changing unsustainable and promoting sustainable patterns of consumption and production and protecting and

> managing the natural resource base of economic and social development are the overarching objectives of and essential requirements for sustainable development.

A second follow up conference, Rio+20, was held in Brazil in 2012. One major theme was how to align environmental and economic goals with a view to achieving sustainable development. This placed the green economy at the center of attention.

During and after the global financial crisis, concepts of green economy or green growth have gained momentum when several United Nations (UN) agencies emphasized the idea of fostering initiatives that promise economic growth, are ecosystem-friendly, and contribute to poverty alleviation at the same time (UNDESA, UNEP, and UNCTAD, 2012: 4). The United Nations Environment Programme (UNEP) claims that the concept of green economy can be understood to be consistent with the more familiar understanding of sustainable development (UNEP, 2011). This means the principle of sustainability comprising protection of the environment, economic, and social development as three basic pillars are also recognized in ideas of a green economy (Khor, 2012: 70). The economic dimension is often put at the forefront (UNEP, 2011: 2). With a view to achieving economic change and development, ecological, environmental, and social harm in a green economy shall be limited (UNEP, 2011: 1).

Sustainable development is now considered an umbrella concept embodying both substantive and procedural components. Substantive components are: the principle of integration, equity (embracing both inter- and intra-generational equity), sustainable use of natural resources, and the right to development. The procedural components are: access to information, participation in the decision-making process, preparation of environmental impact assessment reports, and access to remedies. Many of these components have become binding on states, especially the principle of integration which is usually given effect through the environmental impact assessment process, the principle of equity (embodying both inter- and intra-generational equity aspects), and the procedural components. These are part of international law with binding obligations for states. The Framework Principles on Human Rights and Environment (discussed in Chapter 2) proposed by the former Special Rapporteur on Human Rights and Environment, John Knox, also embody these principles.

From Millennium Development Goals to Sustainable Development Goals

In 2000, the Millennium Summit took place, bringing together leaders from all over the world to adopt the UN Millennium Declaration. In a new global partnership to reduce extreme poverty, a number of measurable targets to be achieved by 2015 were set. The eight Millennium Development Goals included, for instance, the objective to eradicate extreme hunger and poverty (Goal 1). Therefore, the proportion of people whose income is less than 1 USD per day should be halved between 1990 and 2015. Indicators by the World Bank were used to measure this target. The other MDGs comprise achieving universal primary education (Goal 2), promoting gender equality (Goal 3), reducing child mortality (Goal 4), improving maternal health (Goal 5), combatting HIV/AIDS, malaria, and other diseases (Goal 6), ensuring environmental sustainability (Goal 7) and developing a global partnership for development (Goal 8). These goals are very closely related to basic human rights, such as the rights to food, health, shelter, and education.

In 2015, state leaders came together to adopt the post-2015 development agenda of the United Nations. Building on the Millennium Development Goals and with a view to complementing what they have not achieved, UN member states agreed on the Sustainable Development Goals (SDGs) together with 169 targets specifying how to achieve them by 2030. In the UN General Assembly resolution adopting the post-2015 development agenda, there is explicit reference to the realization of human rights and the economic, social, and environmental dimensions as the three pillars of sustainable development. In contrast to the MDGs, the SDGs are further differentiated (17 goals instead of 8 goals) and make more explicit reference to climate change and its consequences by highlighting the need to protect ecosystems, halt biodiversity loss, and address sustainable conservation and management of forests and oceans. Sustainability, inclusiveness, equality, comprehensiveness, and universality belong to the key principles of the 2030 Agenda for Sustainable Development.

Sustainable Development Goals

The SDGs and human rights are closely inter-related (see Table 1.1). Goal 13, climate action, for instance, is consistent and mutually reinforcing with the human right to health, the right to adequate food, and the right to safe drinking water, but also with the right of all peoples to freely dispose of their natural wealth and resources (anchored, among others, in the International Covenant on Economic, Social and Cultural Rights, ICESCR, 1966). Goal 4, quality education, strongly relates to the right to education, the right to work, including technical and vocational training and international co-operation (stipulated in the ICESCR), as well as equal rights of women and girls in the field of education (as emphasized in the Convention on the Elimination of all Forms of Discrimination against Women, CEDAW).

Sustainable development versus sustainability

Given that the word "development" is associated with developing countries that need more economic development, sustainable development tends to be also associated with developing countries, because industrialized countries have already achieved economic development. However, this argument does

Table 1.1 Sustainable Development Goals, 2015

Goal 1	No Poverty
Goal 2	Zero Hunger
Goal 3	Good Health and Well-Being
Goal 4	Quality Education
Goal 5	Gender Equality
Goal 6	Clean Water and Sanitation
Goal 7	Affordable and Clean Energy
Goal 8	Decent Work and Economic Growth
Goal 9	Industry, Innovation and Infrastructure
Goal 10	Reduced Inequalities
Goal 11	Sustainable Cities and Communities
Goal 12	Responsible Production and Consumption
Goal 13	Climate Action
Goal 14	Life Below Water
Goal 15	Life on Land
Goal 16	Peace, Justice and Strong Institutions
Goal 17	Partnerships for the Goals

Source: United Nations, 2015.

not take into account the consumerist lifestyle of the Global North that has a greater impact on the environment and natural resources than poverty and underdevelopment. In fact, a person in a developed country consumes 80% more energy than a person in a developing country. Thus, sustainable development is as much about economic development in the Global South as about decadent consumption in the Global North. Given this scenario, it could be argued that "sustainability" is a better term to use, as it covers both sets of actors and both sets of activities.

Others believe that sustainability is the destination while sustainable development is the journey to get there: "Sustainable development has thus been seen as the transition strategy to reach sustainability" (Washington, 2015). Given that neither term has an exact definition, it can mean anything to anyone. Indeed, as discussed, one of the biggest critiques of sustainable development is that it is broad and vague. Some use the terms "sustainable development" and "sustainability" interchangeably. Others equate sustainable development with sustainable growth. Some consider sustainable development to be a contestable concept, such as democracy or justice, that depend on other norms or principles to achieve them (Dobson, 1999). Washington believes that there is a deep confusion about sustainable development and sustainability in terms of their meaning and purpose:

> 'Sustainability' to my mind has to be a much *broader concept* than 'sustainable development.' It should focus on sustainability in the long-term for all aspects of the human and natural environment. It should certainly not be about growth in numbers, resource use or GDP. As Shiva (1992) predicted,

> 'sustainable development' has been co-opted by neoliberal ideology to continue 'business as usual' growth. However, growth is the cause of escalating degradation of the biosphere. If we are to demystify 'sustainability' then our focus must (first and foremost) center on being sustainable, not on endless growth on a finite planet – an impossibility.
>
> (Washington, 2015: 365, emphasis in original)

Certainly, the current level of growth and consumption is clearly not sustainable. Climate change is a good example of a global problem caused by unchecked development. Despite overwhelming scientific evidence, some countries have refused to take mitigation measures, opting instead to jeopardize the very survival of future generations and the health of the planet itself. Indeed, we human beings now have the dubious distinction of having a geologic era named after us – the Anthropocene – because of the extent to which humans have altered the earth system. Of course, this generalization ignores the overwhelmingly disproportionate contribution of certain countries and certain groups of people, as well as the disproportionate burden shouldered by certain countries and people as a result.

Others such as Agyeman, Bullard, and Evans (2003) have coined the phrase "just sustainabilities." They believe that "unless society strives for a greater level of social and economic equity, both within and between nations, the long term objective of a more sustainable world is unlikely to be secured" (Agyeman, Bullard, and Evans, 2003: 2). In other words, they contend that "a sustainable society must also be a just society, locally, nationally and internationally, both within and between generations and species" (Agyeman, Bullard, and Evans, 2003: 3). They argue that sustainability is not simply an "environmental" concern, even though environmental sustainability is important: "A truly sustainable society is one where wider questions of social needs and welfare, and economic opportunity, are integrally connected to environmental concerns" (Agyeman, Bullard, and Evans, 2003: 2). This indeed is our central thesis as well. No sustainable development programs and policies will be successful if they do not integrate social and environmental justice concerns into that process. In the end, however, whatever the phrase we adopt, we need to accept that the current model of development is totally and utterly unsustainable:

> There is a key need to demystify 'sustainability', to understand that it cannot (on a finite planet) be based on endless growth. This is why sustainability *is different* to sustainable development, as 'development' has for so long been equated to growth in Western society. We need now to focus on a 'sustainability' that heals the damage caused by the endless growth myth. Time to stop beating around the bush. If we break the denial dam, then we can urgently commence the 'Great Work' of healing the planet.
>
> (Washington, 2015: 372, emphasis in original)

Critical voices

Several scholars, however, point to tensions in the pursuit of these combined goals, i.e., the economic, social, and environmental dimension of sustainable development (Kopnina and Shoreman-Ouimet, 2015). The main critical aspect brought forward is that sustainable development is so broadly defined that international organizations and national governments can easily claim that their policies promote sustainability. Furthermore, there is a contradiction in these three objectives, as further economic growth will always take place at the expense of environmental protection (Washington, 2015: 363).

Those arguing from the perspective of an ecologist as opposed to an environmentalist ideology especially question the compatibility of capitalism and ecological sustainability (Brand, 2012). Current unsustainable production and consumption patterns are hindering factors in a transition to green growth (Brand, 2012: 30) and also hamper comprehensive realization of sustainable development. There is a risk of imbalance in realizing the three pillars of sustainability that may come at the expense of the ecological dimension or social development.

Critical scholars emphasize how major international organizations operating within the dominant framework of neoliberalism always depart from the assumption that sustainable development and economic growth are compatible goals. Even more so, economic growth is usually seen as the key driver of sustainable development. By now, however, we should have realized that we are living on a planet with finite resources and that we have exploited our "cumulative carbon budget" (Shue, 2014). Kopnina (2012: 707) emphasizes that "The most fundamental paradox of sustainable development can be summed up in its oxymoronic goal of both promoting development through economic growth and re-distribution of wealth and keeping the health of the ecosystem – including humans – intact."

As noted above, some contend that these contradictory goals can only be overcome with zero growth or de-growth – especially with a view of guaranteeing basic human rights to future generations (Meadows, Randers, and Meadows, 2004). However, concepts such as de-growth are problematic in the context of historic and existing international injustices between developing and developed countries when developing countries need to ensure that the basic needs of their populations are met, especially access to electricity. To achieve this, co-operation between developing and developed countries is imperative.

If we are serious about achieving the true goals of sustainable development, not merely paying lip service, the Global North needs to de-carbonize its economy immediately and reduce its decadent consumption patterns so that those in the Global South can achieve a decent standard of living with access to basic amenities that those in the Global North take for granted. Similarly, the increasingly affluent class of people in the Global South should not emulate their counterparts in the Global North, as this planet cannot simply sustain the level of consumption that we currently see in the Global North.

Other frameworks

Of course, it would be a mistake to assume that these three frameworks are the only frameworks that are relevant for a discussion on the relationship between human rights and environmental protection or in relation to environmental rights. There are many other important frameworks that can be used. Three other frameworks in particular are important, and we will incorporate them into various chapters to the extent possible. These are: traditional knowledge of indigenous peoples, the law and society approach, and the North–South divide.

Traditional knowledge of indigenous peoples is an important framework. Even international law has acknowledged its importance by incorporating it into treaties. The Convention on Biological Diversity is an example. The modern movement on rights of nature is very much based on indigenous cultural practices where nature and its components are considered as sacred.

The Arctic Centre at University of Lapland provides a good definition of traditional knowledge:

> Traditional knowledge refers to the knowledge and practices of indigenous and local communities that have developed over centuries and are traditionally transferred from elders to young people in concrete working and life situations. Traditional knowledge is dynamic, and it can be transferred and expressed orally, through stories, legends, rituals, songs and laws. It can also be preserved in artifacts handed from father to son or mother to daughter.
>
> The indigenous people have knowledge and understanding of their environment and ecosystems and ways how to use and manage them. Often this knowledge is very particular and detailed. In recent years traditional knowledge has been increasingly considered alongside scientific knowledge within the context of research and conservation efforts related to Arctic peoples and nature.
>
> (Arctic Centre, 2018)

Traditional Ecological Knowledge (TEK) is being increasingly threatened by climate change where indigenous and local communities are no longer able to rely on their traditional knowledge due to changing weather patterns which have made weather erratic and unpredictable. Their traditional lands and traditional practices are also being affected. For example, their traditional food sources are no longer available, and they are now forced to rely on store-bought food, leading to health issues. These challenges were included in the petition filed before the Inter-American Commission of Human Rights by a group of Inuit from the US and Canada against the US with regard to climate change. This petition is discussed in more detail in Chapter 9.

Changing cultural norms within these communities where younger generations are no longer relying on traditional knowledge is another threat to its continuity. How to treat TEK alongside scientific knowledge is another challenge where, traditionally, indigenous knowledge has been dismissed as "unscientific." However, the scientific community is slowly waking up to the reality that TEK can be invaluable, especially with regard to adapting to climate change, as many of these communities have intimate knowledge of their land and are used to living in harsh conditions, especially in the Arctic. As adverse consequences associated with climate change get more and more acute, the world can learn a lot from indigenous practices relating to food security, traditional medicines and remedies, and agricultural practices.

The law and society approach is another framework that might be useful. The objective of law and society scholars is to "study living law – the study of how the legal system actually operates" (Friedman, 1986: 764). Friedman identifies two modern ideas inherent in the movement: legal systems are man-made objects, i.e., social creations; and second, the idea of cultural relativity, i.e., "law varies in time and space, according to the conditions of the culture in which it is embedded" (Friedman, 1986: 764). Pointing out that these two ideas contradict pre-modern theories of law according to which law is *not* man-made, he further posits: "Without a conception of law as essentially a human institution, a product of culture, no social science of law is thinkable" (Friedman, 1986: 764). This, of course, leads to the question as to what constitutes law and whether law is a universal aspect of social life. Does every society have a legal system? What do we mean by "law" and "legal system"? Where does one draw the boundaries? "Law" is not limited to formal law. Many societies have informal systems or customary practices that exist alongside formal laws and institutions. Thus, legal pluralism is an area of research, and many scholars have devoted their time to study it. Another area of study is the impact of law and legal institutions themselves. Deterrence, compliance, and non-compliance are all part of issues that have an impact on how law operates in society. As Friedman concludes:

> Basically, the law and society movement has made, and can make, a major contribution, just by insisting on its way of looking at law, together with its box of tools to use in carrying out the work and its simple but powerful ideas for arranging and explaining the work. Its angle is a kind of deliberate detachment: separating *study* of law from the values that are part of its substance. … And at its best, the law and society movement holds up to the law a mirror and a yardstick; it stands for clarity, honesty, and rigor in the study of legal institutions. How many of its rivals can seriously make that claim?
> (1986: 780)

Thus, the law and society approach seeks to understand how law actually operates in practice, and its impact on society. This approach is useful for the study of environmental rights; as we shall see in many chapters of this book, with regard to environmental rights, the law on the books is quite different from how judges have interpreted it.

Another framework to understand the geopolitics surrounding environmental lawmaking is the North–South framework. North–South divisions that have colonial and post-colonial roots underpin many of the global environmental negotiations. It is essential to understand these divisions underlying many of the tensions and compromises during treaty negotiations. Without addressing these divisions, it is impossible to design a proper legal regime, and many environmental negotiations have stalled because these divisions were not properly understood or addressed. Climate change is a good example of divisions along North–South, North–North, and South–South divisions. It has been pointed out:

> North–South conflicts originating in the economic realm have profoundly shaped the evolution of international environmental law and policy. The global North industrialized and developed by exploiting the planet's natural resources without regard for the environmental consequences. Northern consumption patterns which are increasingly emulated by Southern elites, have brought the planet's ecosystems to the brink of collapse and will constrain the development of options of present and future generations, particularly in the global South. Indeed, some observers have argued that the North owes an ecological debt to the South for 'resource plundering, unfair trade, environmental damage, and the free occupation of environmental space to deposit waste'.
>
> (Atapattu and Gonzalez, 2015: 9–10, footnotes omitted)

Structure of the volume

While it would be impossible to cover all aspects of the relationship between human rights and environmental protection and its specialized topics, we plan to cover the most important principles so that the student will be able to apply these principles to issues at hand. The student of environmental rights will also be able to appreciate the intrinsic link between human rights and the environment, as well as the limitations inherent in applying a human rights approach to environmental issues.

The book will be constructed as follows. **Part I (Introduction and evolution)** will consist of four chapters. The **Introduction** will frame the issues. We will use three frameworks in the book: human rights, environmental justice, and sustainable development (including sustainable development goals). We will also differentiate anthropocentric and eco-centric approaches; and address the changing nature of human rights – from individual civil and political rights, and economic, social and cultural rights, to group rights, inter-generational rights, and to emerging rights to nature.

Chapter 2 will trace the evolution of a human right to a healthy environment, including the institutional framework of human rights. It will cover developments since the adoption of the 1972 Stockholm Declaration on the Human Environment which constitutes the foundation of international environmental law, to the Paris Agreement on Climate Change adopted in December 2015 which specifically includes a reference to human rights in the Preamble, the first binding environmental instrument to do so. It will conclude with a survey of Framework Principles on Human Rights and Environment proposed by the UN Special Rapporteur, John Knox, in his final report. **Chapter 3** will discuss the pros and cons of a human rights approach to environmental protection, while **Chapter 4** will highlight developments at the regional level. The relevant case law of the three regional human rights systems will be discussed here.

Part II of the textbook is devoted to a discussion of human rights of special relevance to environmental protection. In keeping with the mapping report of the former UN Special Rapporteur on Human Rights and the Environment, John Knox, we will have a chapter each discussing the **substantive rights** such as right to life, health, food, water, housing, privacy, and an adequate standard of living (**Chapter 5**), and **procedural rights** such as right to information, right to participate in the decision-making process, right to remedies, and environmental/social impact assessments (**Chapter 6**). **Chapter 7** discusses **constitutional developments** relating to environmental rights while **Chapter 8** surveys **selected domestic cases**, as many of the developments with regard to the convergence between human rights and environment have taken place at the national level.

Part III of the textbook is devoted to climate change and human rights. While the link between climate change and human rights may be fairly obvious, climate change poses particular challenges to the human rights framework given its global scope, historic responsibility of industrialized countries, and the disproportionate impact on certain regions and certain groups. Climate change was chosen as one focus area precisely because of these challenges and to highlight the various limitations of the human rights framework vis-à-vis climate change. **Chapter 9** will trace the developments from the adoption of the UN Framework Convention on Climate Change (UNFCCC) in 1992 to the Paris Agreement in 2015. In particular, it will discuss the pros and cons of including a human rights provision in the Paris Agreement, the reference to climate justice, procedural rights and the rights of nature, and the inclusion of a loss and damage mechanism. **Chapter 10** will discuss social movements and the role of civil society, including transnational advocacy networks and the various strategies used inside and outside the UNFCCC negotiations, including litigation. This chapter will highlight the strategies that have worked as well as those that have failed, with a discussion of why they either worked or failed.

Chapter 11 discusses vulnerability and climate change. We will elaborate on climate-induced migration and conflicts over natural resources, the situation of small island states, women, children, disabled people, and indigenous peoples and explain how the human rights framework protects them, and whether a justice framework is better suited.

The final chapter in this section addresses mitigation and adaptation through a human rights lens (**Chapter 12**). Mitigation is discussed through a case study of a hydropower dam in Ethiopia and its impact on indigenous peoples and agro-pastoralist groups. There are examples of mitigation and adaptation measures infringing upon rights of people, particularly vulnerable groups. This chapter discusses the importance of applying a human rights and justice lens to both mitigation and adaptation.

The final part (**Part IV**) of the textbook is devoted to **emerging issues** and issues that pose particular challenges to environmental rights. **Chapter 13** will examine the **extraterritorial application of human rights** which discusses the territoriality principle and the vertical nature of human rights and the fact that the human rights framework does not readily lend itself to transboundary damage caused by environmental issues or damage caused by global issues such as climate change. The chapter surveys the Maastricht Principles on Extraterritoriality, some of the recent decisions of various judicial bodies and what these developments mean for victims of transboundary environmental damage.

Chapter 14 deals with business and human rights and discusses the Guiding Principles on Business and Human Rights (called the "Ruggie Principles"), and the move toward holding corporate actors accountable to protect the rights of people. Many of the violations of rights have taken place at the hands of the corporate sector that is not directly responsible for protecting the rights of people under international human rights law. There is, however, an emerging school of thought, reinforced by the Ruggie Principles that these non-state actors should uphold the basic rights of people in their operations.

The final chapter in this part, **Chapter 15**, discusses rights of nature and inter-generational rights, both of which have entered mainstream discussions on rights and have the potential to revolutionize the traditional notion of "human" rights. While rights of nature are new, the inter-generational equity principle has been recognized by scholars for some time and is included as one of the guiding principles in the UNFCCC. Whether future generations have rights under human rights law has become a hotly debated issue in the context of climate change.

The concluding chapter (**Chapter 16**) will summarize the main trends and offer some ideas for future developments in this field, particularly in relation to climate change. It will briefly explain the relationship between human rights, environmental protection, and good governance. It will discuss whether a rights and justice framework would be useful, particularly in relation to vulnerable communities who are disproportionately affected by environmental issues.

The field of human rights and environment is a rapidly evolving area and we have tried to make the book as up to date as possible. Developments until October 2018 are included.

DISCUSSION QUESTIONS

1. Which of the three main frameworks introduced above do you find most useful for analyzing issues at the intersection of human rights and the environment – and why?
2. How do environmental challenges and climate change exacerbate situations of injustice?
3. In which way are the Sustainable Development Goals (SDGs) and human rights inter-related?
4. Which historical aspects in North–South relations have led to the environmental injustices that we can observe today?
5. Why is Traditional Ecological Knowledge (TEK) relevant in the context of climate change? Can you research an example demonstrating this relevance?

References

Agyeman, J. Bullard, R.D., and Evans, B. (2003). *Just Sustainabilities. Development in an Unequal World*. London: Earthscan.

Alston, P. (1984). Conjuring Up New Human Rights: A Proposal for Quality Control. *American Journal of International Law* 78 (3): 607–21.

American Declaration of the Rights and Duties of Man (1948). Adopted by the Ninth International Conference of American States, Bogota, Colombia, April 1948 [Online]. Available at: www.cidh.oas.org/Basicos/English/Basic2.American%20Declaration.htm [Accessed: 18 October 2018].

Anton, D. and Shelton, D. (2011). *Environmental Protection and Human Rights*. Cambridge: Cambridge University Press.

Arctic Centre (2018): Traditional Knowledge [Online]. Available at: www.arcticcentre.org/EN/communications/arcticregion/Arctic-Indigenous-Peoples/Traditional-knowledge [Accessed: 18 October 2018].

Atapattu, S. (2016). Justice for Small Island Nations: Intersections of Equity, Human Rights, and Environmental Justice. In: Abate, R. (ed.), *Climate Justice: Case Studies in Global and Regional Governance Challenges*. Washington DC: Environmental Law Institute, p. 299

Atapattu, S. (2019). From 'Our Common Future' to Sustainable Development Goals: Evolution of Sustainable Development under International Law. *Wisconsin International Law Journal* (forthcoming).

Atapattu, S. and Gonzalez, C.G. (2015). The North-South Divide in International Environmental Law: Framing the Issues. In: Alam, S., Atapattu, S., Gonzalez, C., and Razzaque, J. (eds.), *International Environmental Law and the Global South*. Cambridge: Cambridge University Press, p.1.

Beckman, L. and Page, E.A. (2008). Perspectives on Justice, Democracy and Global Climate Change. *Environmental Politics* 17 (4): 527–35.

Brand, U. (2012). Green Economy – The Next Oxymoron? No Lessons Learned from Failures of Implementing Sustainable Development. GAIA 21 (1): 28–32.

Brandstedt, E. and Bergman, A. (2013): Climate Rights – Feasible or Not? *Environmental Politics* 22 (3): 394–409.

Bullard, R. P. (2001). Environmental Justice in the 21st Century: Race Still Matters. *Phylon* 49 (3/4): 151–71.

Cahill-Jackson, J. (2012). Mossville Environmental Action Now v. United States: Is a Solution to Environmental Injustice Unfolding? *Pace International Law Review Online Companion* 3 (6): 173–208.

Cancun Agreements (2010). Framework Convention on Climate Change: Report of the Conference of the Parties on its Sixteenth Session, held in Cancún from 29 November to 10 December 2010, FCCC/CP/2010/7/Add.1.

Caney, S. (2006). Cosmopolitan Justice, Rights and Global Climate Change. *Canadian Journal of Law and Jurisprudence* 19 (2): 255–78.

Caney, S. (2011): Climate Change, Energy Rights and Equality. In: Arnold, D.G. (ed.), *The Ethics of Global Climate Change*. Cambridge: Cambridge University Press, pp.77–103.

Cerrell Associates (1984). *Political Difficulties Facing Waste-to-Energy Conversion Plant Siting*. Cerrell Associates, Los Angeles. Prepared for California Waste Management Board, State of California.

Committee on Economic, Social and Cultural Rights (2000). General Comment No. 14 : The right to the highest attainable standard of health, UN doc. E/C.12/2000/4.

Convention Against Torture and Other Cruel, Inhuman or Degrading Treatment or Punishment (1984). [Online]. Available at: www.ohchr.org/en/professionalinterest/pages/cat.aspx [Accessed: 1 September 2018].

Convention on All Forms of Discrimination Against Women (1979). [Online]. Available at: www.un.org/womenwatch/daw/cedaw/text/econvention.htm [Accessed: 20 October 2018].

Convention on the Rights of Persons with Disabilities (2006). [Online]. Available at: www.un.org/development/desa/disabilities/convention-on-the-rights-of-persons-with-disabilities/convention-on-the-rights-of-persons-with-disabilities-2.html [Accessed: 20 October 2018].

Convention on the Rights of the Child (1989). [Online]. Available at: www.ohchr.org/en/professionalinterest/pages/crc.aspx [Accessed: 1 September 2018].

Dawe, N. and Ryan, K. (2003). The Faulty Three-Legged-Stool Model of Sustainable Development. *Conservation Biology* 17 (5): 1458–60.

De Schutter, O. (2012). *International Human Right Law*. Cambridge: Cambridge University Press.

Dobson, A. (1999). *Fairness and Futurity*. Oxford: Oxford University Press.

Dudai, R. (2009). Climate Change and Human Rights Practice: Observations on and around the Report of the Office of the High Commissioner for Human Rights on the Relationship between Climate Change and Human Rights. *Journal of Human Rights Practice* 1 (2): 294–307.

Environmental Justice Network (1991). Principles of Environmental Justice. [Online]. Available at: www.ejnet.org/ej/principles.html [Accessed: 17 December 2018].

Energy Justice Network (2007). Targeting "Cerrell" Communities [Online]. Available at: www.ejnet.org/ej/cerrell.pdf [Accessed: 1 October 2018].

Executive Order No. 12898 (1994). 59 Fed. Reg. 7629, 16 February 1994, President Clinton, Federal Actions to Address Environmental Justice in Minority Populations and Low-Income Populations [Online]. Available at: www.archivesgov/federal-register/executive-orders/pdf/12898.pdf [Accessed: 24 October 2018].

Flowers, N. (1998). A Short History of Human Rights [Online]. Available at: http://hrlibrary.umn.edu/edumat/hreduseries/hereandnow/Part-1/short-history.htm [Accessed: 18 October 2018].

Foster, S. (1998). Justice From the Ground Up: Distributive Inequities, Grassroots Resistance and the Transformative Politics of the Environmental Justice Movement. *California Law Review* 86 (4): 775–841.

Friedman, L. (1986). The Law and Society Movement. *Stanford Law Review* 38 (3): 763–80.

Gonzalez, C. (2012). Environmental Justice and International Environmental Law. In: Alam, S., Bhuiyan, J.H., Chowdhury, T.M.R., and Techera, E. (eds.) *Routledge Handbook of International Environmental Law*. London: Routledge, pp.77–98.

Gonzalez, C. (2015). Environmental Justice, Human Rights, and the Global South. *Santa Clara Journal of International Law* 1: 151–95.

Gordon, R. (2015). Unsustainable Development. In: Alam, S., Atapattu, S., Gonzalez, C., and Razzaque, J. (eds.) *International Environmental Law and the Global South*. Cambridge: Cambridge University Press, pp.401–24.

Hayward, T. (2007). Human Rights Versus Emissions Rights: Climate Justice and the Equitable Distribution of Ecological Space. *Ethics and International Affairs* 21 (4): 421–50.

Hiskes, R.P. (2005). The Right to a Green Future: Human Rights, Environmentalism and Intergenerational Justice. *Human Rights Quarterly* 27 (4): 1346–64.

Hiskes, R.P. (2009). *The Human Right to a Green Future: Environmental Rights and Intergenerational Justice*. Cambridge: Cambridge University Press.

Humphreys, S. (2014). Climate Justice: The Claim of the Past. *Journal of Human Rights and the Environment* 5: 134–148.

International Covenant on Civil and Political Rights (1966). [Online] Available at: www.ohchr.org/EN/ProfessionalInterest/Pages/CCR.aspx [Accessed: 24 October 2018].

International Covenant on Economic, Social and Cultural Rights (1966). [Online]. Available at: www.ohchr.org/Documents/ProfessionalInterest/cescr.pdf [Accessed: 1 October 2018].

International Convention for the Protection of All Persons from Enforced Disappearance (2006). [Online]. Available at: www.ohchr.org/Documents/ProfessionalInterest/disappearance-convention.pdf [Accessed: 1 October 2018].

International Convention on the Elimination of All Forms of Racial Discrimination (1965). [Online]. Available at: www.ohchr.org/en/professionalinterest/pages/cerd.aspx [Accessed: 1 October 2018].

International Convention on the Protection of All Migrant Workers and Members of their Families (1990). [Online]. Available at: www.ohchr.org/en/professionalinterest/pages/cmw.aspx [Accessed: 1 October 2018].

Johannesburg Declaration on Sustainable Development (2002). A CONF.199/20 [Online]. Available at: www.un-document-net/aconf199-20.pdf [Accessed: 24 October 2018].

Khor, M. (2012). Challenges of the Green Economy Concept and Policies in the Context of Sustainable Development, Poverty and Equity. In: UNDESA, UNEP, and UNCTAD (eds.) *The Transition to a Green Economy: Benefits, Challenges and Risks from a Sustainable Development Perspective*. Report by a Panel of Experts to the Second Preparatory Committee Meeting for the United Nations Conference on Sustainable Development. New York, NY: United Nations Publishers, pp.69–97.

Kopnina, H. (2012). Education for Sustainable Development (ESD): The Turn Away from 'Environment' in Environmental Education? *Environmental Education Research* 18 (5): 699–717.

Kopnina, H. and Shoreman-Ouimet, E. (2015): Introduction: The Emergence and Development of Sustainability. In: Kopnina, H. and Shoreman-Ouimet, E. (eds.) *Sustainability: Key Issues*. London: Routledge, pp.3–24.

Kuehn, R. (2000). A Taxonomy of Environmental Justice. *Environmental Law Reporter* 30: 10681–703.

La Asamblea Legislativa Plurinacional (2010). *Law of the Mother Earth*. La Paz: Plurinational Legislative Assembly of Bolivia.

Meadows, D.H., Randers, J., and Meadows, D.L. (2004). *The Limits to Growth: The 30-Year Update*. White River Junction, VT: Chelsea Green Publishing Co.

Mickelson, K. (2015). The Stockholm Conference and the Creation of the South-North Divide in International Environmental Law and Policy. In: Alam, S., Atapattu, S., Gonzalez, C.G., and Razzaque, J. (eds). *International Environmental Law and the Global South*. Cambridge: Cambridge University Press, pp.109–29.

Moellendorf, D. (2011): Common Atmospheric Ownership and Equal Emissions Entitlements. In Arnold, D.G. (ed.) *The Ethics of Global Climate Change*. Cambridge: Cambridge University Press, pp.104–23.

Morrow, K. (2010). Climate Change and Human Rights: The Defining Dilemma of Our Times? *Journal of Human Rights and the Environment* 1 (2): 131–4.

Mossville Environmental Action Now v. United States (2010). Report No. 43/10 (17 March 2010): Petition 242-05, Admissibility [Online]. Available at: http://srenvironment.org/regional-decision/inter-american-commission/ [Accessed: 24 October 2018].

Nussbaum, M. (2006). *Frontiers of Justice: Disability, Nationality, Species Membership*. Cambridge, MA: Harvard University Press.

Newell, P. (2005). Race, Class and the Global Politics of Environmental Inequality. *Global Environmental Politics* 5 (3): 70–94.

Newell, P. (2012): *Globalization and the Environment: Capitalism, Ecology and Power*. Cambridge: Polity Press.

OHCHR (2009). *Report of the Office of the United Nations High Commissioner for Human Rights on the Relationship between Climate Change and Human Rights*, A/HRC/10/61. Geneva: Human Rights Council.

Page, E. (2006). *Climate Change, Justice and Future Generations*. Cheltenham: Edward Elgar.

Page, E. (2008): Distributing the Burdens of Climate Change. *Environmental Politics* 17 (4): 556–75.

Schlosberg, D. (2009). Climate Justice Beyond Equity: The Flourishing of Human and Non-Human Communities. Paper presented at the American Political Science Association Annual Meeting, Toronto, Canada, September 2009.

Shelton, D. (2009): Describing the Elephant: International Justice and Environmental Law. In: Ebbson, J. and Okowa, P. (eds.) *Environmental Law and Justice in Context*. Cambridge: Cambridge University Press, pp.55–75.

Shue, H. (2011): Human Rights, Climate Change and the Trillionth Ton. In: D.G. Arnold (ed.) *The Ethics of Global Climate Change*. Cambridge: Cambridge University Press, pp.292–314.

Shue, H. (2014): *Climate Justice: Vulnerability and Protection*. Oxford: Oxford University Press.

The Social and Economic Rights Action Center and the Center for Economic and Social Rights v. Nigeria (2001). African Commission on Human and Peoples' Rights, Comm. No. 155/96

Statute of the International Court of Justice (1945) [Online]. Available at: www.icj-cij.org/en/statute [Accessed: 1 October 2018].

UNDESA, UNEP, and UNCTAD (2012): *The Transition to a Green Economy: Benefits, Challenges and Risks from a Sustainable Development Perspective*. Report by a Panel of Experts to the Second Preparatory Committee Meeting for the United Nations Conference on Sustainable Development. New York, NY: United Nations Publications.

UNEP (2011). *Towards a Green Economy: Pathways to Sustainable Development and Poverty Eradication*. Nairobi: United Nations Environment Programme.

UN General Assembly (2015). Transforming Our World: The 2030 Agenda for Sustainable Development, A/RES/70/1 [Online]. Available at: www.un.org/ga/search/viewdoc.asp?symbol=A/RES/70/1&Lang=E [Accessed: 24 October 2018].

UNFCCC (1992). United Nations Framework Convention on Climate Change. New York, NY: United Nations, General Assembly.

United Nations (1945). Charter of the United Nations, 1 UNTS XVI.

United Nations (1992). Convention on Biological Diversity, Rio de Janeiro, Earth Summit.

United Nations (1992). Non-legally Binding Authoritative Statement of Principles for a Global Consensus on the Management, Conservation and Sustainable Development of All Types of Forests, Rio de Janeiro, Earth Summit.

United Nations (1992). Rio Declaration on Environmental and Development. New York, United Nations, General Assembly.

United Nations (1995). Copenhagen Declaration on Social Development, Copenhagen, World Summit for Social Development.

United Nations (2002). Johannesburg Declaration on Sustainable Development, Johannesburg, World Summit on Sustainable Development.

United Nations (2015). Sustainable Development Goals [Online]. Available at: www.un.org/sustainabledevelopment/sustainable-development-goals/ [Accessed: 2 October 2018].

United Nations Declaration on the Rights of Indigenous Peoples (2007). A/RES/61/295 [Online]. Available at: www.un.org/en/genocideprevention/documents/atrocity-crimes/Doc.18_declaration%20rights%20indigenous%20peoples.pdf [Accessed: 1 October 2018].

United Nations General Assembly (1948). Universal Declaration of Human Rights, 217 A (III).

Vienna Declaration on Human Rights (1993) [Online]. Available at: www.ohchr.org/Documents/ProfessionalInterest/vienna.pdf [Accessed: 18 October 2018]

Washington, H. (2015). Is 'Sustainability' the Same as 'Sustainable Development'. In: Kopnina, H. and Shoreman-Ouimet, E. (eds.) *Sustainability: Key Issues*, London: Routledge, pp.359–376.

WCED (1987): *Our Common Future*. Report of the World Commission for Environment and Development. Oxford: Oxford University Press.

2 Emergence of a human right to a healthy environment

Introduction

In his separate opinion in the *Case Concerning the Gabcikovo Nagymaros Project* (Hungary v. Slovakia) (International Court of Justice, 1996) Judge Christopher Weeramantry noted:

> The protection of the environment is likewise a vital part of contemporary human rights doctrine, for it is a *sine qua non* for numerous human rights such as the right to health and the right to life itself. It is scarcely necessary to elaborate on this, as damage to the environment can impair and undermine all the human rights spoken of in the Universal Declaration and other human rights instruments.

This clearly elucidates the link between environmental damage and human rights. This has not always been the case, however. The link between human rights and environmental degradation was recognized fairly recently under international law, although contemporary human rights law does not, as yet, recognize a distinct right to a healthy environment.

The first seeds of the link between human rights and environmental degradation were sown by the Stockholm Declaration on the Human Environment itself. This document, adopted in 1972, is considered to be the foundation of international environmental law. Principle 1 of the Declaration stated:

> Man has the fundamental right to freedom, equality and adequate conditions of life, in an environment of a quality that permits a life of dignity and well-being, and he bears a solemn responsibility to protect and improve the environment for present and future generations.
>
> (Stockholm Declaration, 1972: Principle 1)

While falling short of endorsing a distinct human right to a healthy environment, Principle 1 clearly recognized that an environment of a certain quality is necessary for human beings to lead a life of dignity and well-being. In other words, a degraded environment could result in a violation of protected rights. This recognition is important, but even 30 years after the Stockholm Declaration was adopted, international human rights law has not progressed much beyond this recognition. The Paris Agreement on Climate Change was the first time that an environmental treaty incorporated the need to respect human rights in relation to an environmental issue (Knox, forthcoming). However, specific recognition of a right to a healthy environment in international human rights law is confined to three regional treaties: African Charter on Human and People's Rights; Additional Protocol to the American Convention of Human Rights; and UNECE Convention on Access to Information, Public Participation in Decision-making and Access to Justice in Environmental Matters.

Despite this lethargy in articulating a distinct right to a healthy environment, the international community has come a long way since the adoption of the Stockholm Declaration. Moreover, environmental rights have been articulated and refined by regional human rights bodies and judiciaries around the world. This chapter will discuss this evolution. It will first discuss the developments at the United Nations (UN): the appointment of a Special Rapporteur on human rights and environment in 1989 and the reports submitted by her; the appointment of a Special Rapporteur on the adverse effects of the movement and dumping of toxic and dangerous products and wastes on the enjoyment of human rights in 1995, and the appointment of an Independent Expert on Human Rights and the Environment in 2012, whose mandate was extended and converted to a Special Rapporteur position in 2015. This appointment culminated in the submission of the Framework Principles on Human Rights and Environment that were submitted to the UN Human Rights Council. A new Special Rapporteur was appointed in 2018. We will explore how environmental rights have been articulated at the international level and the jurisprudential bases for them. We will then briefly discuss developments at the regional level; a detailed discussion of the cases decided by the regional human rights bodies will appear in Part II of the book. We will also discuss some of the proposals that have been advanced to recognize a distinct right to a healthy environment.

Articulation of environmental rights and jurisprudential bases

Ironically, two developments *at the national level* contributed to the convergence of the two fields of human rights and environmental protection at the international level. The first development is the environmental impact assessment process at the national level, which is subject to public scrutiny and public participation in many countries. These documents are often considered public documents, and legislation in many countries provide for public participation through comments, public hearings, etc. This process has incorporated the procedural rights that are recognized under international human rights law – access to information, public participation, and access to remedies (Sands and Peel, 2012: 648). Access to environmental information is now firmly entrenched in international law and is closely connected to participatory rights in environmental impact assessment procedures and with the development of procedural rights in human rights law (Sands and Peel, 2012: 648). Referred to as "environmental democracy" or "access rights," these three rights have been incorporated in international environmental law. Principle 10 of the Rio Declaration on Environment and Development 1992 incorporated these three pillars for the first time in the context of sustainable development. They are considered to constitute the procedural components of sustainable development. They are now embodied in a regional environmental treaty adopted under the auspices of the UN Economic Commission for Europe (ECE). Latin American and Caribbean countries followed suit, adopting their own P-10 treaty which was opened for signature in September 2018.

The second development is the environmental justice movement which originated in the United States as a direct response to the practice of locating polluting and hazardous activities in areas of low-income and minority communities. These polluting activities had a huge impact on the lives of these people and impinged on their rights. The practice is by no means over, despite the Executive Order issued by President Clinton in 1994. There are many examples of low-income and minority communities in the United States that are being disproportionately affected by pollution, environmental degradation, including climate change, and severe weather events: Hurricane Katrina, the villages of Kivalina and Shishmaref in Alaska, and the water crisis in Flint, Michigan are a few examples. A petition alleging environmental racism is pending before the Inter-American Commission of Human Rights in the *Mossville* case.

Another area where the convergence between human rights and environment has taken place is in relation to indigenous peoples. Indigenous peoples have an intimate relationship with the land they inhabit and their culture is closely related to land. Environmental degradation has a severe impact, not only on their protected rights, but also on their culture, and sometimes their very survival. They have an intimate knowledge of their environment, and this knowledge has been passed from generation to generation. In the Inuit petition discussed in Chapters 4 and 9, the petitioners claimed, *inter alia*, that they could no longer rely on their traditional knowledge due to the unpredictability of a changing climate.

Over the years, the articulation of environmental rights has taken many forms and approaches. These fall into three broad categories: (a) "greening" human rights law by using existing rights – both civil and political rights, and economic, social, and cultural rights – to articulate environmental rights. Under this approach, substantive rights recognized under international human rights have been invoked in relation to environmental issues; (b) using procedural rights recognized under human rights law to lead to better outcomes relating to projects that can have a significant impact on the environment, or to provide relief to victims; (c) recognition of a distinct right to a healthy environment as a stand-alone right (Collins, 2015). This development has taken place only at the regional level.

The relationship between human rights and environment is two-fold: the enjoyment of human rights can be directly affected by environmental degradation. Moreover, human beings depend on nature for the services it provides – air purification, water, decomposition of waste, etc. – which are referred to as "ecosystem services," even though that term itself has been critiqued as being too anthropocentric. Similarly, there could be ecological limitations to human rights. This refers to "the idea that individual freedom is not only determined by a social context – but also by an ecological context as determined by principles such as ecological integrity or planetary boundaries" (Burdon, 2015).

Scholars have put forward several jurisprudential bases for environmental rights. The first basis is that they are embedded in legal systems through common law principles such as nuisance and, therefore, are part of contemporary international law through general principles of law; second, some contend that they are a logical outcome of other rights, such as the right to life and the right to property; third, it can be argued that these rights are becoming part of customary international law given that over 100 countries in the world have environmental rights embodied in their constitutions (Boyd, 2012). Finally, it can be argued that, given its importance, a right to a "healthy" (or "safe," "satisfactory," "good," or whatever adjective is suitable) environment should be incorporated through treaties (Burger, 2003). Many scholars seem to lean toward greening existing rights, rather than adopting a substantive right to environment. Some even go to the extent of arguing against its adoption, on the ground that it will dilute existing rights (Alston, 1984). While a stand-alone right to a healthy environmental has not yet been recognized under international human rights law, it is clear that the international community has come a long way since the Stockholm Conference and is possibly at the brink of recognizing such a right at the international level. Several proposals and initiatives certainly seem to be leaning towards that. Not only are we closer to recognizing such a right, we have advanced even further by recognizing even rights of nature. Before we turn to a discussion of the developments at the UN, we will provide a brief overview of human rights institutions at the international level.

An overview of human rights institutions

International human rights law has largely developed along two tracks: treaty-based mechanisms and charter-based mechanisms. Treaty-based mechanisms include several specialized regimes governing various vulnerable groups: women, children, migrant workers, and disabled people. Charter-based mechanisms include special procedures: special rapporteurs, independent experts and fact-finding missions, and Universal Periodic Review (UPR). In addition to international mechanisms, there are three regional mechanisms that are discussed in Chapter 4.

Treaty-based mechanisms

Eleven treaty-based mechanisms currently exist under international human rights law. These are the mechanisms under the following treaties: ICCPR, ICESCR, Convention on the Rights of the Child (CRC), Convention on the Elimination of All Forms Discrimination against Women (CEDAW), Convention on the Elimination of Racial Discrimination (CERD), Convention on Migrant Workers and Their Families (CMW), Convention against Torture (CAT), Convention on Disabled People (CRPD), and the Convention on Enforced Disappearances (CED).

Treaty-based bodies adopt several mechanisms to address the issues covered under their treaties: submission of country reports (the practice of submitting shadow reports by civil society groups has become quite entrenched now although there is no formal mechanism for their submission); concluding observations on country reports; general comments on various issues; and an individual complaints procedure if the treaty in question provides for that. There are three mechanisms by which complaints can be brought before these human rights bodies: individual communications; state-to-state complaints; and inquiries (OHCHR). Currently, seven treaties accept individual communications: ICCPR, CERD, CAT, CEDAW, CRPD, CED, and ICESCR (OHCHR). For two treaty bodies – those under the CRC and CMW – the individual communications procedure has not entered into force. Together, these reports, general comments, concluding observations, and pronouncements on individual petitions have contributed to the development of international human rights law which have occasionally involved environmental issues.

Charter-based mechanisms

The UN Charter itself provides the legal basis for these mechanisms (United Nations, 1945). Acting pursuant to Article 65 of the UN Charter, the Economic and Social Council has established various bodies to address human rights. These include the Human Rights Council (which succeeded the Commission on Human Rights), the complaints mechanism, universal periodic review, and special procedures that address either specific country situations or thematic issues. These are called "special rapporteurs," or "independent experts," or they can be a working group (De Schutter, 2010: 881). Of these, the Special Rapporteur on human rights and the environment and the Independent Expert on human rights and the environment are relevant to the present discussion. In addition, Universal Periodic Review (UPR) is a unique process and has proven to be an important mechanism to evaluate the human rights situation in all UN member countries. This includes environmental obligations to the extent that they impact human rights. It is "a State-driven process, under the auspices of the Human Rights Council, which provides the opportunity for each State to declare what actions they have taken to improve the human rights situations in their countries and to fulfill their human rights obligations" (UN Human Rights Council, n.d.).

Office of the High Commissioner for Human Rights

The High Commissioner for Human Rights is the principal human rights official of the UN and heads the Office of the High Commissioner for Human Rights (OHCHR) and spearheads the UN's work on human rights:

> OHCHR's thematic priorities are strengthening international human rights mechanisms; enhancing equality and countering discrimination; combating impunity and strengthening accountability and the rule of law; integrating

> human rights in development and in the economic sphere; widening the democratic space; and early warning and protection of human rights in situations of conflict, violence and insecurity.
>
> (OHCHR)

In addition, it supports UN human rights treaty bodies and special procedures, thus bridging the gap between treaty-based mechanisms and charter-based mechanisms.

Developments at the UN: the creation of special mandates

As discussed briefly in Chapter 1, the World Commission on Environment and Development (WCED) appointed by the UN General Assembly in 1983 issued its report in 1987. Called "Our Common Future," this report had a huge impact on the development of international environmental law and its future trajectory. It called upon states to integrate environmental protection into the economic development process and advanced the notion of sustainable development as the new policy framework for states. Defined as "development that meets the needs of the present without compromising the ability of future generations to meet their needs" (WCED, 1987), sustainable development made a direct reference to human beings. Since then, sustainable development has evolved to encompass three pillars with a definite human rights dimension: environmental protection, economic development, and social development (Copenhagen Declaration on Social Development, 1995; Johannesburg Declaration on Sustainable Development, 2002). The social pillar of sustainable development has received less attention, but recognizes that without addressing social inclusion, social justice, and equality, sustainable development cannot be realized.

The draft legal principles for environmental protection and sustainable development proposed by the WCED Experts Group on Environmental Law articulated a fundamental right to a healthy environment: "All human beings have the fundamental right to an environment adequate for their health, and well-being" (Draft Principles, 1994: Principle 1). These draft principles, unfortunately, did not proceed much further and continue to gather dust in the UN archives.

Appointment of a Special Rapporteur on human rights and the environment[1]

The UN Sub-Commission on Prevention of Discrimination and Protection of Minorities (as it was then called) of the UN Commission on Human Rights appointed a Special Rapporteur in 1990 to study the relationship between human rights and the environment. Welcoming this development, the UN General Assembly adopted a resolution endorsing that "all individuals are entitled to live in an environment adequate for their health and well-being" (UNGA resolution, 1991). It further noted that "a better and healthier environment can help contribute to the full enjoyment of human rights by all" (UNGA resolution, 1991). While this acknowledged the link between a healthy environment and the enjoyment of rights, it did *not* acknowledge a human right to a healthy environment.

Ms. Fatma Zohra Ksentini was appointed as the UN Special Rapporteur on Human Rights and the Environment in 1990. She submitted four reports including a note during her tenure as Special Rapporteur. In addition, a set of draft articles on human rights and the environment was appended to the final report. An overview of her reports is given here. In her final report, the Special Rapporteur noted:

> For the particular purposes of this study of human rights and the environment, it is equally important to establish the legal framework for pursuing what have become the essential demands of this century, in order to take up the legitimate concerns of our generation, to preserve the interests of future generations and mutually to agree upon the components of a right to a healthy and flourishing environment.
>
> Providing these various agents with the relevant legal framework and means of expression will reinforce the channels for dialogue which in turn would lead to the further refinement of this right as well as its harmonious application.
>
> (Ksentini, 1994: para. 5)

In her report, the Special Rapporteur identified several principles that have become binding on states in the environmental field: assessing the consequences of an activity before engaging in it; informing other states of potential transboundary impacts of activities; consulting with potential victims where there are transboundary impacts; urgently informing other states and providing mutual assistance in the event of accidents; and providing access to remedies to citizens of the victim state on the basis of non-discrimination (para. 5). Many of these are procedural in nature and are general international law principles that are also applicable to environmental issues. Evaluating the impact of development projects on the environment, as well as its transboundary impact where applicable, is a well-established principle of both domestic law and international law.

The Special Rapporteur discussed international human rights law provisions relevant to environmental protection, including the Universal Declaration of Human Rights (UDHR), the Convention on the Rights of the Child, the Convention on the Elimination of all Forms of Discrimination Against Women, and the International Covenant on Economic, Social and Cultural Rights (ICESCR), as well as the regional treaties that specifically embody a right to environment – the African Charter on Human and People's Rights and the Additional Protocol to the American Convention on Human Rights.

The report made specific reference to Principles 1 and 21 of the Stockholm Declaration and concluded as follows:

> The relationships established by the Stockholm Declaration between the environment, development, satisfactory living conditions, dignity,

> well-being and individual rights, including the right to life, constitute recognition of the right to a healthy and decent environment, which is inextricably linked, both individually and collectively, to universally recognized fundamental human rights standards and principles, and which may be demanded as such by their beneficiaries, i.e. individuals alone or in association with others, communities, associations and other components of civil society, as well as peoples.
>
> (Ksentini, 1994: para. 31)

It is not clear how the Special Rapporteur came to this conclusion. While it is true that the Stockholm Declaration makes a link between the environment, development, dignity, individual rights, and satisfactory living conditions, that by itself does not lead to the conclusion that there exists a separate right to a healthy and decent environment. In fact, almost 20 years after the compilation of the report, a distinct right to a healthy and decent environment has yet to be recognized under international human rights law.

Linking environment, sustainable development, and participatory rights, the report points out that the right to participation has both individual and collective dimensions, and that without the right to participation, sustainable development would be totally without substance (Ksentini, 1994: para. 70). The importance of the right to participation was reiterated in the context of environmental protection by the adoption of Principle 10 of the Rio Declaration. In addition to these general aspects, the Special Rapporteur discussed indigenous peoples and the environment, protection of the environment during armed conflict, and the link between the environment and international peace and security, as well as the impact of environmental degradation on vulnerable people (Ksentini, 1994: para. 70).

The draft principles on human rights and the environment which were appended to the final report specifically recognize that "sustainable development links the right to development and the right to a secure, healthy and ecologically sound environment" (Draft Principles, 1994) and that human rights violations lead to environmental degradation, and environmental degradation leads to human rights violations. The draft principles recognized a substantive right to a healthy environment: "All persons have the right to a secure, healthy and ecologically sound environment. This right and other human rights, including civil, cultural, economic, political and social rights, are universal, interdependent and indivisible" (Draft Principles, 1994: Part I, Article 2).

Other substantive rights included in the draft principles were: a right to freedom from pollution, the right to protection and preservation of air, soil, water, sea-ice, flora, fauna, and essential processes and areas necessary to maintain biodiversity and ecosystems, a right to the highest attainable standard of health, a right to safe and healthy food and water, a right to a healthy working environment, and a right to adequate housing. While rights such as health, food, water, and adequate housing are established rights under international law and their inclusion here only reinforces them, the other "rights" are clearly *lex ferenda*.

The proposed right to be free from pollution would be included in the right to a healthy environment, but the inclusion of a right to preservation of air, soil, water, sea-ice, and areas necessary to maintain biodiversity and ecosystems clearly goes beyond the parameters of traditional human rights law. These can be the indirect result of a human right to environment, not its main focus. One of the drawbacks of a human rights approach to environmental protection is that it cannot protect other species or ecosystems, except indirectly.

The draft articles also endorsed the right of indigenous peoples to control their lands, territories and natural resources and to maintain their traditional way of life. In addition, they have the right to protection against any action that may result in the destruction of their territories, "including land, air, water, sea-ice, wildlife and other resources" (Draft Principles, 1994: Part II, Article 14). The UN Declaration on the Rights of Indigenous Peoples (UNDRIP) adopted in 2007 includes a similar provision that requires states to provide for effective mechanisms for the prevention of and redress for, *inter alia*, "any action which has the aim or effect of dispossessing them of their lands, territories or resources" (Article 8(2)).

Procedural rights are also included in the draft articles – these include the customary international law rights of information, effective public participation, the right to hold opinions, freedom of association, and effective remedies. The right to environmental and human rights education is also included among procedural rights, although it seems to be more suited to be included among substantive rights.

The draft articles are noteworthy for the inclusion of duties of states and people: they stress that all persons have the duty to protect and preserve the environment. Furthermore, they elaborate on the steps states have to take to ensure that the right to a secure, healthy, and ecologically sound environment is fulfilled. In this regard, states have to adopt administrative, legislative, and other measures to effectively implement the rights in the Declaration with the aim of preventing environmental harm. The Declaration contains an elaborate list of measures that states must take, including: collection and dissemination of environmental information; prior assessment of activities for their environmental impact; ensuring public participation in decision-making; provision of effective remedies; monitoring and management of natural resources; adopting measures to reduce wasteful processes of production and consumption patterns; and licensing and regulation of substances and activities that are harmful to the environment. Moreover, states are required to ensure that transnational corporations carry out their duties of environmental protection, sustainable development, and respect for human rights. The Declaration calls upon all international organizations and agencies to observe the rights and duties in the Declaration.

The draft Declaration is unprecedented, as it extends its reach to cover activities of transnational corporations. Activities of transnational corporations pose significant challenges for both international law and national law, as they transcend both systems and hence are subject to neither. Many corporations have been responsible for gross violations of human rights and severe environmental destruction in developing countries (Kravchenko and Bonine, 2008: chapter 9).

Despite several attempts to regulate their activities, the UN has not succeeded in adopting hard law on the subject. What remains are several soft law instruments. The *Guiding Principles on Business and Human Rights* proposed by John Ruggie (2011) is the latest attempt in this regard. These principles, based on the premise that the root causes lie in the governance gaps created by globalization (Simons, 2012: 5) adopt the same typology that is used in the present volume: the obligations to protect, respect, and fulfill. The obligations are framed in relation to states as duty-bearers as all human rights are, but it is clear that business enterprises are the main object of the guiding principles. The guiding principles impose an explicit obligation on business enterprises to respect human rights – at a minimum, those rights that are embodied in the international bill of rights, including international labor standards. If a right to a healthy environment is recognized under international law, then these guiding principles would extend to cover that too. Simons argues that in order to address corporate impunity, one needs to address the root causes of governance gaps and that "one of the most significant impediments to corporate human rights accountability is the structure of the international legal system itself" and that "Ruggie's approach to addressing corporate human rights impunity was misconceived" (Simons, 2012: 5).

Unfortunately, the draft Declaration or at least a watered-down version of it, was never adopted. It seems destined to gather dust in the UN archives, and the hard work that went into the reports and the draft Declaration seems to be appreciated only by scholars interested in the nexus between human rights and the environment. It is hoped that the appointment of an independent expert on human rights and the environment, discussed later, will not meet the same fate.

Appointment of a Special Rapporteur on the implications for human rights of the environmentally sound management and disposal of hazardous substances and wastes

Although this topic has not attracted much attention in scholarly circles, the topic of hazardous substances also implicates human rights. The original title read as follows: "Special Rapporteur on the adverse effects of the illicit movement and dumping of toxic and dangerous products and wastes on the enjoyment of rights" (Commission on Human Rights, 1995). The first resolution on the topic was adopted in 1995. Commission on Human Rights resolution 1995/81 affirmed that the illicit traffic and the dumping of toxic and dangerous products and wastes constitutes a serious threat to the human rights to life and health.

In creating the mandate, the Commission on Human Rights (as it was then called), recognized that:

> The illicit traffic and dumping of toxic and dangerous products and wastes pose a serious threat not only to the environment, but also to the enjoyment of internationally-protected human rights – the right to life, the enjoyment of the highest attainable standard of physical and mental health, the rights to clean water, food, adequate housing and safe and healthy working

conditions, the right to information, the right to participation and freedom of association, and other human rights enshrined in the Universal Declaration of Human Rights and other international instruments.

(1995)

The mandate of the Special Rapporteur was threefold:

(a) outline the elements of the problem and conduct a general survey of issues involving the human rights of victims, with special reference to Africa and other developing countries
(b) identify, investigate, and monitor actual situations and complaints received by the Special Rapporteur
(c) produce annually a list of countries and transnational corporations engaged in illicit traffic of toxic and dangerous products and wastes to developing countries

The mandate was strengthened and expanded in 2011 to cover not only the illicit movement but also the whole life-cycle of hazardous materials (HRC resolution 18/11). On the basis of this resolution, the Special Rapporteur was tasked "to monitor the adverse effects that the generation, management, handling, distribution and final disposal of hazardous substances and wastes may have on the full enjoyment of human rights, including the right to food, adequate housing, health and water" (Pallemaerts, 2013). In 2012, the mandate was expanded to cover the protection of human rights defenders, given the increasing incidence of their victimization (Pallemaerts, 2013).

Five Special Rapporteurs have been appointed to date under this mandate. In 2011, the Human Rights Council decided to strengthen and expand the mandate, affirming that the way hazardous substances and wastes were managed throughout their life-cycle may have an adverse impact on the full enjoyment of human rights (HRC Resolution 18/11). The mandate was expanded to include information on: human rights issues raised by transnational corporations and other business enterprises in relation to hazardous substances and wastes; the scope of national legislation relating to the human rights implications of such substances; ambiguities in international instruments that allow the movement and dumping of such substances; support and assistance for victims of human rights violations; and human rights abuses committed against human rights defenders working in this area (Report of Special Rapporteur, Baskut Tuncak, 2014).

The previous Special Rapporteur, Okechukwu Ibeanu, noted in his report that developing countries accept hazardous products and toxic wastes due to poverty, but lack disposal facilities, placing the human rights of local populations at risk. Moreover, women, children, and youth are particularly at risk because they are often among the poorest and therefore likely to work in polluting industries and scavenge dumps for reusable materials. They also have limited access to information. The Special Rapporteur called for greater global attention to the gender and age dimensions of illicit movement and dumping of toxic products and wastes.

The report identified rights to life, health, adequate food, work, and the right to a remedy as being at risk. He also identified the rights to information and participation as being central to the issue. He pointed to the general trend that states, corporations, and other private entities do not share vital information about the potential effects of pollution and irreversible damage to the environment until an incident occurs. Even then, they are reluctant to divulge information and often give false information misleading the public.

The Special Rapporteur considered that the rights to information and participation are both rights in themselves, and also essential to exercise other rights. They play an important role in the context of illicit movement and dumping of toxic products. He pointed out that the rights to information and participation in relation to both human rights and environmental protection are well reflected in the international legal framework under both areas of law. These include the UDHR, the ICCPR, the Rio Declaration on Environment and Development, the Rotterdam Convention on Prior Informed Consent Procedure for Certain Hazardous Chemicals and Pesticides in International Trade, the Stockholm Convention on Persistent Organic Pollutants, the Basel Convention on the Control of Transboundary Movements of Hazardous Wastes and their Disposal, and most importantly, the Aarhus Convention (Ibeanu, 2009).

During the brief period that Professor Marc Pallemaerts was the Special Rapporteur before his untimely death in 2014, he submitted one report in which he identified the regulatory gaps in international law.

The current Special Rapporteur, Baskut Tuncak, noted in his preliminary report that his mandate reflects current issues in both environmental and human rights laws (Human Rights Council, 2014) and that modern laws "increasingly embraced a life-cycle approach to chemicals management, helping to reduce the human rights implications of hazardous substances and wastes" (Human Rights Council, 2014). This mandate overlaps with that of the Independent Expert (now Special Rapporteur) on Human Rights and Environment, to which we now turn.

Appointment of an Independent Expert on Human Rights and the Environment

The Human Rights Council, by resolution 19/10, established the mandate for an Independent Expert on Human Rights and the Environment in March 2012 (Human Rights Council, 2012). John Knox, Henry C. Lauerman Professor of International Law at Wake Forest Law School, was appointed as the Independent Expert. He held an initial consultation to discuss his mandate in October 2012 and the present author was invited to participate. Since then, the Independent Expert has had several consultations with various stakeholders and produced three reports. The mandate of the Independent Expert includes: studying the human rights obligations relating to the enjoyment of a safe, clean, healthy, and sustainable environment; identifying best practices relating to the use of human rights obligations in relation to environmental protection; and preparing a compendium of best practices.

In his first report submitted to the Human Rights Council in December 2012, the Independent Expert noted that environmental rights are a latecomer to the human rights field. At the time the main human rights treaties were adopted, the environment had not emerged as a concern, but due to technological developments and with the realization that human beings can cause immense destruction, "environmental concerns have moved from the periphery to the centre of human efforts to pursue economic and social development" (Knox, 2013a). He noted further that the close relationship between human rights and the environment has resulted in two main developments: (a) the recognition of an explicit right to a healthy environment; and (b) increased attention paid to the environment in relation to already recognized human rights, such as the right to life and the right to health (Knox, 2013a).

With regard to the former, he noted that many constitutions in the world adopted in recent years incorporate some form of environmental right, while two regional human rights treaties do so as well. Globally, no such attempt has been made, despite the appointment of a Special Rapporteur in the 1990s and the draft principles proposed by her. The Independent Expert believes that if the UDHR were to be drafted today, it is most likely to embody an environmental right, but the UN did not take several opportunities it had to proclaim such a right, starting with the Stockholm Declaration. At the global level, more attention seems to have been paid to "greening" existing rights, rather than adopting an explicit right to a healthy environment. This effort has resulted in identifying two sets of rights: "(a) rights whose enjoyment is particularly vulnerable to environmental degradation; and (b) rights whose exercise supports better environmental policymaking" (Knox, 2013a: para. 36). The Independent Expert categorized the former set of rights as substantive rights and the latter as procedural rights.

After a brief survey of the evolution of environmental rights under international law, the Independent Expert pointed out that some aspects of the relationship between human rights and the environment are firmly established: first, environmental degradation can and does negatively impact the enjoyment of certain protected rights, most notably the rights to life, health, food, and water; second, procedural rights can lead to better decisions concerning the environment which, in turn, will protect the human rights affected by the environmental issue. He emphasized, however, that the obligations that human rights law imposes in relation to environmental protection are less clearly understood and the mandate given to him by the Human Rights Council is precisely to study the nature of those obligations and offer clarification. In order to do so, the Independent Expert identified several issues to be studied:

> the relationship between human rights obligations and best practices; the connections between substantive and procedural rights and duties; vulnerable groups and non-discrimination; human rights obligations relating to transboundary and global environmental harm; the application of human rights norms to non-State actors; and the relationship between a right to a healthy environment and other human rights.
>
> (Knox, 2013a: para. 36)

He concluded by noting that while many developments have taken place with regard to the link between human rights and the environment during the last two decades, some aspects of this relationship need further clarity. Thus, the human rights obligations relating to a safe, healthy, and sustainable environment need to be clarified so that states will be in a position to fulfil those obligations fully. It is hoped that the Independent Expert will succeed in clarifying the parameters of these obligations, and that the UN will endorse those recommendations, rather than sweep them under the rug this time too. As the Independent Expert himself observed in his first report, "the United Nations has not taken advantage of subsequent opportunities to recognize a human right to a healthy environment" (Knox, 2013a: para. 14), and it is hoped that the appointment of an Independent Expert by the Human Rights Council will provide the impetus to do so. The international community can no longer ignore the impact of environmental degradation on the enjoyment of human rights, and the time has come to recognize this important right at the international level. Climate change may well be the catalyst for change in this area.

In his second report (called the "Mapping Report"), the Independent Expert mapped the human rights obligations relating to the environment with an extensive review of global and regional sources (Knox, 2014). He concluded that human rights law includes obligations relating to the environment, both substantive and procedural, as well as those applicable to vulnerable groups. Procedural obligations – which are now more or less established – include the duty to assess environmental impacts on human rights, the obligation to provide information and facilitate participation, and the duty to provide access to remedies. Substantive obligations include the duty to adopt legal and institutional frameworks "that protect against environmental harm that interferes with the enjoyment of human rights, including harm caused by private actors" (Knox, 2014: para. 84). In addition to the general requirement of non-discrimination in applying environmental laws, states may have additional obligations to protect particularly vulnerable groups – women, children, and indigenous groups. While human rights obligations relating to the environment are still evolving, they are already clear enough to provide guidance to states, and the Independent Expert urged states to take human rights obligations into account in the development and implementation of environmental policies. His third report has a compilation of good practices, elaborating on the obligations identified in the mapping report. The term "practice" was defined broadly to include legislation, policies, case law, administrative practices, and projects, and included a wide range of actors (Knox, 2015).

The Independent Expert and UNEP identified criteria to guide the identification of good practices, rather than best practices, as follows:

> To be a "good" practice, the practice should integrate human rights and environmental standards, including through the application of human rights norms to environmental decision-making and implementation or the use of environmental measures to define, implement and (preferably) exceed minimum standards set by human rights norms. The practice should be exemplary

> from the perspectives of human rights and of environmental protection, and there should be evidence that the practice is achieving or working towards achieving its desired objectives and outcomes.
>
> (Knox, 2015: para. 13)

The good practices identified include procedural obligations: make environmental information public; facilitate public participation in environmental decision-making; protect rights of expression and association; and access to legal remedies, and examples from across the world were provided for each obligation. Similarly, for substantive obligations, specific reference was made to the Ruggie Principles with regard to non-state actors, obligations relating to transboundary environmental harm, including preparation of transboundary environmental impact assessments, providing relief to foreign victims of environmental harm, and obligations relating to vulnerable groups that included women and children. The Independent Expert concluded by identifying two thematic areas that need further clarification and study:

> Many areas require further attention in this respect. The Independent Expert continues to emphasize the importance of two areas in particular: the importance of clarifying and implementing human rights obligations relating to transboundary environmental harm, in particular with regard to the global harm caused by climate change; and the urgent need to take more effective steps to protect the rights of environmental human rights defenders.
>
> (Knox, 2015: para. 106)

It is heartening that the Independent Expert's mandate was extended for a further period of three years and the position was converted to that of a Special Rapporteur (OHCHR, n.d. b). This bears testimony to the success of the work done by the Independent Expert, and the emphasis he placed on the need to take the link between human rights and the environment more seriously. The UN human rights bodies finally seem to be taking environmental rights seriously.

Special Rapporteur on Human Rights and Environment

As noted, in 2015, the mandate of the Independent Expert on Human Rights and Environment was converted to a Special Rapporteur position and extended for a further period of three years. As Special Rapporteur, John Knox submitted four reports to the Human Rights Council – on climate change (2016), biodiversity (2017), children's rights (2018) and the final report with Framework Principles on Human Rights and the Environment appended (2018). While his report on climate change will be explored in greater detail in Chapter 9, his report on biodiversity and human rights is significant. It is probably the first time that an explicit link was made between biodiversity and human rights.

In his report on climate change, the Special Rapporteur noted that the relationship between climate change and human rights has received increased attention from various stakeholders in recent years. Pursuant to the Human Rights Council resolution 7/23, the HRC expressed concern about the impact of climate change on the enjoyment of rights, and requested the Office of the High Commissioner for Human Rights to prepare a report exploring the link between climate change and human rights. This report, while acknowledging the link between climate change and human rights, did not readily accept that climate change violated human rights law. In other words, the obligations that states have with regard to damage caused by climate change are not very clear-cut. One of the challenges is the human rights structure itself – because human rights operate vertically between states as duty-bearers and people as rights-holders, if the pollution originates from a source outside the territory, then it is not clear to what extent the state is liable under human rights law.

In his report on biodiversity and human rights, the Special Rapporteur noted:

> The full enjoyment of human rights, including the rights to life, health, food and water, depends on the services provided by ecosystems. The provision of ecosystem services depends on the health and sustainability of ecosystems, which in turn depend on biodiversity. The full enjoyment of human rights thus depends on biodiversity, and the degradation and loss of biodiversity undermine the ability of human beings to enjoy their human rights.
>
> (Knox, 2017: para. 5)

He was, however, careful to note that biodiversity has intrinsic value that may not be captured within a human rights framework. He pointed out that while the importance of a healthy environment is widely recognized, the relationship between human rights and biodiversity is not well understood. Human beings depend on the environment for their very survival – the food they eat, the air they breathe, the water they drink, and the medicine they take, and for decomposition of waste, aeration of soil, and virtually every function provided by nature without which human beings cannot survive. Increasingly, unfortunately, biodiversity and ecosystems are coming under threat, which in turn has an impact on human beings and their rights. The Special Rapporteur identified rights that are directly affected by biodiversity loss and noted: "biodiversity contributes to particular ecosystem services that directly support the full enjoyment of human rights" (Knox, 2017: para. 10). The rights that are at risk of being infringed include the rights to life, health, food, water, culture, an adequate standard of living, and non-discrimination. In addition, the report points out that loss of biodiversity often results from and reinforces existing patterns of discrimination, and some communities are disproportionately affected:

> Although everyone depends on ecosystem services, some people depend on them more closely than others. For indigenous peoples, forest-dwellers, fisherfolk and others who rely directly on the products of forests, rivers, lakes and oceans for their food, fuel and medicine, environmental harm can and often does have disastrous consequences. This is true not only because of

> their close relationship to nature, but also because they typically have little economic and political power within their countries, so they cannot easily obtain substitutes for their lost natural resources. Their marginalization means that they have limited or no access to decision-making processes or legal remedies. Their legal rights to the territory and resources on which they depend may not even be recognized by their Governments.
>
> (Knox, 2017: para. 22, footnotes omitted)

The report also discussed human rights obligations relating to the conservation and sustainable use of biodiversity. These are procedural and substantive in nature. Procedural obligations are the duties (a) to assess impacts and make environmental information public; (b) to facilitate public participation in the decision-making process; and (c) to provide access to remedies in the event of any harm. The substantive obligations are to adopt legal and institutional frameworks that protect against environmental harm that interferes with the enjoyment of rights. While states have the discretion to strike a balance between environmental protection and other legitimate societal needs, the balance must be reasonable and should not result in unjustified, foreseeable infringements of human rights. States' duty to safeguard biodiversity in order to protect rights from infringement extends to environmental harm from private actors, and businesses have a responsibility to respect the rights relating to biodiversity as well.

The report also highlights the principle of cooperation in protecting biodiversity and ecosystem services:

> The effective protection of biodiversity, like the effective mitigation of climate change, is possible only with international cooperation as States have often recognized. Many of the components of biodiversity, the threats to biodiversity and benefits biodiversity provides have transboundary or global dimensions.
>
> (Knox, 2017: para. 36)

Framework Principles on Human Rights and Environment

Sixteen principles are included in the Framework Principles, as well as commentary on each principle. Before these were finalized, draft guidelines on human rights and environment were published and circulated among various stakeholders requesting written comments. An expert seminar and a public consultation were also held before the principles were finalized. Soliciting broad participation by many stakeholders in this endeavor allowed many voices to be heard and many concerns to be addressed. The Framework Principles were appended to the final report of the Special Rapporteur and were submitted to the UN Human Rights Council in March 2018.

The report was careful to point out that the 16 principles identified in the Framework Principles do not create new obligations. Rather, they reflect the application of existing human rights obligations in relation to environmental issues.

While not all states may have accepted all the obligations included therein, they reflect the broad consensus or, at a minimum, best practices:

> The coherence of these interpretations, however, is strong evidence of the converging trends towards greater uniformity and certainty in the understanding of human rights obligations relating to the environment. These trends are further supported by State practice, including in international environmental instruments and before human rights bodies. *As a result, the Special Rapporteur believes that States should accept the framework principles as a reflection of actual or emerging international human rights law.* He is confident that, at a bare minimum, States will see them as best practices that they should move to adopt as expeditiously as possible.
>
> (Knox, 2018: para. 9, emphasis added)

The Principles recognize the intertwined nature of human rights, environmental protection and sustainable development:

> Human beings are part of nature, and our human rights are intertwined with the environment in which we live. Environmental harm interferes with the enjoyment of human rights, and the exercise of human rights helps to protect the environment and to promote sustainable development.
>
> (Knox, 2018: Annex, para. 1)

When he submitted the report to the Human Rights Council, the Special Rapporteur stated that

> it is high time that the international community recognized a human right to a healthy environment … A healthy environment is necessary for the full enjoyment of many human rights, including the rights to life, health, food, water, and development. At the same time, the exercise of other freedoms, including the rights to information, participation and remedy, is vital to the protection of the environment.
>
> (Knox, 2018)

He called upon the Human Rights Council to support the recognition of this right in a global instrument. Quoting Victor Hugo, he said that the right to a healthy environment is an idea whose time has come.

While the Framework Principles are a major step forward (even though they have not yet been formally adopted), it is a regrettable that many of these principles are framed in soft language. If, as the Special Rapporteur points out, these principles are already part of international law, then there is no reason why they should be written in soft language (using the word "should" as opposed to "shall" or that "states have an obligation to … "). This could have the effect of diluting already existing obligations and states treating them lightly, and not as obligatory. Certainly, the procedural aspects and the principle of non-discrimination are binding on states.

The procedural obligations are also incorporated in two regional treaties – the Aarhus Convention (1998) and the most recent Latin American and Caribbean Agreement which was modeled after the Aarhus Convention. John Knox's term as Special Rapporteur ended in 2018, and David Boyd was appointed the new Special Rapporteur by the Human Rights Council whose term began in August 2018. An expert's consultation is scheduled for October 2018 to discuss his mandate.

Other convergences

The convergence between human rights and environment is also seen in treaty bodies established under various human rights treaties, especially the Human Rights Committee and the Committee on Economic, Social and Cultural Rights. Some of these complaints are discussed in Chapters 5 and 6.

Universal Period Review (UPR) is another mechanism that has seen this convergence. The Human Rights Council has urged "states to take human rights into consideration when developing their environmental policies" (2011: Resolution 16/11). As discussed in Chapter 9, the Human Rights Council has adopted several resolutions on climate change and human rights.

According to a report compiled by the Independent Expert on Human Rights and the Environment, a total of 45 states discussed their constitutional right to a healthy environment during their UPR review (Knox, 2013b). Mongolia, Indonesia, and the Solomon Islands discussed the impacts of climate change on the enjoyment of a right to a healthy environment, while Peru discussed the impact of mining on the right to a healthy environment. Many countries referred to the impact of environmental harm on human health. The sources of harm included toxic waste dumping (Ivory Coast), nuclear waste (Belarus, Kyrgyzstan), mining operations (Ghana, Brazil, Mongolia, Nauru, and Bolivia), natural disasters (Greece), oil exploration (Nigeria), gold mining (Mongolia), and acid rain from volcanic explosions (Vanuatu).

Proposals that have been advanced

Draft International Covenant on the Human Right to the Environment

The *Centre International de Droit Comparé de l'Environnement* (International Centre of Comparative Environmental Law), which has consultative status with ECOSOC, published a draft International Covenant on the Human Right to the Environment in 2017, modeled after the two main Covenants on Civil and Political, and on Economic, Social and Cultural Rights. It points out that 69 years after the adoption of the UDHR (now 70 years) and 51 years after the adoption of the two Covenants, the time has come to devote a third Covenant to the human right to the environment. The draft recognizes that environmental damage on human rights can have a greater impact on autochthonous and local communities because of their greater dependence on nature. The draft Preamble reaffirms that all human rights *including the right to environment*, are universal, indivisible, and interdependent, and it is the responsibility of the states, individuals and other public, and *private entities* to protect and promote these rights (2017, emphasis added).

Draft Article 1 provides for the right to a healthy environment:

> 1 Everyone, including future generations, has the right to live in an ecologically balanced environment capable of assuring his or her health, security, and wellbeing.
> 2 To that end, every person has the obligation to protect the environment and to contribute to its improvement.
> 3 The State Parties undertake to effectively implement the rights and duties in the present Covenant.
>
> (2017)

Other rights included in the Draft Covenant are: the right to a heightened level of protection of the environment and non-retrogression (Article 2); the right to precautionary measures (Article 3); the right to prevention (Article 4); the right to environmental assessment, including the assessment of transboundary impacts (Article 5); the right to reparation of environmental damage (Article 6); the right to education (Article 7); the right to freedom of opinion and expression (Article 8); the right to information (Article 9); the right to participation (Article 11); and the right to recourse (Article 11). In addition, other rights such as right to water, food, and rights of indigenous peoples are included. Of particular note are the inclusion of rights of persons in disaster situations (Article 15) and rights of environmental refugees and internally displaced persons (Article 16).

The Draft Covenant includes principles such as fairness and solidarity (Article 17), non-discrimination (Article 18), and sustainable protection (Article 19). Part II of the Draft Covenant is devoted to international cooperation, while Part III addresses monitoring and enforcement. The Committee on Economic, Social and Cultural Rights is entrusted with the task of monitoring the implementation of the draft covenant.

This is an important step forward in recognizing environmental rights. However, it is not clear why it was felt necessary to include already established rights such as rights to education, food, and water, while other rights such as the right to housing and sanitation are not included. Including some established duties of states under environmental law – such as environmental impact assessments – as rights of people is certainly a novel approach. While clearly considerable effort has gone into drafting this Covenant, it is unlikely that states will accept it without a fight. Conspicuously absent are obligations relating to investing in developing countries and the protection of human rights of people in the host state. While transboundary environmental damage is addressed in the Covenant, it does not address the complex issue of human rights violations in neighboring states due to such transboundary environmental damage.

Many of the provisions included in the Draft Covenant are already part of international law, including the preventive principle; assessment of impacts of projects, programs, and plans, including their transboundary impact; provision of information; and rights of indigenous people. New additions include rights of persons in disaster situations and rights of environmental refugees. One omission is in relation to extraterritorial damage. Individual petitions are limited to those against the petitioner's state of nationality. While states must evaluate the transboundary impact of their activities and plans, and indeed the ICJ held in the *Pulp Mills* case that this is now a customary international law principle, it is not clear what remedies are available to victims of transboundary harm, or at least whether they are entitled to information of potential transboundary harm. The Draft Covenant was submitted to the Human Rights Council by the Centre in February 2017. The outcome of these efforts remains to be seen.

Global Pact for the Environment

In another development, the French President Macron spearheaded efforts to adopt a Global Pact on the Environment. Drafted by Le Club des Juristes, the preliminary draft of the Global Pact refers to both SDGs and the urgent need to address climate change. The Preamble acknowledges "the growing threats to the environment and the need to act in an ambitious and concerted manner at the global level" and "determined to promote a sustainable development that allows each generation to satisfy its needs without compromising the capability of future generations to meet theirs, while respecting the balance and integrity of the Earth's ecosystem" (Global Pact for the Environment, 2017: Preamble).

The Preamble also refers to human rights and is remarkably similar to the Preamble of the Paris Agreement on Climate Change adopted in 2015:

> Conscious of the need to respect, promote and consider their respective obligations on human rights, the right to health, the rights and knowledge of indigenous peoples, local communities, migrants, children, persons with disabilities and people in vulnerable situation, under their jurisdiction.

Draft Article 1 acknowledges that "every person has the right to live in an ecologically sound environment adequate for their health, well-being, dignity, culture and fulfillment" (Global Pact for the Environment, 2017). Article 2 lays down the duty of every state, international organization, and every person (natural/legal/private/public) to take care of the environment, while Article 3 calls upon parties to integrate environmental protection into planning and implementation of policies, especially in relation to climate change, protection of oceans, and maintenance of biodiversity. The draft incorporates most principles and rights that are recognized under international human rights law and international environmental law. These include sustainable development, inter-generational equity, the principle of prevention, the precautionary principle, the "polluter pays" principle, and the need to ensure adequate remediation of environmental harm. The procedural rights, education and training, research and innovation, cooperation, non-regression, and protection of the environment during armed conflicts are also included. Of particular interest is the provision relating to the compliance mechanism. The Pact proposes the establishment of a Committee of Independent Experts to monitor compliance focusing on facilitation in a transparent, non-adversarial, and non-punitive manner. It is not, however, clear who should be appointed to the Committee and by whom.

It also refers to the vital role of women in sustainable development (similar to the Rio Declaration) and the need to promote gender equality and empowerment of women. It also endorses the inter- and intra-generational equity principle. The document endorses many of the existing principles of international environmental law identified above and also include adopting effective environmental norms, cooperation, and the special needs of developing countries to be given special attention and education and training.

A summit on the Pact was held in September 2017 and in May 2018, the UN General Assembly adopted resolution 72/277 titled "Towards a Global Pact for the Environment." In its resolution, the General Assembly requested the Secretary-General to submit a technical and evidence-based report that identifies and assesses possible gaps in international environmental law and environment-related instruments to strengthen the implementation of these treaties. It also decided to establish an *ad hoc* open-ended working group to consider the report and discuss possible options to address these gaps and make recommendations which may include convening an inter-governmental conference to adopt an international instrument in 2019. This is a significant development. Together with the Framework Principles proposed by the former Special Rapporteur on Human Rights and Environment, John Knox, this may signal a move towards adopting, at long last, a human right to a healthy environment.

Despite their weaknesses, these two documents are remarkably similar with many "rights" overlapping. While the Global Pact recognizes the intrinsic value of the environment especially preserving biodiversity, it falls short of recognizing rights of nature, even though the document is called the *Global Pact for the Environment.* It does, however, affirm a human right to a healthy environment. There is a definite momentum towards recognizing this right and the time seems to be opportune to do so.

Conclusion

The topic of environmental rights (or human rights and the environment) has come a long way since it was first articulated by scholars in the early 1990s. In some respects, the topic has come a full circle – from a Special Rapporteur on the topic who submitted a set of draft articles which went nowhere to another Special Rapporteur two decades later who has submitted another set of principles on the topic to the Human Rights Council. We urge the international community to adopt these principles. The time has come for the international community to recognize a stand-alone right of every person to a healthy (or safe, secure, sustainable) environment. There is sufficient evidence of state practice and developments at regional and national levels, including recognition in over 100 national constitutions, to do so.

DISCUSSION QUESTIONS

1. Do you think the international community should adopt a substantive right to a healthy environment?
2. If so, what is the best way to adopt it? UN General Assembly resolution? A binding treaty? Other options?
3. What are the pros and cons of a soft law approach versus a hard law approach?
4. What are the implications of the failure to adopt a substantive right to a healthy environment?
5. How would you support the current momentum towards accepting such a right?

Note

1 This section draws from the author's book: Atapattu, S. (2016). *Human Rights Approaches to Climate Change: Challenges and Opportunities*. New York, NY: Routledge.

References

Additional Protocol to the American Convention of Human Rights in the Area of Economic, Social and Cultural Rights (1988). [Online]. Available at: www.oas.org/juridico/english/treaties/a-52.html [Accessed: 10 October 2018].

African Charter on Human and People's Rights (1981) [Online]. Available at: www.achpr.org/files/instruments/achpr/banjul_charter.pdf [Accessed: 18 October 2018].

Alston, P. (1984). Conjuring Up New Human Rights: A Proposal for Quality Control. *American Journal of International Law* 28: 607.

Anton, D.K. and Shelton, D. (2011). *Environmental Protection and Human Rights*. Cambridge: Cambridge University Press.

Atapattu, S. (2006). *Emerging Principles of International Environmental Law*. New York, NY: Transnational.

Atapattu, S. (2016). *Human Rights Approaches to Climate Change, Challenges and Opportunities*. London: Routledge.

Boyd, D. (2012). *The Environmental Rights* Revolution: *A Global Study of Constitutions, Human Rights, and the Environment*. Vancouver: UBC Press.

Burdon, P. (2015). Environmental Human Rights: A Constructive Critique. In: Grear, A. and Kotze, L. (eds.), *Research Handbook on Human Rights and the Environment*, Cheltenham: Edward Elgar, p.61.

Bryner, N. (2017). France Pushes for Global Pact on the Environment. [Online]. Available at: http://legal-planet.org/2017/06/27/france-pushes-for-global-pact-on-the-environment/ [Accessed: 18 October 2018].

Burger, M. (2003). Bipolar and Polycentric Approaches to Human Rights and the Environment. *Columbia Journal of Environmental Law* 28: 371.

CELDF (2018). Rights of Nature [Online]. Available at: https://celdf.org/rights/rights-of-nature/ [Accessed: 18 October 2018].

Collins, L. (2015). The United Nations, Human Rights and the Environment. In: Grear, A. and Kotze, L. (eds.), *Research Handbook on Human Rights and the Environment*, Cheltenham: Edward Elgar, p.219.

Commission on Human Rights Resolution 1995/81 (1995). Adverse effects of the illicit movement and dumping of toxic and dangerous products and wastes on the enjoyment of human rights. [Online]. Available at: www2.ohchr.org/english/issues/environment/waste/[Accessed: 10 October 2018].

Convention against Torture and Other Cruel, Inhuman or Degrading Treatment or Punishment (1984). [Online]. Available at: www.ohchr.org/en/professionalinterest/pages/cat.aspx [Accessed: 1 September 2018].

Convention on All Forms of Discrimination against Women (1979). [Online]. Available at: www.un.org/womenwatch/daw/cedaw/text/econvention.htm [Accessed: 20 October 2018].

Convention on the Rights of Persons with Disability (2006). [Online]. Available at: www.un.org/development/desa/disabilities/convention-on-the-rights-of-persons-with-disabilities/convention-on-the-rights-of-persons-with-disabilities-2.html [Accessed: 20 October 2018].

Convention on the Rights of the Child (1989). [Online]. Available at: www.ohchr.org/en/professionalinterest/pages/crc.aspx [Accessed: 1 September 2018].

Copenhagen Declaration on Social Development. (1995). A/CONF.166/9. [Online]. Available at: www.un-documents.net/cope-dec.htm [Accessed: 24 October 2018].

Declaration of the United Nations Conference on the Human Environment (1972) [Online]. Available at: www.un-documents.net/unchedec.htm [Accessed: on 18 October 2018].

Dellinger, M. (2012). Ten Years of the Aarhus Convention: How Procedural Democracy is Paving the Way for Substantive Change in National and International Environmental Law. *Colorado Journal of International Environmental Law & Policy* 23: 309.

de Schutter, O. (2010). *International Human Rights Law*. Cambridge: Cambridge University Press.

Draft International Covenant on the Human Right to the Environment (2017) [Online]. [Online]. Available at: https://cidce.org/wp-content/uploads/2016/08/Draft-of-the-International-Covenant-on-the-Human-Right-to-the-Environment_15.II_.2017_EN.pdf [Accessed: 18 October 2018].

Draft Principles on Human Rights and Environment (1994). E/CN.4/Sub.2/1994/9, Annex I. [Online]. Available at: http://hrlibrary.umn.edu/instree/1994-dec.htm [Accessed: 10 October 2018].

Draft Global Pact for the Environment (2017) [Online]. Available at: http://pactenvironment.emediaweb.fr/wp-contet/uploads/2017/07/Global-Pact-for-the-Environment-projec-24-June-2017.pdf [Accessed: 18 October 2018].

Gonzalez, C. (2012). Environmental Justice and International Environmental Law. In: Alam, S., Bhuiyan, J.H., Chowdhury, T.M.R., and Techera, E. (eds.), *Routledge Handbook of International Environmental Law*. London: Routledge.

Executive Order No. 12898 (1994). 59 Fed. Reg. 7629, 16 February 1994, President Clinton, Federal Actions to Address Environmental Justice in Minority Populations and Low-Income Populations [Online]. Available at: www.archivesgov/federal-register/executive-orders/pdf/12898.pdf [Accessed: 24 October 2018].

Framework Principles on Human Rights and Environment (2018). [Online]. Available at: www.ohchr.org/EN/Issues/Environment/SREnvironment/Pages/FrameworkPrinciples Report.aspx [Accessed: 10 October 2018].

Human Rights Council (2008). Resolution 7/23, Human Rights and Climate Change. [Online]. Available at: http://ap.ohchr.org/documents/e/hrc/resolutions/a_hrc_res_7_23.pdf [Accessed: 10 October 2018].

Human Rights Council (2011). Resolution 16/11, Human rights and the environment, A/HRC/RES/16/11. [Online]. Available at: www2.ohchr.org/english/bodies/hrcouncil/docs/16session/A.HRC.RES.16.11_en.pdf [Accessed: 10 October 2018].

Human Rights Council (2011). Resolution 18/11, Mandate of the Special Rapporteur on the implications for human rights of the environmentally sound management and disposal of hazardous substances and wastes A/HRC/RES/18/11. [Online]. Available at: www.ohchr.org/Documents/HRBodies/SP/Nominations/EritreaBelarusEnvironment/HRC.RES.18.11_toxic_waste.pdf [Accessed: 10 October 2018].

Human Rights Council (2012). Human Rights and the Environment. Resolution A/HRC/RES/19/10 [Online]. Available at: http://daccess-dds-ny.un.org/doc/RESOLUTION/GEN/G12/131/59/PDF/G1213159.pdf?OpenElement [Accessed: 18 October 2018].

Human Rights Council (2014). Report of the Special Rapporteur on the implications for human rights of the environmentally sound management and disposal of hazardous substances and wastes, Baskut Tuncak (2014), A/HRC/27/54. [Online]. Available at: https://documents-dds-ny.un.org/doc/UNDOC/GEN/G14/150/07/PDF/G1415007.pdf?OpenElement [Accessed: 10 October 2018].

Human Rights Council (2016). Report of the Special Rapporteur on the issue of human rights obligations relating to the enjoyment of a safe, clean, healthy and sustainable environment, Climate Change, A/HRC/31/52. [Online]. Available at: https://documents-dds-ny.un.org/doc/UNDOC/GEN/G16/015/72/PDF/G1601572.pdf?OpenElement [Accessed: 10 October 2018].

Ibeanu, O. (2009). Report of the Special Rapporteur on the adverse effects on the movement and dumping of toxic and dangerous products and wastes on the enjoyment of human rights. A/HRC/5/5, UN Human Rights Council [Online]. Available at: www2.ohchr.org/english/issues/environment/waste/index.htm [Accessed: 18 October 2018].

International Convention for the Protection of All Persons from Enforced Disappearance (2006). [Online]. Available at: www.ohchr.org/Documents/ProfessionalInterest/disappearance-convention.pdf [Accessed: 1 October 2018].

International Convention on the Elimination of All Forms of Racial Discrimination (1965). [Online]. Available at: www.ohchr.org/en/professionalinterest/pages/cerd.aspx [Accessed: 1 October 2018].

International Convention on the Protection of All Migrant Workers and Members of their Families (1990). [Online]. Available at: www.ohchr.org/en/professionalinterest/pages/cmw.aspx [Accessed: 1 October 2018].

International Court of Justice (1996). *Case Concerning the Gabcikovo Nagymaros Project (Hungary v. Slovakia)*.

International Court of Justice (2010). *Pulp Mills on the River Uruguay (Argentina v. Uruguay)*.

International Covenant on Civil and Political Rights (1966). [Online]. Available at: www.ohchr.org/Documents/ProfessionalInterest/ccpr.pdf [Accessed: 1 October 2018].

International Covenant on Economic, Social and Cultural Rights (1966). [Online]. Available at: www.ohchr.org/Documents/ProfessionalInterest/cescr.pdf [Accessed: 1 October 2018].

Johannesburg Declaration on Sustainable Development (2002). A/CONF.199/20. [Online]. Available at: www.un-documents.net/jburgdec.htm [Accessed: 24 October 2018].

Knox, J.H. (2013a). Report of the Independent Expert on the issue of human rights obligations relating to the enjoyment of a safe, clean, healthy and sustainable environment. A/HRC/22/43, Human Rights Council [Online]. Available at: www.ohchr.org/Documents/HRBodies/HRCouncil/RegularSession/Session22/A-HRC-22-43_en.pdf [Accessed: 18 October 2018].

Knox, J.H. (2013b). Mapping Human Rights Obligations Relating to the Enjoyment of a Safe, Clean, Healthy and Sustainable Environment, Individual Report on the UN General Assembly and the Human Rights Council, including the Universal Periodic Review Process, Report No. 6. A/HRC/25/53. Geneva: Office of the High Commissioner for Human Rights (OHCHR).

Knox, J. H. (2014). Report of the Independent Expert on the issue of human rights obligations relating to the enjoyment of a safe, clean, healthy and sustainable environment. A/HRC/22/43, Human Rights Council [Online]. Available at: www.ohchr.org/EN/Issues/Environment/IEEnvironment/Pages/Annualreports.aspx[Accessed: 10 October 2018].

Knox, J. H. (2015). Report of the Independent Expert on the issue of human rights obligations relating to the enjoyment of a safe, clean, healthy and sustainable environment. A/HRC/22/43, Human Rights Council [Online]. Available at: www.ohchr.org/EN/Issues/Environment/IEEnvironment/Pages/Annualreports.aspx [Accessed: 1 October 2018].

Knox, J.H. (2018). Human Rights Council Report of the Special Rapporteur on the issue of human rights obligations relating to the enjoyment of a safe, clean, healthy and sustainable environment, Framework Principles on Human Rights and the Environment, A/HRC/37/59. [Online]. Available at: https://documents-dds-ny.un.org/doc/UNDOC/GEN/G18/017/42/PDF/G1801742.pdf?OpenElement [Accessed: 10 October 2018].

Knox, J. (forthcoming). The Paris Agreement as a Human Rights Treaty. In: Akande, D., Kuosmanen, J., McDermott, H., and Roser, D. (eds.), *Human Rights and 21st Century Challenges: Poverty, Conflict, and the Environment*. Oxford: Oxford University Press.

Ksentini, F.Z. (1994). Review of Further Developing in the Fields with which the Sub-commission Has been Concerned Human Rights and the Environment: Final Report Prepared by Mrs. Fatma Zohra Ksentini, Special Rapporteur. E/CN.4/Sub.2/1994/9, UN Economic and Social Council [Online]. Available at: www1.umn.edu/humanrts/demo/HRandEnvironment_Ksentini.pdf [Accessed: 18 October 2018].

Kravchenko, S. and Bonnie, J. (2008). *Human Rights and the Environment: Cases, Law & Policy*. Durham, NC: Carolina Academic Press.

Kuehn, R.R. (2000). A Taxonomy of Environmental Justice. *Environmental Law Reporter* 30: 10681.

Mosville Environmental Action Now v. USA (2010). Petition 242-05, Report No 43/10, Admissibility (March 2010). [Online]. Available at: www.worldcourts.com/iacmhr/eng/decisions/2010.03.17_Mossville_Environmental_Action_Now_v_United_States.pdf [Accessed: 1 September 2018].

Need to ensure a healthy environment for the well-being of individuals (1991). UN GA Res 45/94, UN GAOR, 45th Sess., at 2, UN Doc.A/RES/45/94.

Office of the High Commissioner for Human Rights (OHCHR) (n.d. a). Human Rights Bodies – Individual Complaints [Online]. Available at: www.ohchr.org/EN/HRBodies/TBPetitions/Pages/HRTBPetitions.aspx [Accessed: 18 October 2018].

Office of the High Commissioner for Human Rights (OHCHR) (n.d. b). Special Rapporteur on Human Rights and the Environment [Online]. Available at: www.ohchr.org/EN/Issues/Environment/IEEnvironment/Pages/IEenvironmentIndex.aspx [Accessed: 18 October 2018].

Office of the High Commissioner for Human Rights (OHCHR) (n.d. c). Overview of the Mandate [Online]. Available at: www.ohchr.org/EN/Issues/Environment/IEEnvironment/Pages/Overview.aspx [Accessed: 10 October 2018].

Office of the High Commissioner for Human Rights (OHCHR) (n.d. d) [Online]. Available at: www.ohchr.org/EN/AboutUs/Pages/WhoWeAre.aspx [Accessed: 18 October 2018].

Our Common Future (1987). *Report of the World Commission on Environment and Development*. Oxford: Oxford University Press.

Pallemaerts, M. (2013). Preliminary Report [Online]. Available at: www.ohchr.org/EN/HRBodies/HRC/RegularSwssions/Swssion24/Documents/A-HRC-24-39_en.pdf [Accessed: 18 October 2018].

Payne, C.R. (2018). A Global Pact for the Environment. *ASIL Insights* 22 (12): September [Online]. Available at: www.asil.org/insights/volume/22/issue/12/global-pact-environment [Accessed: 18 October 2018]

Petition to the Inter-American Commission on Human Rights seeking relief for violations resulting from global warming caused by acts and omissions of the US (2005). [Online]. Available at: www.inuitcircumpolar.com/inuit-petition-inter-american-commission-on-human-rights-to-oppose-climate-change-caused-by-the-united-states-of-america.html [Accessed: 1 September 2018].

Regional Agreement on Access to Information, Public Participation and Justice in Environmental Matters in Latin America and the Caribbean ("P-10 Agreement") (2018). [Online]. Available at: https://repositorio.cepal.org/bitstream/handle/11362/43583/1/S1800428_en.pdf [Accessed: 24 October 2018].

Rio Declaration on Environment and Development (1992). A/CONF.151/26 (Vol. I) [Online]. Available at: www.un.org/documents/ga/conf151/aconf15126-1annex1.htm [Accessed: 10 October 2018].

Ruggie, J. (2011): Report of the Special Representative of the Secretary-General on the issue of human rights and transnational corporations and other business enterprises, A/HRC/17/31, Human Rights Council [Online]. Available at: www.ohchr.org/documents/issues/business/A.HRC.17.31.pdf [Accessed: 18 October 2018].

Sand, P.H. (2015). The Right to Know: Freedom of Environmental Information in Comparative and International Law. Reproduced in Peter H. Sand (ed.), *The History and Origin of International Environmental Law*. Cheltenham, U.K.: Edward Elgar, p.203.

Sands, P. and Peel, J. (2012). *Principles of International Environmental Law*, 3rd ed. Cambridge: Cambridge University Press, p.648.

Simons, P. (2012). International Law's Invisible Hand and the Future of Corporate Accountability for Violations of Human Rights. *Journal of Human Rights and the Environment* 3 (1): 5.

UN OHCHR (2018). UN expert calls for global recognition of the right to safe and healthy environment (5 March 2018). [Online]. Available at: www.ohchr.org/EN/NewsEvents/Pages/DisplayNews.aspx?NewsID=22755&LangID=E [Accessed: 24 October 2018].

UNECE Convention on Access to Information, Public Participation in Decision-Making and Access to Justice in Environmental Matters (Aarhus Convention) (1998)

[Online]. Available at: www.unece.org/fileadmin/DAM/env/pp/documents/cep43e.pdf [Accessed: 18 October 2018].

United Nations (1945). Charter of the United Nations, 1 UNTS XVI.

United Nations General Assembly (2007). UN Declaration on the Rights of Indigenous Peoples. Adopted by the UN General Assembly in resolution 61/295 [Online]. Available at: www.un.org/esa/socdev/unpfii/documents/DRIPS_en.pdf [Accessed: 18 October 2018].

United Nations General Assembly (1991). Need to Ensure a Healthy Environment for the Well-being of Individuals. GA Res 45/94, p.2 [Online]. Available at: www.un.org/en/ga/search/view_doc.asp?symbol=A/RES/45/94&Lang=E&Area=RESOLUTION [Accessed: 18 October 2018].

UN Human Rights Council (n.d.). *Universal Periodic Review* [Online]. Available at: www.ohchr.org/en/hrbodies/upr/pages/uprmain.aspx [Accessed: 18 October 2018].

Wates, J. (2005). The Aarhus Convention: A New Instrument Promoting Environmental Democracy. In Segger, M. C. and Weeramantry, C.G. (eds.), *Sustainable Justice: Reconciling Economic, Social and Environmental Law*. Leiden, Netherlands: Martinus Nijhoff, p.393.

Weissbrodt, D. and Mahling, W. (n.d.) Highlights of the 46th Session of the Sub-Commission [Online]. Available at: www1.umn.edu/humanrts/demo/subrept.htm [Accessed: 18 October 2018]

World Resources Institute (n.d.). The Access Initiative (TAI). [Online]. Available at: www.wri.org/our-work/project/access-initiative-tai [Accessed: 18 October 2018].

3 Pros and cons of a human rights-based approach to environmental protection

Introduction

As we discussed in Chapter 2, victims of environmental degradation are increasingly resorting to the human rights framework to seek redress. They have been able to obtain relief by way of compensation and other remedies. The human rights framework has been successful in providing relief to victims, especially at the regional level, when they have not been able to get any relief from their national institutions and judiciaries. We will discuss these cases in Chapters 4 and 8. But is human rights a good framework to seek relief for environmental degradation? What are the advantages and what are the drawbacks? In this chapter, we will discuss these as well as the limitations inherent in the human rights framework itself.

Contradictory or complementary?

For most part, international protection of the environment and international protection of human rights developed in different fora (Anton and Shelton, 2011), without much interaction until the late 1980s. Despite this, their close relationship was recognized by states, as was acknowledged at the Stockholm Conference in 1972 (Stockholm Declaration: Principle 1).

Some believe that despite having distinct origins, "the protection of the environment and the promotion of human rights are increasingly seen by many as intertwined, complementary goals" (Anton and Shelton, 2011: 119). Others, however, reject this connection, arguing that they are basically incompatible. According to them, "human rights and environmental protection are based on fundamentally different and ultimately irreconcilable value systems; as such they are much more likely to conflict with each other than to be complementary or mutually reinforcing" (Anton and Shelton, 2011: 119). Environmental lawyers, on the one hand, argue that a human rights focus will detract attention from other species and will reduce all environmental values to an instrumental use for humanity. Human rights lawyers, on the other hand, argue that linking human rights and environment "diminishes the importance and focus on protection of more immediate human rights concerns" (Anton and Shelton, 2011: 119). Many believe that human rights violations are more serious than damage caused by environmental degradation.

There are several reasons for these divergent views. Human rights violations are often more visible than environmental degradation. They are also more immediate than environmental damage, except where accidents cause severe human suffering, as is the case in Bhopal and Chernobyl. Moreover, not all human rights violations are the result of environmental degradation and not all environmental damage results in human rights violations. However, to the extent that environmental degradation or pollution results in human rights violations, then the human rights framework becomes a useful tool to the victims.

Some argue that existing rights can be interpreted expansively to include environmental rights. Indeed, this is how many courts have articulated environmental rights. Others insist that international law must adopt a substantive right to a healthy environment. As we saw in Chapter 2, there is a definite momentum towards accepting environmental rights even though there is also resistance to the idea. This trend is most visible at national and regional levels. The Paris Agreement adopted in 2015 achieved the honor of being the first environmental treaty to include a specific provision on human rights, even though the provision itself could have been better drafted and the language has watered down the typology of human rights found in human rights treaties (UNFCCC, 2015).

Scholars believe that human rights law and environmental protection interrelate in four different ways. First, when drafting environmental instruments, relevant human rights guarantees are emphasized. In this situation, victims have to rely on weak enforcement mechanisms in environmental treaties. The second approach is where existing human rights law is invoked where human rights are threatened by environmental degradation. This approach relies on the existing sophisticated redress mechanisms of human rights law. The third approach is to recognize a distinct right to an ecologically-balanced and sustainable environment. This formulation recognizes that the environment should be safe, not just for humans, but for other species as well. This avoids the criticisms of a purely anthropocentric approach to environmental protection but "strict environmentalists continue to object to the anthropocentrism inherent in taking a human rights approach to environmental protection" (Anton and Shelton, 2011: 131). The fourth approach addresses the issue as a matter of human responsibilities, rather than as rights.

It is important to recognize that, despite the overlap between them, human rights and environmental protection remain distinct areas of law. It is equally important to recognize their inter-relationship. An environment of a particular quality is necessary to enjoy many of the rights recognized under human rights law. In a degraded or polluted environment, it becomes difficult to enjoy protected rights. Indeed, human beings need services provided by nature to survive. According to UN Water, between 5–10 million people die every year due to polluted drinking water, most of them women and children living in poverty. Freshwater shortages are already raising concerns about the potential for conflicts and exacerbating existing ones. Climate change is predicted to cause conditions to deteriorate further. Air pollution, soil contamination, and loss of biodiversity, food crops, agricultural lands, and traditional lands result in a violation of protected rights. Thus, examining the link between environmental degradation and human rights is vital.

Furthermore, human rights can strengthen environmental protection. Access to information and participation in decision-making enhance environmental protection. Unlike human rights violations, private actors are usually responsible for environmental degradation, and strengthening compliance mechanisms plays an important role. These participatory rights (or access rights) in turn strengthen civil society participation and its watchdog role. States become responsible for their failure to monitor these activities or to prevent damage to citizens. Many of these rights overlap with principles of good governance. Thus, ensuring these rights also leads to beneficial results by ensuring that governmental entities are held accountable for their decisions.

Because international human rights law does not yet recognize a distinct right to a healthy environment, victims have resorted to using existing rights to articulate environmental claims. Some scholars believe that, given developments at regional and national levels, the time is ripe to recognize a stand-alone right to a healthy environment. Critics, on the other hand, claim that this right cannot be properly defined and, in any event, efforts to do so may divert attention from other, more established, rights (Anton and Shelton, 2011: 132). However, laws often respond to social problems and threats to the enjoyment of rights. They also reflect emerging social values. Indeed, "as environmental protection comes to be perceived as fundamental to human dignity and well-being, it moves towards the requisite acceptance. The growing awareness of the breadth and depth of environmental crises can be seen in increasing recourse to rights language" (Anton and Shelton, 2011: 132). Indeed, the threat that climate change poses to the enjoyment of rights may well be the turning point.

Proponents argue that the judiciary should be given the opportunity to define the parameters of the right which would be neither non-derogable nor inalienable. Courts will have to decide whether, and in what circumstances, individual rights will be superseded by the greater good of society or the benefit of the activity in question for the national economy. They will also consider whether government entities should be given a margin of appreciation and whether their decisions should be given deference. The courts can use existing standards developed by international organizations, such as the World Health Organization, when deciding whether the environment has become unhealthy for the inhabitants. The principles that have evolved through the jurisprudence of regional systems of human rights can provide guidance to international bodies.

However, there is no denying that the human rights framework has limitations vis-à-vis environmental damage. The main limitations are its vertical nature and territorial application. Human rights operate essentially within the area where states exercise territorial sovereignty, and not beyond. Thus, where there is transboundary environmental damage, the reach of human rights law becomes rather limited. In other words, victims of transboundary pollution cannot rely on human rights law, as the perpetrator of the damage is not the victim's national state. An example will illustrate this point.

States A and B are neighboring states. There is an industry operated by a private company near the border with State B which emits a large amount of fumes untreated to the atmosphere, resulting in air pollution and respiratory problems for the citizens of State B. The question that arises is whether there is any recourse open to the citizens of State B under international human rights law against State A. Because human rights law operates vertically between states as duty-bearers and their citizens as rights-holders, State A does not have any obligations under international human rights law towards citizens of State B. Because State B is not the perpetrator, it bears no responsibility under human rights law towards its citizens, although it may have to take temporary measures to protect them and provide with medical treatment. If the pollution is severe and affects a whole region, State B might take up the issue with State A through diplomatic channels. This issue will be explored in Chapter 13.

On the other hand, human rights can give a human face to a particular problem and bring attention to the victims. Many environmental issues lack a human face and, therefore, the public finds it difficult to relate to these environmental issues, as they seem rather abstract. Climate change is a good example of an environmental problem that did not have a human face until recently. The Inuit petition succeeded in giving climate change a human face.

What are the advantages and disadvantages of a human rights-based approach to environmental protection, and of strengthening the human rights regime in the context of contemporary environmental challenges? We will discuss some contrasting perspectives in this chapter. It is important to realize that, while a human rights-based approach offers legal guidance, helps define concrete obligations and the respective duty-bearers, there are also some disadvantages when employing a human rights framework to environmental protection. These mainly relate to the nature of human rights themselves. In many contexts, human rights are still contested. Developing countries with their colonial history fear that liberal (predominantly Western) states will interfere in their internal affairs under the pretext of protecting human rights. They believe that this could come hand-in-hand with conditionalities for development cooperation or trade agreements. Moreover, it makes the policy field of the environment much more complex, and can lead to situations in which actors have to make decisions that are not in their area of expertise. We will now consider various arguments for and against a human rights-based approach to environmental protection.

Advantages

Acknowledging complex interlinkages

Environmental challenges are already undermining a broad range of human rights and it becomes increasingly difficult to deny the complex relationship between environmental challenges, environmental policies, and human rights. On the one hand, the consequences of environmental change have adverse effects on the enjoyment of human rights. On the other hand, environmental policies, such as those relating to the Green Economy, as well as climate mitigation and adaptation endeavors, can lead to the infringement of rights of local communities, in particular of indigenous peoples.

In the face of climate change, all three categories of human rights – civil and political rights; economic, social, and cultural rights; and collective rights – can be at risk. Environmental impacts, including heat waves, floods, storms, droughts, and exceptional weather events can – in extreme cases – threaten civil and political rights, like the right to life. It is codified in the International Covenant on Civil and Political Rights (ICCPR, 1966) and the Convention on the Rights of the Child (CRC, 1989). Sea levels rise, temperature increases, and changes in precipitation can also negatively affect the right to food, the right to water, the right to health, and the right to adequate housing. All of these social rights are part of the International Covenant on Economic, Social and Cultural Rights (ICESCR, 1966). At the same time, they are also anchored in several core treaties of the UN Human Rights System, such as the 1979 Convention on the Elimination of All Forms of Discrimination against Women (CEDAW), the 1989 CRC, and the 2006 Convention on the Rights of Persons with Disabilities (CRPD). Cultural rights, for example, are also affected by the implications of environmental degradation. Human rights arguments in this area are raised to strengthen international cooperation for adaptation and mitigation measures to protect people and world heritage, like historic buildings, archeological sites, or unique landscapes, from adverse consequences such as unpredictable weather conditions, temperature increases, floods, extreme weather events, and droughts (Maus, 2014).

Environmental degradation and climate change also affect collective rights, such as the right to self-determination. It is anchored in the ICESCR and the ICCPR, and holds that peoples should be free to determine their political status as well as their economic, social, and cultural development. The right to self-determination can be threatened by rising sea levels and extreme weather events endangering the territorial existence of low-lying small island states. It can also be at risk if indigenous peoples lose their traditional land as a source of subsistence and culture. When human beings have to abandon their territory due to sea levels rising or severe weather events, their legal status and protection in the international system is unsettled (OHCHR, 2009; Atapattu, 2015).

Moreover, environmental and climate impacts adversely affect the collective right to a healthy environment. Such a right is not yet anchored in any international human rights treaty. Nevertheless, it included in regional conventions, such as the 1981 African Charter on Human and Peoples' Rights, the 2003 Protocol to the African Charter on Human and Peoples' Rights on the Rights of Women in Africa, the 2004 Arab Charter on Human Rights, and the 1988 Additional Protocol to the American Convention on Human Rights. Thus, it can be considered to be a binding human right in several world regions. In addition, there is a global trend to adopt environmental rights in national or federal constitutions (May and Daly, 2014).

On the other hand, there are empirical studies demonstrating that the implementation of environmental and climate policies can also lead to human rights violations. This concerns mitigation measures adopted under the climate regime such as Reducing Emissions from Deforestation and Forest Degradation (REDD+) programs (Faustino and Furtado, 2014), Clean Development Mechanism (CDM) projects (Schade and Obergassel, 2014), and Green Economy (GE) strategies (Bratman, 2015). It has been highlighted that the transition process to renewable energy and a low-carbon economy can lead to human rights violations of local communities, indigenous peoples, and (agro)-pastoral groups.

Infringements of rights pertaining to environmental and climate policy implementation often occur in the context of conflicts around property, land, and natural resources. Thus, in extreme cases, violations of the right to life have been reported, particularly when violent relocations were undertaken, or where people resisted governmental resettlement decisions (Human Rights Watch, 2012). Furthermore, the right to property specified in Article 17 of the Universal Declaration of Human Rights (UDHR, 1948), and the right to development recognized in the Declaration on the Right to Development (Orellana, 2010), are often neglected. Similarly affected is the right to self-determination and the right to dispose of one's own means of subsistence, as well as related social rights to adequate housing, to food, to water, and to health (ICESCR, 1966). Thus, human rights are at risk due to severe environmental impacts, and they can be infringed by the implementation of environmental policies if a rights-based approach is not adopted. Recognizing the intrinsic relationship between human rights and the environment is critical to the adoption of sound laws and policies.

From environmental justice claims to just environmental practices

Human rights can bridge the divide between normative environmental justice claims and just environmental practices. Human rights constitute a normative concept in political theory and telling us how something "should" be, in an ideal world. But at the same time, human rights instruments, like the International Covenant on Economic, Social and Cultural Rights, also entail concrete implementation measures. This means they prescribe detailed steps states have to take in order to implement the obligations in the human rights instrument they have ratified. These include adopting enabling legislation, amending existing legislation and harmonizing it with international standards, and adopting policies to realize the rights accepted. These could entail, for example, education policies, health policies, or social protection schemes. In this way, human rights can be understood to be both normative and empirical. If we think back to the normative claims in relation to environmental justice we discussed in Chapter 1, we can understand human rights as a yardstick telling us how environmental justice claims can be put into practice through respecting, protecting, and fulfilling human rights.

Since the adoption of the Universal Declaration of Human Rights, recognition for equal and inalienable rights has been understood as the foundation for justice (UDHR, 1948: Preamble). Today, scholars increasingly link existing human rights obligations to the environment (Knox, 2015a), to climate change (Knox, 2015b), and to climate justice (Humphreys, 2014). John Knox, the former UN Special Rapporteur on Human Rights and Environment, provides examples of such linkages, emphasizing that states already have duties, such as providing access to environmental information and protecting the right of free expression and right of association in relation to environmental issues, and they need to provide for the right to participation in environmental decision-making (Knox, 2015a). At the 2015 Paris Climate Conference, John Knox also reminded the negotiating state parties that: "States' human rights obligations also encompass climate change," and that "they must ensure that all of their actions comply with their human rights obligations. That includes their actions relating to climate change" (Knox, 2015b).

In fact, there is no single state in the world that has not committed to at least one of the UN core conventions on human rights (Liese, 2006; Knox, 2015b). This means every government has to respect, protect, and fulfill certain core standards, also in the face of environmental challenges. Maintaining and complying with these core human rights standards is the first step to establishing just environmental practices. However, human rights are not static, but dynamic, and international, as well as regional, human rights architectures will further develop in the face of severe environmental challenges, such as climate change. We can already observe some of these new developments. These include discussions on extraterritorial state obligations relating to the responsibilities of states for transboundary environmental harm, the institutionalization of a collective right to a healthy environment at the regional level, e.g., in the African Charter of Human and Peoples' Rights, but also increasing adoption of environmental rights in national constitutions and legislation.

Making use of the powerful language of human rights

The language of human rights is strong and emphasizes the need for immediate political action – it cannot be easily ignored in contemporary politics (Hiskes, 2009: 2). State actors from liberal and democratic countries, where human rights are anchored in constitutions, national legislation, and political discourse, consider them to be an integral part of their identity (Risse, Ropp, and Sikkink, 1999: 23). This means that framing environmental concerns as a human rights issue emphasizes the urgency of the matter: if we want to protect and provide for basic rights standards, we need to address environmental challenges swiftly and thoroughly.

Authoritarian regimes and some developing states, or states like the United States, may take a different perspective. Some of them are rather skeptical when it comes to viewing environmental and climate change through a human rights lens. The United States fear historic responsibilities and costly obligations in relation to past, present, and future emissions. Other countries fear that questions concerning the environment could be used to intervene in their internal affairs disregarding their state sovereignty. They also suspect that this could lead to new forms of conditionalities, i.e., imposing conditions to democratize countries, in order to receive development aid. There are some models in international relations scholarship, however, suggesting the ways in which human rights can be used by transnational advocacy networks and domestic opposition groups to initiate a change towards adopting basic rights standards in these countries. The Boomerang Pattern or the Spiral Model of Human Rights Change emphasize the way human rights norms become adopted and internalized if there is pressure from above (i.e., from transnational human rights networks) and from below (i.e., from domestic opposition groups) that can be exerted and upheld. This particularly applies to repressive, less democratic states (Risse, Ropp, and Sikkink, 1999, 2013). These models could be relevant when applying a human rights-based approach to environmental challenges. Human rights change within a country can help a state's citizens to access information regarding the environment, to participate in environmental decision-making, or to exhaust legal remedies when they are in any way negatively affected by an environmental or climate policy.

Embedded in a human rights discourse, environmental change is framed a matter of immediate human urgency, as opposed to a technocratic matter of scientific policy solutions or inter-state negotiations. It emphasizes a "pressingly relevant" need for action employing the "muscular language of human rights" (Hiskes, 2009: 3–5), and it can unleash argumentative, persuasive, or pressure mechanisms exerted on state actors to change their policies and make them more human-centered. The employment of pressure and persuasion mechanisms is more likely to succeed when international human rights NGOs, like Human Rights Watch and Amnesty International, take up issues like climate change together with international organizations, like the Office of the High Commissioner for Human Rights (Human Rights Watch, 2012, 2015; Amnesty International and Human Rights Watch, 2015; Greenpeace and Amnesty International, 2015).

Placing human beings at the center of political action

By employing a human rights-based approach to the environment, the plight of individual and collective rights-holders becomes central to the discussion. Simon Caney has suggested that human rights standards are helpful in defining a minimum threshold of what is acceptable, for example, when it comes to the impact of climate change on human beings (Caney, 2005, 2006). Thus, human rights agreements have a hybrid function; they create standards for states to be applied in relation to individuals, i.e., they define norms that are relevant within societies for the respective rights-holders (Risse, 2004: 228). Human rights define the relationship between a state government and its citizens (Schmitz and Sikkink, 2013: 827). This means they also emphasize responsibilities states have in protecting their citizens affected by environmental challenges, such as pollution, floods, droughts, or water contamination. This focus on the rights-holders is also important if one takes the transboundary character of environmental change and its unequal implications on certain community groups and individuals within society into consideration. Particularly vulnerable groups, or those that are historically marginalized, such as women, children, and the elderly, disabled, or indigenous peoples, and standards for their protection move to the center of attention and policy solutions.

This is also increasingly important in the area of climate change, where state negotiations have for a long time merely focused on international responsibilities and complex technical solutions although vulnerable groups, especially those in developing countries, are already experiencing negative consequences of climate change. Even though states will not always come to a satisfactory agreement during climate negotiations, immediate political action needs to be taken to protect the societal groups that are already suffering. An advantage of a human rights-based approach is that it does not only point to vulnerability, but supports an understanding of affected individuals and communities as active rights-holders who can make claims towards the government and the international community. The concept of rights also embraces recognition of human dignity and equality, as stipulated in the Universal Declaration of Human Rights, building the foundation for peace and justice. To uphold human dignity in the context of a degrading environment means to respect, protect, and fulfill human rights.

Assigning obligations

Moreover, a human rights-based approach can be fruitful for assigning obligations. In many cases, these can be derived from the required implementation measures of the existing human rights treaties and instruments to which state parties have committed. Obligations exist between a ratifying state government and its citizens, but also between ratifying capable states and other, less capable, countries. In the Framework Principles on Human Rights and the Environment, for instance, John Knox specifies that the obligation to international cooperation does not mean that every state takes exactly the same action, but that commitments take respective capabilities into account. Thus, there are different requirements for states in different economic situations, including technical and financial assistance from developed states to other states, in order to protect the basic human rights of citizens from developing countries. These state obligations – known as the duty to cooperate, which forms one of the foundational principles of international law – have been anchored in the UN human rights edifice since the inception of international human rights. References to it can be found in the Charter of the United Nations, the Universal Declaration of Human Rights, and in the International Covenant on Economic, Social and Cultural Rights (ICESCR) (de Schutter et al., 2012: 1091–1104). Obligations of international cooperation are also stipulated in specific treaties, like the Convention on the Rights of the Child (1989), the Convention on the Elimination of All Forms of Discrimination Against Women (1979), and other conventions containing economic, social and cultural rights. The 2011 Maastricht Principles on Extraterritorial State Obligations constitute a recent attempt to strengthen extraterritorial obligations of states and international organizations, as well as state responsibility for the acts and omissions of non-state actors and transnational corporations (ETO Consortium, 2013: 7). These principles are discussed in Chapter 13.

Closely related to assigning obligations are mechanisms of accountability. Human rights assessments can help to identify accountable duty-bearers or just procedures to attribute accountability (ICHRP, 2008: 7). These are key questions in the environmental area to which mechanisms and actors of the human rights regime can contribute substantially.

Initiating cooperation programs

Human rights can build the constitutive basis for international cooperation programs. These programs can be based on obligations for international cooperation and assistance as stipulated in the International Covenant on Economic, Social and Cultural Rights (ICESCR, 1966) and other social rights instruments and reaffirmed in the Maastricht Principles (even though these are not binding). Article 2.1 of ICESCR reads:

> Each State Party to the present Covenant undertakes to take steps, individually and through international assistance and co-operation, especially economic and technical, to the maximum of its available resources, with a view to achieving progressively the full realization of the rights recognized in the present Covenant by all appropriate means, including particularly the adoption of legislative measures.

Cooperation in the progressive realization of human rights highlights the responsibilities, not only of individual states, but also of the international community in relation to economic, social, and cultural rights. Their implementation can be severely hampered in a situation of environmental challenge, such as severe weather events, changes in precipitation, or pollution. These situations can be a barrier to fully realizing the rights to health, food, water, adequate housing, or education. Thus, economic and technical resources through bilateral or multilateral cooperation programs may be needed in order to fulfill these human rights. This means cooperation in the environmental policy field can be based on states' commitment to progressively realize economic, social, and cultural rights and to offer assistance to other states if resources are available, as outlined in Article 2.1 of the ICESCR.

Another side of the story is that in the environmental and climate policy field, cooperation programs have sometimes led to an infringement of human rights, especially of indigenous peoples. One important example is the Barro Blanco project in Panama, a hydroelectric dam project that was temporarily suspended in 2015 and deregistered as a Clean Development Project (CDM) in 2016 due to strong opposition by the indigenous Ngöbe and Buglé communities. Protests arose because these groups had not been properly consulted before dam-building commenced, despite the fact that there were plans to flood riverbanks along the townships where these communities live. This meant the livelihood of 5,000 Ngöbe farmers were endangered with this project, as they could not use the fertile lands of the riverbeds for agriculture or to grow medical plants. This had an impact on their right to means of subsistence, including food and water. As some sacred places were planned to be flooded, the cultural rights of these indigenous peoples were also endangered. Programs under the CDM are cooperation projects that allow countries with a commitment to reduce greenhouse gas emissions, i.e., developed countries, to implement an emission reduction project in a developing country to earn certified emission reduction credits to meet their respective target. In the 2015 Paris Agreement, states agreed that:

> Parties should, when taking action to address climate change, respect, promote and consider their respective obligations on human rights, the right to health, the rights of indigenous peoples, local communities, migrants, children, persons with disabilities and people in vulnerable situations and the right to development, as well as gender equality, empowerment of women and intergenerational equity.
>
> (UNFCCC, 2015: Preamble)

In order to design climate action and other environmental programs consistent with human rights standards, states are increasingly resorting to the introduction of procedural rights and safeguards. In 2010, safeguards were introduced to Reducing Emissions from Deforestation and Forest Degradation (REDD+) programs. For projects under the Clean Development Mechanism (CDM) and the newly established Sustainable Development Mechanism (SDM) under the Paris Agreement, they are still under consideration. Procedural rights are of particular importance in environmental law. They establish a link between the state and civil society by fostering transparency and participation in environmental decision-making (Gupta, 2008). The most important procedural rights are the right to information, the right to participation, and the right to justice, the last-mentioned right usually meaning access to judicial and administrative recourse procedures. All of these rights are anchored in the 1948 Universal Declaration of Human Rights, the 1966 International Covenant on Civil and Political Rights, and other human rights instruments. Of far more influence in environmental matters, however, is the Convention on Access to Information, Public Participation in Decision-Making and Access to Justice in Environmental Matters, also known as the Aarhus Convention, which was adopted in 1998. Procedural rights are discussed in Chapter 6. At this point, it is important to be aware that "a human rights-based approach to climate change negotiations, policies and measures" (OHCHR, 2010) can help vulnerable states in cooperation with capable states and international organizations to face the consequences of climate change, to design just climate policies, and to implement the human rights of their citizens at the same time. This will be true for other environmental challenges as well.

Building an analytical umbrella

A human rights framework can build an analytical umbrella around many relevant issues pertaining to the environment. It accommodates various aspects and problems that are relevant in the face of environmental challenges, ranging from questions of human security, to migration and the protection of climate refugees, to gender equality and development concerns. The Office of the High Commissioner for Human Rights, for instance, has elaborated a human rights-based approach to development (OHCHR, 2006a) and a human rights-based approach to climate change (OHCHR, 2010). An important advantage of an approach based on human rights is that it uses universally recognized standards as a yardstick for political action. This allows for consistent policy-making in a world of complexity in which many policy fields interact with the environment and human rights. Another advantage is that within the human rights system, policy-makers can revert to established standards, implementation measures, and monitoring and evaluation procedures, as well as experienced actors and networks. A broad-ranging human rights framework accommodates a variety of issues relating to environmental justice (that cannot merely be treated in isolation from each other) with a view to protecting affected vulnerable groups and individuals as rights-holders.

Redress for victims of environmental abuse

Finally, obtaining redress for victims of environmental abuse is one of the main advantages of a human rights approach. Because the environmental framework lacks redress mechanisms akin to the human rights framework, victims often resort to the human rights framework to seek redress, provided other issues like causation, jurisdiction, and local remedies are satisfied. In addition, litigation can have a wider impact beyond the immediate parties to the case. As the Inuit petition before the Inter-American Commission of Human Rights showed us, litigation (even failed litigation) can draw attention to a particular problem and its victims, and also influence international negotiations and civil society efforts relating to climate change (Hunter, 2012; Atapattu, 2015).

Disadvantages

While a human rights approach holds a certain appeal in relation to environmental issues, the two areas do not necessarily fit well together, and some equate it to trying to fit square pegs in round holes. Let us now turn to some of the disadvantages or weaknesses that a human rights-based approach to environmental protection entails.

Enforcement

The first drawback of a human rights approach is characteristic of the human rights regime in general: it is usually confronted with enforcement and compliance challenges. Although some human rights conventions, such as the Convention on the Rights of the Child (CRC) and the Convention on the Elimination of All Forms of Discrimination Against Women (CEDAW), have near universal ratification (only the United States has not ratified the CRC), key gaps in implementation prevail in many countries (OHCHR, 2006b). In international relations scholarship on norms, this "compliance gap" (Dai, 2013) has been grasped in conceptual differentiations between "norm adoption" and the often-lacking "norm internalization" (Finnemore and Sikkink, 1998: 902–9), or between "prescriptive status" and defective "norm-consistent behavior" (Risse, Ropp, and Sikkink, 1999: 29–33). The progressive realization of economic, social, and cultural rights is especially a key challenge in international and regional human rights systems (which lack legal commitment by important states, like the United States, for instance). At the same time, rights such as water, food, health, and adequate housing are most severely endangered in situations of flooding, environmental pollution, changes in precipitation, or droughts. Many states, even liberal democracies, neglect certain human rights because action towards compliance is too costly or conflicts with other political goals that are being prioritized by respective governments. Developing countries lack the resources, infrastructure, and social services to implement economic, social, and cultural rights, and heavily depend on international assistance and cooperation to progressively realize them. Despite an advanced monitoring and reporting system, there are hardly any enforcement or sanctioning mechanisms at the international level in place. This means norm-violating states do not have to expect any punishments apart from "naming and shaming," a strategy that places governments that infringe human rights at the center of negative attention in public discourses. The chronic under-fulfillment of human rights objectives can – in the end – lead to skepticism towards adopting a human rights-based approach to environmental protection (Woods, 2010; Dudai, 2009).

Establishing cause and effect

Scholars and practitioners have pointed to the impossibility of establishing cause and effect when it comes to environmental implications and their effects on human rights (OHCHR, 2009; Dudai, 2009). Environmental destruction, pollution, and anthropogenic climate change have mainly been caused in industrialized states, mostly by private companies, but have devastating effects, predominantly among vulnerable societal groups in developing countries that often lack the resources to adapt. According to human rights law, the ratifying states are primarily responsible for respecting, protecting, and promoting the rights of those citizens living in their own territories. This means that developing states (who do not bear the main responsibility for climate change) have the obligation to respect, protect, and fulfill their citizens' human rights, whether or not they were the main culprits. If they lack the capacity to do so, other Member States and the international community have a duty to cooperate and provide international assistance. However, it is difficult to establish human rights responsibilities across borders (ICHRP, 2008: 4). Even if one particular country or private company within a state has demonstrably caused a serious amount of greenhouse gas emissions, we need to ask ourselves the following questions: which damage has this caused, can we clearly establish cause and effect, and where and whom would they have to compensate for this? This complex constellation of cause and effect over time, with public and private actors involved in different world regions can make identifying duty-bearers and assigning (extraterritorial) obligations an ambiguous task.

Overstretching the human rights catalogue?

Human rights scholars like Philip Alston, have raised concerns about potentially over-expanding the catalogue of international human rights, which comes at the risk of devaluing core human rights (Dudai, 2009: 304; Alston, 1984). Since 1998, the UN Commission on Human Rights, and its successor, the Human Rights Council have passed a number of resolutions on the relationship between the environment and human rights, the first one addressing dumping of toxic and dangerous waste. By 2017, six resolutions adopted by the Human Rights Council have focused entirely on interlinkages between climate change and human rights. The question is whether, next to non-binding resolutions and declarations, a new convention on the same level as the core UN human rights treaties (like the CRC or CEDAW) can evolve, potentially embracing an international human right to a healthy environment. There could be strong arguments for rights concerns in the context of environmental change to be covered under the ICESCR – and even adding another optional protocol to it can be regarded as a fairly ambitious endeavor (Dudai, 2009: 305). There are many arguments in favor of environmental rights, especially the urgent need for political action and for assigning responsibilities. Historically, human rights have been of a dynamic nature, and could be regarded as a response to current challenges and dominant political discourses. But the prevailing skepticism around economic, social, and cultural rights, and even more so regarding collective rights, will make the formulations and adoption (and implementation) of environmental human rights a difficult task to accomplish. However, there are some promising developments, as discussed in Chapter 2.

Human rights conflicting with environmental and climate policies

There is increasing evidence that environmental and climate policies have the potential to conflict with human rights and indigenous peoples' rights (see also, ICHRP, 2008: 5–6). Important examples are climate mitigation policies, such as Reducing Emissions from Deforestation and Forest Degradation (REDD+), and projects under the Clean Development Mechanism (CDM) or Green Economy (Green Growth) policies. These policies are a political response to environmental degradation and climate change. They are initiated to protect human beings on a large scale from the implications of a changing climate by reducing net emissions of greenhouse gases through sustainable forest management, as well as conserving and enhancing forest carbon stocks (REDD+), or through, for instance, fostering alternative energy sources (CDM, Green Economy policies). REDD+ projects, however, have in the past severely affected the rights of indigenous peoples who have used forests for their subsistence for generations. This is why procedural safeguards were introduced to REDD+ implementation during the 2010 climate negotiations in Cancun. However, there is still empirical evidence that human rights are neglected in the realization of these climate mitigation programs. The establishment of hydroelectric dams constitutes another important example in this area. Dams are often built under the framework of CDM or Green Economy policies in order to create renewable energy sources. However, they often lead to situations in which local population groups are (sometimes violently) relocated, their access to water is denied, and they cannot sustain their livelihoods or practice their culture and traditions on sacred lands. Affected groups are very often indigenous peoples and (agro-)pastoralists. This, again, leads to complex situations of injustice: those who have contributed the least to climate change suffer disproportionately from political actions taken in this regard, although they have the fewest resources to adapt. As long as climate policies are not in accordance with human rights standards, policy-making in this area is inconsistent and problematic, but many states are still opposed as standards can make climate policies more complex and costlier.

Decision-making for future generations

In Chapter 1, we introduced the concept of inter-generational justice as an important dimension of environmental justice. Inter-generational justice issues are difficult to implement in practice by current decision-makers. So far, they have mostly been addressed by scholars in political theory, philosophy, and law (Hiskes, 2005, 2009; Caney, 2006; Shue, 2014). Although some scholars argue that constitutionally guaranteeing a fundamental human right to an adequate and healthy environment today would enable favorable conditions for rights protection in the future (Hayward, 2007; Hiskes, 2009), several concerns remain. Who exactly can speak for future generations, how can they be represented, and in what way can we assess their interests (Dudai, 2009: 302)? Can we convince today's policy-makers to take future interests into account? Will the current electorate agree with this, despite different immediate interests? Or are environmental problems of the next decades too far away in time and space to discuss them today? Protecting the rights of future generations in a changing environment means reshaping current political systems and institutions (Beckman and Page, 2008: 529), which is quite an ambitious endeavor in a world that is already confronted with complex problems and a multitude of diverging interests. Questions like democratic representation of future generations and guaranteeing a healthy environment to our descendants are often too disconnected from today's political realities and agendas. Sustainable development also requires that we take into account the interests of future generations when taking decisions today. Some countries have established institutions to represent the rights of future generations: Hungary's Ombudsman for future generations is a good example (FDSD, n.d.).

Ideological and disciplinary tensions

In addition to the reasons discussed above, there are ideological tensions between human rights and environmental movements. Whereas the human rights movement has an anthropocentric orientation, placing human beings in the center of decision-making, ecologists prefer to follow an eco-centric approach, focusing on the entire ecosystem. Ecologists criticize those with an anthropocentric orientation because they claim that this has led to severe environmental destruction and the devastating climatic challenges we are facing today. They argue that human beings are only one component of the ecosystem, and protection of the entire ecosystem should have priority in political decision-making. Often, ecologists are also opposed to capitalism as the main principle of economic order in most countries today. They argue that capitalism – which started to dominate the societies of many developed countries during times of industrialization – can be understood as one of the fundamental causes of all problems relating to the environment. Capitalism, however, is very much accepted among most human rights practitioners, because it usually goes hand in hand with liberal democratic political systems that respect human rights. This tension can lead to different priorities when it comes to the formulation of policy programs, specifically on issues like development, economic growth, or population control (e.g., Dudai, 2009), and can lead to a situation in which a human rights framework may lack the necessary support of environmental groups. The discourse on the right to development is a good example of this tension. Even though sustainable development and the right to development originated almost at the same time in the UN, it took the international community decades to narrow the gap between these two and incorporate sustainable development into the right to development discourse, which ultimately culminated in the adoption of Sustainable Development Goals.

In addition to these ideological tensions, there are also disciplinary differences (see also ICHRP, 2008: 3). Today, environmental science is an interdisciplinary academic subject, but it has its origins in the natural sciences. A human rights-based approach originating in law and social sciences might simply not be acceptable for everyone. Each discipline will, of course, also work with different theoretical assumptions, research questions, and methods, and a human rights perspective might not be considered as relevant for all scholars and practitioners working on environmental problems. In the area of climate change, for instance, negotiations for a considerable period of time focused on technological and scientific solutions. Many negotiators were opposed to taking up different issues, such as those based on justice and human rights concerns, because the debates were so complex already and a consensus was difficult to find. They argued that human rights would just further complicate the matter (Wallbott and Schapper, 2017). Therefore, references to rights norms have only slowly entered the climate change negotiations and can be found in the Cancun Agreement (2010) and the Paris Agreement (UNFCCC, 2015). But still, not everyone concerned with environmental problems, especially those working in different disciplinary fields, agrees with the need for a human rights-based approach to the environment.

The contested nature of a collective human right to a healthy environment

The human right to a healthy environment as a collective right is severely contested. Collective rights belong to the third generation of human rights and, except for the right to self-determination, are not anchored in the UN human rights edifice. Many states, in particular liberal democracies, question the universality of collective rights, although they are part of a regional human rights instruments, like the African Charter on Human and Peoples' Rights (Banjul Charter, 1981). The Banjul Charter states that: "All peoples shall have the right to a general satisfactory environment favourable to their development" (Banjul Charter, 1981: Article 24). Thus, environmental rights have found entry into regional conventions and many national constitutions but there is resistance against the establishment of an internationally recognized human right (above national law). Many Western states are rather careful when it comes to formulating or agreeing to the adoption of collective rights, as they do not conform to their commitment to liberalism as a political ideology. Liberalism emphasizes individualism and individual freedoms. Human rights and liberties are promoted to limit the power of the state. Limited governmental interference plays a crucial role in liberalism, and such interferences is only acceptable if an individual harms others who need to be protected by the state. Another important aspect is compensation. Western industrialized countries are mainly responsible for causing severe damage to the environment. If there were a collective right to a healthy environment, these states would have to agree to binding legal commitments towards their own citizens that could be quite costly. This is why many states are hesitant to accept such commitments. Although there is an intensive academic discourse in political theory and philosophy encompassing normative claims for a right to ecological space (Hayward, 2007), environmental rights (Hiskes, 2009), climate rights (Brandstedt and Bergman, 2013), or concrete formulations for a substantive human right for the protection of the environment (Turner, 2014), its development in practice is still controversial.

Conclusion

This chapter discussed the pros and cons of a human rights approach to environmental issues. What is interesting about the convergence between human rights and the environment is that both civil and political rights, and economic, social, and cultural rights, as well as procedural rights have been invoked to seek redress for damage caused by environmental harm, blurring the traditional distinction between different generations of rights (Atapattu, 2015: 106).

To summarize the disadvantages here – environmental issues encompass a much broader category of species than would be covered by a human rights approach. Species other than human beings obviously do not enjoy human rights, although it can be argued that by ensuring a healthy environment for humans, we are indirectly protecting the environment and other species. Moreover, there is an increasing tendency to adopt rights of nature, which is discussed in Chapter 15.

Secondly, environmental problems often affect a large number of people, whereas human rights violations are generally individual in nature (except, of course, massive violations of human rights such as genocide, apartheid, and war crimes, and rights which can only be exercised collectively, such as freedom of association). Moreover, environmental issues affect future generations and are sometimes transboundary in nature. Furthermore, human rights claims usually require a prior violation, whereas the main focus of environmental protection regimes is to prevent future environmental degradation. Furthermore, environmental issues do not fall neatly within the existing generations of rights; civil and political rights, as well as economic, social, and cultural rights, and procedural rights have been invoked in relation to environmental issues, while some contend that environmental rights are better categorized as third-generation rights (Marks, 1980–81; cf., Alston, 1982). The traditional remedies available under the human rights framework, such as compensation, may not be suitable for communities that are losing their traditional lands, and even their territories, due to climate change. Finally, in the absence of a specific right to a clean environment, the claimants have to establish a causal link between the environmental issue and the protected right. If this link cannot be established, the claim will fail. Given these drawbacks, it can be argued that a human rights approach may not be best suited for many kinds of environmental issues.

Despite these drawbacks, the human rights approach holds a certain appeal for victims of environmental damage. It gives a human face to environmental issues and draws attention to the victims. Moreover, victims of environmental damage have the opportunity to obtain relief for the harm they suffered. In addition, such litigation strategies can draw political attention to the issue in question, leading to action at the domestic level and further negotiations at the international level (Hunter, 2009). However, it must be stressed that adopting a human rights approach to environmental issues should not be taken as endorsing an anthropocentric approach to environmental issues. Human rights should complement the eco-centric approach, not supplant it.

DISCUSSION QUESTIONS

1. Which arguments are stronger from your perspective – those in favor or those opposed to a human rights-based approach to environmental protection?
2. Do you think a human right to a healthy and sustainable environment would "overstretch" the human rights catalogue?
3. What are the advantages of framing environmental degradation and climate change as a human rights concern?
4. Why are some states opposed to adopting a human rights-based approach to environmental protection? What are the political reasons behind this?
5. Can a human rights-based approach to environmental protection diminish inter-generational injustice?

References

African Charter on Human and Peoples' Rights (Banjul Charter) (1981) [Online]. Available at: www.achpr.org/instruments/achpr/ [Accessed: 20 October 2018].

African Union (2003). Protocol to the African Charter on Human and Peoples' Rights on the Rights of Women in Africa. Maputo, Mozambique.

Alston, P. (1982). A Third Generation of Solidarity Rights: Progressive Development or Obfuscation of International Human Rights Law? *Netherlands International Law Review* 29 (3): 307–32.

Alston, P. (1984). Conjuring Up New Human Rights: A Proposal for Quality Control. *American Journal of International Law* 78 (3): 607–21.

Amnesty International and Human Rights Watch (2015). Joint Statement: Protection of Human Rights from Climate Change Requires Urgent Shift to 100% Renewable Energy for All [Online]. Available at: www.amnesty.org/en/documents/pol30/3033/2015/en [Accessed: 17 June 2016].

Atapattu, S. (2015): The Role of Human Rights Law in Protecting Environmental Rights in South Asia. In: Haglund, L. and Stryker, R. (eds.), *Closing the Rights Gap: From Human Rights to Social Transformation.* Oakland, CA: University of California Press, p.105.

Beckman, L. and Page, E. (2008). Perspectives on Justice, Democracy and Global Climate Change. *Environmental Politics* 17 (4): 527–35.

Brandstedt, E. and Bergman, A.K. (2013). Climate Rights – Feasible or Not? *Environmental Politics* 22 (3): 394–409.

Bratman, E. (2015). Passive Revolution in the Green Economy: Activism and the Belo Monte Dam. *International Environmental Agreements: Politics, Law and Economics* 15 (1): 61–77.

Caney, S. (2005). Cosmopolitan Justice, Responsibility and Global Climate Change. *Leiden Journal of International Law* 18 (4): 747–75.

Caney, S. (2006). Cosmopolitan Justice, Rights and Global Climate Change. *Canadian Journal of Law and Jurisprudence* 19 (2): 255–78.

Convention on the Rights of the Child (CRC) (1989) [Online]. Available at: www.ohchr.org/en/professionalinterest/pages/crc.aspx [Accessed: 20 October 2018].

Dai, X. (2013). The "Compliance Gap" and the Efficacy of International Human Rights Institutions. In: Thomas, R., Ropp, S., and Sikkink, K. (eds.), *The Persistent Power of Human Rights: From Commitment to Compliance*. Cambridge: Cambridge University Press, pp.85–102.

Dudai, R. (2009). Climate Change and Human Rights Practice: Observations on and Around the Report of the Office of the High Commissioner for Human Rights on the Relationship between Climate Change and Human Rights. *Journal of Human Rights Practice* 1(2): 294–307.

de Schutter, O., Eide, A., Khalfan, A., Orellana, M., Salomon, M.E., and de Siderman, I. (2012). Commentary to the Maastricht Principles on Extraterritorial Obligations of States in the Area of Economic, Social and Cultural Rights. *Human Rights Quarterly* 34 (4): 1084–169.

ETO Consortium (2013). Maastricht Principles [Online]. Available at: www.etoconsortium.org/nc/en/main-navigation/library/maastricht-principles/?tx_drblob_pi1[downloadUid]=23 [Accessed: 20 October 2018].

Faustino, C. and Furtado, F. (2014). The Green Economy, Forest Peoples and Territories: Rights Violations in the State of Acre, Fact-finding and Advocacy Mission. Preliminary Report. Rio Branco, Uruguay: World Rainforest Movement.

FDSD (n.d.). Hungarian Parliament's Deputy Commissioner for Fundamental Rights, the Ombudsman for Future Generations [Online]. Available at: www.fdsd.org/ideas/the-hungarian-parliaments-ombudsman-for-future-generations/ [Accessed: 20 October 2018].

Finnemore, M. and Sikkink, K. (1998). Norm Dynamics and Political Change. *International Organization* 52(4): 887–917.

Greenpeace and Amnesty International (2015). Joint Statement: Protection of Human Rights from Climate Change Requires Urgent Shift to 100% Renewable Energy for All [Online]. Available at: www.amnesty.org/en/documents/pol30/3033/2015/en/. [Accessed: 17 June 2016].

Hayward, T. (2007). Human Rights Versus Emissions Rights: Climate Justice and the Equitable Distribution of Ecological Space. *Ethics and International Affairs* 21 (4): 431–50.

Hiskes, R.P. (2005). The Right to a Green Future: Human Rights, Environmentalism and Intergenerational Justice. *Human Rights Quarterly* 27 (4): 1346–64.

Hiskes, R.P. (2009). *The Human Right to a Green Future: Environmental Rights and Intergenerational Justice*. Cambridge: Cambridge University Press.

Human Rights Watch (2012). *What Will Happen if Hunger Comes? Abuses Against the Indigenous Peoples of Ethiopia's Lower Omo Valley*. Research Report. New York, NY: Human Rights Watch.

Human Rights Watch (2015). *There is No Time Left: Climate Change, Environmental Threats, and Human Rights in Turkana County, Kenya*. Research Report. New York, NY: Human Rights Watch.

Hunter, D. (2009). The Implications of Climate Change Litigation: Litigation for International Environmental Law-Making. In: Burns, W.C.G. and Osofsky, H.M. (eds.), *Adjudicating Climate Change: State, National, and International Approaches*. Cambridge: Cambridge University Press, pp.357–74.

ICHRP (2008). *Climate Change and Human Rights: A Rough Guide*. Geneva: International Council on Human Rights Policy.

International Covenant on Civil and Political Rights (ICCPR) (1966). Adopted and Opened for Signature, Ratification and Accession by General Assembly Resolution 2200 A (XXI) of 16 December 1966.

International Covenant on Economic, Social and Cultural Rights (ICESCR) (1966). Adopted and opened for signature, ratification and accession by General Assembly resolution 2200 A (XXI) of 16 December 1966.

Knox, J.H. (2015a). Human Rights, Environmental Protection, and the Sustainable Development Goals. *Washington International Law Journal* 24 (517) [Online]. Available at SSRN: https://ssrn.com/abstract=2660392 [Accessed: 20 October 2018].

Knox, J.H. (2015b). COP21: 'States' Human Rights Obligations Encompass Climate Change' – UN Expert [Online]. Available at: www.ohchr.org/EN/NewsEvents/Pages/DisplayNews.aspx?NewsID=16836&LangID=E [Accessed: 20 October 2018].

League of Arab States (2004). *Arab Charter on Human Rights*. Tunis, Tunisia.

Liese, A. (2006). *Staaten am Pranger. Zur Wirkung internationaler Regime auf innerstaatliche Menschenrechtspolitik*. Wiesbaden: VS Verlag für Sozialwissenschaften.

Maus, S. (2014). Hand in Hand against Climate Change: Cultural Human Rights and the Protection of Cultural Heritage. *Cambridge Review of International Affairs* 27(4): 699–716.

May, J.R. and Daly, E. (2014). *Global Environmental Constitutionalism*. Cambridge: Cambridge University Press.

Marks, S.P. (1981). Emerging Human Rights: A New Generation for the 1980s? *Rutgers Law Review* 33 (2): 435–52

OHCHR (2006a). *Frequently Asked Questions on a Human Rights-Based Approach to Development Cooperation*. Geneva: United Nations, Office of the United Nations High Commissioner for Human Rights.

OHCHR (2006b). *Annual Report* 2006. Geneva: United Nations, Office of the United Nations High Commissioner for Human Rights.

OHCHR (2009). *Report of the Office of the United Nations High Commissioner for Human Rights on the Relationship between Climate Change and Human Rights*. A/HRC/10/61, Human Rights Council, 15 January 2009. Geneva: Office of the High Commissioner for Human Rights.

OHCHR (2010). *Applying a Human Rights-Based Approach to Climate Change Negotiations, Policies and Measures*. Geneva: Office of the High Commissioner for Human Rights.

Orellana, M.A. (2010). Climate Change and the Millennium Development Goals: The Right to Development, International Cooperation and the Clean Development Mechanism. *SUR: International Journal on Human Rights* 7 (12): 145–71.

Organization of American States (1988). Additional Protocol to the American Convention on Human Rights in the Area of Economic, Social and Cultural Rights ("Protocol of San Salvador"). San Salvador, El Salvador.

Risse, T. (2004). Menschenrechte als Grundlage der Weltvergemeinschaftung? Die Diskrepanz zwischen Normanerkennung und Normeinhaltung. In: Volker, R. (ed.), *Weltpolitik heute: Grundlagen und Perspektiven*. Baden-Baden: Nomos, pp.223–44.

Risse, T., Ropp, S.C., and Sikkink, K. (eds.) (1999). *The Power of Human Rights: International Norms and Domestic Change*. Cambridge: Cambridge University Press.

Risse, T., Ropp, S.C. and Sikkink, K. (eds.) (2013). *The Persistent Power of Human Rights: From Commitment to Compliance*. Cambridge: Cambridge University Press.

Schade, J. and Obergassel, W. (2014). Human Rights and the Clean Development Mechanism. *Cambridge Review of International Affairs* 27 (4): 717–35.

Schmitz, H.P. and Sikkink, K. (2013). International Human Rights. In: Walter, C., Risse, T., and Simmons, B.A. (eds.), *Handbook of International Relations*. Thousand Oaks, London and New Delhi: SAGE, pp.827–51.

Shue, H. 2014. *Climate Justice: Vulnerability and Protection*. Oxford: Oxford University Press.

The Petition to the Inter American Commission on Human Rights Seeking Relief from Violation Resulting from Global Warming Caused by Acts and Omissions of the United States, Summary of the Petition (2005). The Inuit petition. [Online]. Available at: http://earthjustice.org/sites/default/files/library/legal_docs/petition-to-the-inter-american-commission-on-human-rights-on-behalf-of-the-inuit-circumpolar-conference.pdf.

Turner, S.J. (2014). *A Global Environmental Right*. Abingdon: Earthscan.

UN Water [Online]. Available at: www.unwater.org/.

UNECE Convention on Access to Information, Public Participation in Decision-Making and Access to Justice in Environmental Matters (Aarhus Convention) (1998) [Online]. Available at: www.unece.org/fileadmin/DAM/env/pp/documents/cep43e.pdf [Accessed: 18 October 2018].

United Nations (1979). *Convention on the Elimination of All Forms of Discrimination against Women*. New York: United Nations General Assembly.

United Nations (2006). *Convention on the Rights of Persons with Disabilities*. New York: United Nations General Assembly.

United Nations Framework Convention on Climate Change (UNFCCC) (2015). *Paris Agreement*. Adopted at the 21st Conference of the Parties (COP), 12 December 2015. United Nations Framework Convention on Climate Change: Paris.

Universal Declaration of Human Rights (UDHR) (1948). Adopted by the United Nations General Assembly G.A. res. 217A (III), U.N. Doc A/810 at 71 (1948). United Nations General Assembly: Paris.

Wallbott, L. and Schapper, A. (2017). Negotiating By Own Standards? The Use and Validity of Human Rights Norms in UN Climate Negotiations. *International Environmental Agreements: Politics, Law and Economics* 17(2): 209–28.

Woods, K. (2010). *Human Rights and Environmental Sustainability*. Cheltenham and Northampton: Edward Elgar Press.

4 Regional systems of human rights

Introduction

Much of the convergence between human rights and the environment has taken place at the regional level. Currently, there are three regional human rights systems in the world: the European Court of Human Rights (the oldest) established under the auspices of the Council of Europe; the Inter-American Commission and Court of Human Rights established under the auspices of the Organization of American States; and the African Commission; and more recently, the Court of Human and Peoples' Rights (the youngest) established under the auspices of the Organization for African Unity. These three systems have been at the forefront of articulating environmental rights and have developed a rich jurisprudence on the subject. Attempts to establish other regional systems have been made, but currently, a large part of the world remains outside the reach of a regional human rights institution.

This chapter discusses the jurisprudence generated by the three regional human rights systems. After a brief discussion of the regional bodies, the chapter will discuss seminal cases decided by them. It will also refer to some of the important cases pending before them.

Regional systems of human rights

One of the significant developments since the Second World War has been the proliferation of international institutions dealing with a multitude of global challenges, ranging from health to education, and from refugees to development. Institutions dealing with human rights have also proliferated, leading to a sophisticated network of international and regional institutions. The United Nations (UN) takes the pride of place as the apex international body established by states with the main task of maintaining international peace and security. Promoting respect for human rights is another important purpose of the UN. The European Union is the only true supra-national, albeit regional, body we have today. Not only have these international organizations flourished and influenced the development of international law, some of them have also acquired international legal personality (*Reparations case*, 1949).

Amidst the proliferation of international institutions, the establishment of regional human rights systems is an important development. Apart from the three regional systems referred to earlier, there are examples of other similar efforts emerging. The Arab Charter on Human Rights was adopted by the League of Arab Nations in 1994 and was revised in 2004. It established a commission which can review state reports (Anton and Shelton, 2011). The Charter includes a right to a healthy environment as a component of the right to an adequate standard of living: "Every person has the right to an adequate standard of living for himself and his family which ensures their well-being and a decent life, including food, clothing, housing, services and the right to a healthy environment" (Arab Charter, 2004). Another example is the establishment of the ASEAN Inter-Governmental Commission on Human Rights by the Association of South East Asian Nations (ASEAN) in 2009. At present the Commission has consultative status but cannot receive individual complaints. In addition, the ASEAN Commission on the Promotion and Protection of the Rights of Women and Children was established in 2010 (Anton and Shelton, 2011). Interestingly, the ASEAN Human Rights Declaration embodies a provision on the right to a healthy environment under economic, social and cultural rights: "Every person has the right to an adequate standard of living for himself or herself and his or her family including: the right to a safe, clean and sustainable environment" (ASEAN Declaration, 2012: Article 284). It also includes the rights to adequate and affordable food, freedom from hunger, access to safe and nutritious food, and to safe drinking water and sanitation. Furthermore, it links good governance and democracy with human rights. The Preamble reaffirms "adherence to the purposes and principles of ASEAN as enshrined in the ASEAN Charter, in particular the respect for and promotion and protection of human rights and fundamental freedoms, as well as the principles of democracy, the rule of law and good governance" (ASEAN Declaration, 2012: Preamble).

The Asian Human Rights Commission, on the other hand, is an independent non-governmental body and functions as a watchdog in the region. It has succeeded in drawing attention to various human rights abuses in the region, but it is neither an inter-governmental body nor an adjudicatory body. The Asian Human Rights Charter was adopted on the occasion of the 50th anniversary of the Universal Declaration of Human Rights in 1998. It was drafted by a group of scholars headed by Professor Yash Ghai, a prominent scholar in Asia, and many NGOs participated in the drafting process. The Charter has been endorsed by various organizations, including NGOs and peoples' organizations, and was adopted to create a popular culture of human rights in the region. The Charter contains an explicit provision on sustainable development and protection of the environment:

> Economic development must be sustainable. We must protect the environment against the avarice and depredations of commercial enterprises to ensure that the quality of life does not decline just as the gross national product increases. Technology must liberate, not enslave human beings. Natural resources must be used in a manner consistent with our obligation to future

> generations. We must never forget that we are merely temporary custodians of the resources of nature. Nor should we forget that these resources are given to all human kind, and consequently we have a joint responsibility for their responsible, fair and equitable use.
>
> (Asian Human Rights Charter, 1998: Article 2.9)

In addition, it includes a right to a healthy environment within the right to life provision:

> Foremost among rights is the right to life, from which flow other rights and freedoms. The right to life is not confined to mere physical or animal existence but includes the right to every limb or faculty through which life is enjoyed. It signifies the right to live with basic human dignity, the right to livelihood, the right to a habitat or home, the right to education and *the right to a clean and healthy environment for without these there can be no real and effective exercise or enjoyment of the right to life*. The state must also take all possible measures to prevent infant mortality, eliminate malnutrition and epidemics, and *increase life expectancy through a clean and healthy environment* and adequate preventative as well as curative medical facilities. It must make primary education free and compulsory.
>
> (Asian Human Rights Charter, 1998: Article 3.2, emphasis added)

Although it called upon Asian states to establish regional or sub-regional institutions for the promotion and protection of human rights with the adoption of an inter-state Convention, this has so far not materialized.

Like the Asian Human Rights Commission, the Asia Pacific Forum (n.d.) is not an inter-governmental body. It is a coalition of National Human Rights Institutions (NHRIs) in the region. It provides support to the NHRIs to become powerful advocates for human rights in their countries. A proposal for a Pacific Human Rights mechanism has also been floated, but nothing concrete has so far taken place (Hay, 2009).

We will now turn to a discussion of the three regional human rights bodies and the jurisprudence generated by them on environmental rights.

The European human rights system

The European Convention for the Protection of Human Rights and Fundamental Freedoms was adopted in 1950 under the auspices of the Council of Europe (Pedersen, 2008). Originally, it established a two-tier system of human rights – a commission and a court – and individuals did not have direct standing before the court. Only the Commission could refer cases to the court. The Inter-American system, discussed next, followed this model. The two institutions were later merged into one, and now there is only a European Court of Human Rights before which individuals have direct standing, provided local remedies have been exhausted. Judge Buergenthal summarizes the significant role played by the Court:

> Over time, the European Court of Human Rights for all practical purposes has become Europe's constitutional court in matters of civil and political rights. Its judgments are routinely followed by the national courts of the state parties to the Convention, their legislatures, and their national governments. The Convention itself has acquired the status of domestic law in most of the state parties and can be invoked as such in their courts.
>
> (Buergenthal, 2006: 792)

The European Convention does not recognize a distinct right to a healthy environment. This is not surprising given that the Convention was adopted in 1950, a time when environmental protection was not a priority. However, when the Council of Europe discussed whether the Convention should be amended to include a right to a healthy environment, the consensus seemed to be that this was not necessary as the Court has used existing rights to articulate a right to a healthy environment (Council of Europe, 2012). Shelton calls this a whiplash against environmental rights:

> Also, at the regional level, the preambles to European Community directives often state their aim "to protect human health and the environment." Yet, the Council of Europe's Committee of Ministers has repeatedly rejected proposals from the Parliamentary Assembly to add a protocol to the European Convention on Human Rights that would provide a right to a safe and healthy environment to the regional guarantees. The European Social Charter mentions only the right to a safe working environment.
>
> (Shelton, 2015: 336)

To be fair, the Court has used existing rights creatively to adjudicate environmental issues. This section will discuss the jurisprudence generated by the European Court in articulating environmental rights. The Council of Europe's *Manual on Human Rights and Environment* summarizes the Court's jurisprudential basis for environmental rights as follows:

> The Court has already identified in its case-law issues related to the environment which could affect the right to life (Article 2), the right to respect for private and family life as well as the home (Article 8), the right to a fair trial and to have access to a court (Article 6), the right to receive and impart information and ideas (Article 10), the right to an effective remedy (Article 13) and the right to the peaceful enjoyment of one's possessions (Article 1 of Protocol No. 1).
>
> Likewise, the Committee has interpreted the right to protection of health (Article 11) under the European Social Charter as including the right to a healthy environment.
>
> (Council of Europe, 2012)

The first case that reflected the convergence between human rights and environmental protection to reach the Court was *Powell v. United Kingdom* in 1990. It dealt with the impact of excessive noise levels from Heathrow Airport on citizens living in the vicinity. The applicants complained that the noise pollution from Heathrow Airport violated their right to privacy under Article 8 of the European Convention on Human Rights. The Court held that although there was an interference with the local residents' quality of life, it was justified by the greater economic interest of the community at large. In coming to this conclusion, the Court adopted a two-prong test: (a) whether the state had a positive duty to protect the applicants' right to privacy; and (b) whether a fair balance has been struck between competing interests:

> In both contexts regard must be had to the fair balance that has to be struck between the competing interests of the individual and of the community as a whole; and in both contexts the State enjoys a certain margin of appreciation in determining the steps to be taken to ensure compliance with the Convention.
>
> (Anton and Shelton, 2011: 487)

A similar decision was reached by the court in *Hatton v. United Kingdom*. In *Lopez Ostra v. Spain*, the Court found in favor of the complainant. Mrs. Ostra and family lived in a town in Spain that had a heavy concentration of leather industries. Due to fumes and smells emanating from tanneries, many people developed health problems, including Mrs. Ostra's family. Expert evidence submitted in court showed that the symptoms displayed by Mrs. Ostra's daughter were consistent with living in a highly polluted area. The daughter's pediatrician recommended that she be moved from the area. The Court found a violation of Article 8 of the European Convention on Human Rights on the right to private and family life (Atapattu, 2018).

Since 1990, the European Court of Human Rights has developed a significant body of jurisprudence on environmental rights. The Court has not only addressed instances of actual damage to people, it has also dealt with situations that have posed an unacceptable risk to people in the vicinity. While most cases dealt with damage that has already taken place, in *Taskin v. Turkey*, the Court "found a violation with respect to the operation of a gold mine, despite the absence of any accidents or incidents with the mine. The mine was deemed to present an unacceptable risk." The Court further held that the procedural guarantees available to the applicants have not been fulfilled (Anton and Shelton, 2011: 502).

Through its jurisprudence, the Court has developed several rules in relation to environmental rights: a fair balance should be struck between competing interests of the individual and the community as a whole; states enjoy a certain margin of appreciation; and courts will generally give deference to the decisions taken by public authorities. The Court has also applied principles of international environmental law, including procedural rights and referred to judgments of other regional courts and international bodies, including the ICJ (Atapattu, 2018).

Interestingly, in *Florea v. Romania* (2010), Article 3 of the Convention, which prohibits inhuman or degrading treatment, was invoked in relation to passive smoking. In two previous cases on passive smoking – *Aparicio Benito v. Spain* (No. 2) (2006) and *Stoine Hristov v. Bulgaria* (2009) – the applicants had invoked Article 2 (the right to life) and Article 8 (the right to respect for family life), not the prohibition on torture. However, the Council of Europe believes that "at present there is not sufficient case-law to be able to draw up any clear principles on environmental protection at the European level" in relation to passive smoking (Council of Europe, 2012: 8).

The latest case to be filed before the European Court relates to climate change and involves all 47 Member States. The London-based Global Legal Action Network (GLAN) plans to bring legal action against all signatories of the European Convention of Human Rights on behalf of six children from Portugal who were affected by the fires in the summer of 2017. Scientists believe that climate change is a contributing factor to these fires, which killed more than 60 people. According to the journalist writing on the proposed legal action:

> GLAN, representing the six Portuguese children, says European countries are not sufficiently cutting greenhouse gas emissions to cut the risk of global warming-related disasters and protect human life. The group wants to file suit against all 47 signatories to the European Convention on Human Rights simultaneously and wants the court to establish emissions cuts for each country and require all of them to cut the use of fossil fuels.
>
> (Magill, 2017)

According to the GLAN webpage, it has launched a crowdfunding campaign "to help bring a potentially ground-breaking case." They plan to argue that the countries that signed the European Convention must significantly strengthen their emissions reduction policies and commit to keeping their fossil fuel reserves in the ground. They plan "to build on the recent impact of cases like the Urgenda (Netherlands) and Our Children's Trust (United States) cases, which have proven the value of legal action as a tool in the fight against climate change" (GLAN webpage). It will be very interesting to see how the Court handles this petition. Exhaustion of local remedies is one of the requirements that complainants have to fulfill before they can approach the Court. Given that climate change is a global problem, and without the cooperation of states this issue cannot be addressed, in some respects, there are no local remedies to be exhausted for mitigation measures. Mitigation measures by all states are required if this situation is to be addressed in any meaningful manner. In that respect, this case may break new ground. In any event, bringing legal action against all 47 parties to the Convention is itself unprecedented.

In general, the Court allows a wide margin of appreciation to the state (see *Hatton and others*, *Buckley v. United Kingdom*) and will not question the decision of the state officials unless they had failed to take action to protect the rights of citizens. Likewise, states are allowed to weigh the rights of individuals on the one hand,

and the greater good of society, on the other. Usually, the latter will supersede the former (*Hatton v. United Kingdom, and Powell and Rayner v. United Kingdom*).

A case that will be filed in the European Court of Human Rights imminently involves a coal mine in Kosovo, the villagers of Hade, and the World Bank. The community leaders plan to take their dispute to the European Court after a World Bank decision failed to end 12 years of displacement caused by a coal mine expansion. Although the case is not directly related to climate change, because the case involves a coal mine, it can be taken as indirectly related to climate change and could have implications for how relocations involving coal mines are treated by human rights bodies. The Word Bank Inspection Panel decided that these villagers had been left "in limbo" for more than a decade after being relocated many times, but the government had failed to provide new housing and services. Ironically, the Bank is considering underwriting another coal power plant in the same area. This will lead to more displacement of people, not to mention the emission of more greenhouse gases contributing to climate change. If the Bank were serious about addressing climate change, it should stop funding any more coal power plants around the world (Atapattu, 2018).

The Inter-American system

The Inter-American Commission and Court of Human Rights were established under the American Convention on Human Rights, concluded in San Jose in 1969 and adopted within the Organization of American States (OAS). The American Declaration of the Rights and Duties of Man, adopted in 1949, preceded the Convention. While the Convention included only civil and political rights, other rights have been added through protocols. The Additional Protocol to the American Convention includes a right to a healthy environment:

1 Everyone shall have the right to live in a healthy environment and to have access to basic public services.
2 The States Parties shall promote the protection, preservation, and improvement of the environment (Article 11).

By ratifying the Convention, states are considered to have automatically accepted the jurisdiction of the Commission to hear cases brought against them by individuals (Anton and Shelton, 2011: 340, referring to Buergenthal). Similar to the European system at the time when it was originally adopted, the Inter-American system adopts a two-tier approach: only the Commission can refer petitions to the Court, and individuals have no standing before the Court. Over the years, the Inter-American Court has developed considerable jurisprudence on environmental rights, and it was the first regional body to recognize property rights of indigenous groups. It also developed the free, prior, and informed consent principle in relation to indigenous peoples. More recently, the Court issued a groundbreaking Advisory Opinion recognizing a stand-alone, justiciable right to a healthy environment.

Mayagna (Sumo) Awas Tingni Community v. Nicaragua was the first time that the collective property rights of indigenous peoples were recognized. In the case of *Mary and Carrie Dann v. US*, the Inter-American Commission held that any determination with regard to indigenous lands must "be based on fully informed consent of the whole community, meaning that all members be fully informed and have the chance to participate." In *Maya Communities of the Toledo District v. Belize* the Commission held that Belize violated property rights of indigenous peoples by granting concessions with the lands "without effective consultations with and the informed consent of the Maya people" (Ward, 2011). The Commission stressed that the duty to consult is a fundamental component of the state's obligations with regard to communal property rights, and that consultations should be held with the goal of obtaining consent.

The case of *Saramaka People v. Suriname* (2007) involved resource concessions that the Suriname government had granted to private companies within the territories of Saramaka people without obtaining their consent or even consulting them. The Court held that Suriname had violated the rights of Saramaka people to judicial protection and property rights and had failed to have effective mechanisms to protect them from acts that violate their rights to property. However, the Court noted that these property rights are not absolute, and the State has the right to restrict property rights in the interests of society. This is similar to the jurisprudence developed by the European Court. These restrictions must be: previously established by law; necessary; proportionate; and with the aim of achieving a legitimate objective in a democratic society. In addition, such restrictions cannot violate the right of indigenous peoples to survival. In order to do so, the Court prescribed a series of safeguards: (a) states must ensure effective participation of the affected parties; (b) states must guarantee that the affected people will receive a reasonable benefit from the project; and (c) prior to granting the concession, environmental and social impacts must be carried out to mitigate any negative effects.

Furthermore, the Court said that participation must be in line with their customs and traditions; states have a duty to disseminate and receive information; and consultations must be in good faith, culturally appropriate, and have the intent to reach an agreement. In the case of large-scale development projects that could impact the survival of indigenous people, states must obtain their free, prior, and informed consent (FPIC). Ward believes that this case clearly sets a precedent within the Inter-American system (Ward, 2011). This case endorses very important principles and sheds light on what "consultation" entails. The Court limited the application of FPIC to large-scale development projects that may threaten the survival of indigenous people. The language of the UN Declaration on the Rights of Indigenous Peoples (UNDRIP, 2007) in relation to FPIC, however, is not so restrictive. The need for consultations and the application of the principle of free, prior, and informed consent was endorsed by the Court again in *Kichwa People of Sarayaku v. Ecuador* (2012).

La Oroya mine in Peru is dubbed the most polluted place on earth. It has put the health of 35,000 people living near the lead smelter at serious risk.

After being cited for several environmental violations, the company closed the smelter in 2009. Having suffered for many years, a group of three NGOs filed a petition in the Inter-American Commission on behalf of a group of affected people in La Oroya. In *Community of La Oroya v. Peru* (2009), the Inter-American Commission held that the case was admissible with regard to several rights enshrined in the American Convention of Human Rights. Although the right to a healthy environment is included in the Additional Protocol to the American Convention, it is not justiciable (however, this position changed in the Advisory Opinion, discussed below). The Commission also held a public hearing on the issue.

The Inter-American Commission of Human Rights was the first regional institution to specifically recognize property rights of indigenous peoples. It also received the first complaint on climate change. Although the Inuit petition, discussed below, was dismissed, the Inter-American Commission held a hearing on the subject in 2007, and has since then explicitly recognized the link between climate change and human rights.

The Inuit petition[1]

The petition brought by the Inuit Circumpolar Conference on behalf of the Inuit of the United States and Canada against the United States in the Inter-American Commission of Human Rights is the first case that utilized the human rights framework to litigate climate change. Even though the case itself was dismissed, it was successful in other respects. It succeeded in giving a human face to the problem, as well as highlighting that the consequences of climate change were already taking place and were affecting real communities and their traditional way of life. The petition helped dispel the belief held by many that climate change would affect some unknown community in a distant future. Because climate change originated as an environmental problem, people still tend to consider it such. Recognizing that there are actual victims of climate change was important to highlight that climate change was already having a significant adverse impact on peoples' lives.

In their petition to the Inter-American Commission, the petitioners alleged that the United States as the biggest contributor to greenhouse gases is responsible for the damage that is being caused to the Arctic. The changes in the weather in the Arctic region are adversely affecting the petitioners' traditional way of life which, in turn, is affecting their human rights. The full petition, 175 pages in length, was complete with photographs, scientific information, and data from the 2004 Arctic Climate Impact Assessment and signed by over 60 Inuit people. The petition alleged that the United States is by far the largest emitter of greenhouse gases, and "thus bears the greatest responsibility among nations for causing global warming" (Inuit Petition, 2005).

The petition summarized the climate change impacts on the Inuit people as follows: because of the changes in the climate, it has become harder for the Inuit to predict weather, which is important for their hunting practices, travel, and food preservation methods. The Inuit culture, which is intrinsically linked to snow and ice, is being increasingly threatened by the changing climate:

> Like many indigenous peoples, the Inuit are the product of the physical environment in which they live. The Inuit have fine-tuned tools, techniques and knowledge over thousands of years to adapt to the arctic environment. They have developed an intimate relationship with their surroundings, using their understanding of the arctic environment to develop a complex culture that has enabled them to thrive on scarce resources. The culture, economy and identity of the Inuit as an indigenous people depend upon the ice and snow.
> (Inuit Petition, 2005)

The changing climate has given rise to many changes in the Arctic environment. Sea ice, which is a critical resource for the Inuit for travel, harvesting, and communication, has become thinner, freezing later and thawing early, thus affecting the everyday life of the Inuit. Moreover, the quantity, quality, and timing of snowfall have also changed. Lack of igloo quality snow is causing problems to travelers and is also resulting in loss of igloo-building knowledge, which is an important component of Inuit culture.

Humans are not the only species being affected by a changing climate. Marine species that depend on sea ice are likely to decline, with some even facing extinction. Other species that are important for harvesting are moving to new locations, exacerbating travel problems for the Inuit people. In addition, higher temperatures and sun intensity have increased the risk of health problems such as sunburn, skin cancer, and cataracts. Moreover, it is predicted that new diseases like West Nile Virus will pose health risks. The petition noted that while current impacts in the Arctic are severe, the projected impacts are expected to be much worse.

The petitioners noted that several principles of international law are relevant here. The United States is a member of the Organization of American States (OAS) and has signed the American Declaration of the Rights and Duties of Man (1948). Moreover, it has ratified the International Covenant on Civil and Political Rights and has signed the International Covenant on Economic, Social and Cultural Rights. In addition, the United States is bound by international environmental law obligations, including the obligation not to cause transboundary harm, and is a party to the United Nations Framework Convention on Climate Change (UNFCCC, 1992) under which it has committed to implementing policies to return to its 1990 levels of greenhouse gas (GHG) emissions.

The petitioners argued that the impacts of climate change, caused by acts and omissions by the United States, violate the Inuit's human rights protected by the American Declaration. These include the rights to culture, property, health, life, physical integrity, security, means of subsistence, residence, freedom of movement, and the inviolability of the home. Despite the ratification of the UNFCCC, the United States has repeatedly declined to take steps to regulate its GHG emissions. The petitioners alleged that the United States was the largest contributor to GHG emissions in the world and with full knowledge that its action is radically transforming the Arctic environment, "the United States has persisted in permitting the unregulated emission of greenhouse gases from within its jurisdiction into the atmosphere" (Inuit Petition summary, 2005: 7).

The petitioners requested the Commission to: (a) make an onsite visit to investigate the harm suffered by the Inuit; (b) hold a hearing; (c) prepare a report declaring that the United States is internationally responsible for violating the rights in the American Declaration; (d) recommend that the United States adopt mandatory measures to limit its emissions; (e) establish and implement a plan, in coordination with the affected Inuit, to protect Inuit culture and resources; (f) establish and implement a plan to provide the assistance necessary to adapt to the impacts of climate change that cannot be avoided; and (g) provide any other relief that the Commission considers appropriate and just (Inuit Petition, 2005).

The Commission dismissed the petition without prejudice, stating that the information provided was insufficient to make a determination. However, at the petitioners' request, the Commission held a hearing to address matters relating to Global Warming and Human Rights (Earthjustice, 2007a, b). Sheila Watt-Cloutier of the Inuit Circumpolar Conference, Martin Wagner of Earthjustice, and Daniel Magraw of Center for International Environmental Law attended the hearing. Although the Commission did not give a decision on the merits of the case, the petition broke new ground in international law.

The petition also exemplified the challenges inherent in this type of litigation. Given that climate change is a global problem, it involves multiple actors in every state, so establishing causation is challenging. Moreover, given the disproportionate contribution of states to causing the problem, the current state responsibility principles are ill-suited to apply to climate change. Another issue is the time lag between emissions and the consequences. In addition, damage to future generations cannot be captured within a human rights framework. Finally, what would an appropriate remedy be in this situation? For example, how can one compensate for loss of culture? (Atapattu, 2016). It is thus not very surprising that the Inuit petition was dismissed by the Inter-American Commission.

At the hearing held by the Commission on climate change and human rights, questions raised by the Commissioners offer "perhaps the best indication of the challenges that future litigation over human rights violations as consequence of climate change will face" (Chapman, 2010: 38): (a) how to attribute or divide responsibility among states in the region and even those who are not in the region; (b) how the rights violations suffered by the Inuit could be linked to acts or omissions of specific states; (c) whether petitioners had exhausted domestic remedies; and (d) are there any examples of good practice undertaken by states that could guide the Commission? (Atapattu, 2016).

Of course, there is no logical reason why we cannot base state responsibility on the common but differentiated responsibility principle (CBDR) (Atapattu, 2016). After all, the international community was willing to base their obligations relating to climate change on this principle – a huge breakthrough in international law (UNFCCC and Kyoto Protocol, 1997: Article 3). If obligations are based on the CBDR principle, the legal regime that comes into play when a violation of that obligation has taken place (called "state responsibility") should, in principle, be based on that principle. While this makes perfect sense in a perfect world, given how controversial the CBDR principle has become, it is highly unlikely that states will be willing to accept it in relation to liability as well.

Athabaskan petition

Closely resembling the Inuit petition, the Arctic Athabaskan Peoples' petition to the Inter-American Commission argues that black carbon pollution from Canada is harming the Arctic environment and the ecosystems of the Arctic region. Black carbon is a component of sooty fine-particle pollution. According to the petition, Canada emits roughly 98,000 tons of black carbon annually, and the major sources of the emissions are diesel emissions and the burning of biomass in agriculture and other sectors. Black carbon is a potent climate-warming agent. The Arctic is warming twice as fast as the rest of the world and black carbon emissions from Canada are a major contributor to this. The petitioners argued that because black carbon is a potent climate-warming agent, and is emitted near the Arctic, it has a significantly higher climate-warming impact. Arctic warming is dramatically changing the Arctic environment and damaging its natural resources. This, in turn, is having a huge impact on Athabaskan peoples and their way of life. As one Special Rapporteur concluded in 2005: "the effects of global warming and environmental pollution are particularly pertinent to the life chances of Aboriginal people in Canada's North, a human rights issue that requires urgent attention at the national and international levels" (Athabaskan petition, 2013: 3).

The petitioners argued that "[T]he rapid warming and melting in Athabaskan Lands, caused in significant part by Canada's failure to reduce black carbon emissions, violates a number of the Arctic Athabaskan peoples' fundamental rights" (Athabaskan petition, 2013: 3). These included the right to culture, property, means of subsistence, and health. The petition argued that while black carbon is considered a "short-lived" pollutant it is also identified as a particularly potent pollutant in regions of ice and snow. There are many steps that Canada can take to reduce its black carbon emissions which could have a significant impact on reducing the warming in the region. Arctic warming and melting is adversely affecting this indigenous group's ability to pass their cultural knowledge to future generations. This knowledge has been developed over millennia, but due to the warming and melting, the weather is no longer predictable, and the behavior of wildlife has become so erratic that elders are no longer confident about teaching the younger generation their traditional ways.

The petitioners further noted that the American Declaration guarantees their right to culture which has been affirmed by the Inter-American Court of Human Rights in several cases – the failure to prevent environmental damage to indigenous lands can cause "catastrophic damage" to indigenous peoples. For indigenous peoples, land is closely linked to their cultural practices, customs, and language. It has been repeatedly recognized that the relationship of indigenous peoples with their territory is "crucial for their cultural structures and requires special measures under international human rights law in order to guarantee their physical and cultural survival" (Athabaskan petition, 2013: 3–4). They further alleged that their right to subsistence is being affected by changing weather patterns, forest fires, and unpredictable storms, and other severe weather events. These events are also affecting their right to property guaranteed under the American Declaration. Moreover, the warming has led to a loss of traditional food which has adversely affected the health of Athabaskan peoples; as they can no longer obtain food through traditional hunting, fishing, and gathering, they have to supplement these foods with store-bought food which is less healthy leading to an increased prevalence of chronic diseases. Their water quality has also been affected and the likelihood of disease and injury due to dangerous conditions has caused psychological stress.

The petitioners requested the Commission to: (a) investigate and confirm the harms suffered by the Athabaskan people as a result of Arctic warming and melting; (b) prepare a report declaring that Canada's failure to take steps to substantially reduce its black carbon emissions violates the rights recognized in the American Declaration of Rights and Duties of Man; and (c) recommend that Canada take steps to limit black carbon emissions and protect Athabaskan culture and resources from the effects of Arctic warming and melting. This petition is currently pending before the Commission.

This case is very similar in scope to the Inuit petition. It remains to be seen how the Inter-American Commission will handle the petition. Unlike the Inuit petition, because the Athabaskan petition requests Canada to take action domestically to reduce black carbon which will have a noticeable impact in the Arctic region, there is a possibility that the Commission will consider the petition favorably. Moreover, unlike in 2007 when the Inuit petition was considered, there is much more attention paid to climate change by the international community, as well as by the Inter-American Commission itself. Scientists are warning of even more dire consequences than predicted before, so it would be hard for the Commission to turn a blind eye to this petition.

Advisory Opinion on human rights and environment

A groundbreaking Advisory Opinion on environmental rights was issued by the Inter-American Court of Human Rights in early 2018, not just for the recognition of a justiciable right to a healthy environment, but also with regard to extraterritorial human rights obligations.[2] On 14 March 2016, Colombia requested an Advisory Opinion from the Inter-American Court regarding the scope of the American Convention in the context of alleged "severe degradation of the human and marine environment of the Wider Caribbean Region that may result from the acts and/or omissions of States with coasts on the Caribbean Sea in the context of the construction of major new infrastructure projects." Colombia asked the Court for an opinion as to whether: (a) a person outside a state's territory falls under the jurisdiction of that state if the person is within the area of application of an environmental agreement to which that state is a party; (b) serious harm to the marine environment by another state would be incompatible with the obligations contained in the American Convention; and (c) the American Convention implies an obligation to respect the norms of international environmental law, including those pertaining to environmental impact assessment and cooperation with potentially affected states.

The Court gave affirmative answers to the questions submitted by Colombia. It highlighted that the proceedings constitute one of the first opportunities this Court has had to address the state obligations that arise from the need to protect the environment under the American Convention. It affirmed the right to a healthy environment as a right in itself, that multiple human rights are vulnerable to environmental degradation, and that, as a result, a series of environmental obligations lie with states for the fulfillment of their obligations to respect and guarantee of these rights.

The Court emphasized that the right to a healthy environment is an *autonomous right* which protects elements of the environment, such as forests, rivers, seas, and other such elements as legal interests in and of themselves, even in the absence of certainty or evidence of risk to individual persons. The reason is the environment's importance for the other living organisms with whom the planet is shared, also deserving protection in and of themselves. In an attempt at fleshing out the content of the right to a healthy environment, the Court took note of five components identified by the working group in charge of monitoring the implementation of the Protocol of San Salvador: (a) to guarantee to every person, without discrimination, a healthy environment; (b) to guarantee to every person, without discrimination, basic public services; (c) to promote the protection of the environment; (d) to promote the preservation of the environment; and (e) to promote the improvement of the environment. The Court also observed that different substantive and procedural obligations flow from the duty to protect other human rights as they relate to the environment, and focused the bulk of its opinion on the rights to life and personal integrity. There is no doubt this Advisory Opinion will contribute to the development of the law in relation to environmental rights, extraterritoriality (discussed in Chapter 13), and climate change and human rights (discussed in Chapter 9), as well as rights of nature (discussed in Chapter 15).

An important case currently pending before the Inter-American Commission is the *Mossville* case. This case was deemed admissible by the Commission in 2010. It involves 14 polluting factories in Mossville, Louisiana, the United States, which have been spewing noxious substances into the environment, polluting air and waterways and affecting the health of residents close by. The area is predominantly inhabited by African-American people, and the petition revolves around environmental racism, environmental justice, human rights, and environmental rights. The issue also caught the attention of the UN Special Rapporteur on Racism when he visited the United States in 1995. In *Mossville Environmental Action Now v. United States*, the petitioners alleged that the Mossville residents, who are predominately African-American, are subject to a disproportionate pollution burden, which they refer to as "environmental racism," in breach of the American Declaration of the Rights and Duties of Man. They further alleged that the State is responsible for the violation of Mossville residents' rights to equality, life, health, and privacy guaranteed under the American Declaration of Human Rights. This is the first case that has been brought against the United States before an international forum on the basis of environmental justice and, in particular, environmental racism. The remedies that the petitioners are seeking include medical services for residents suffering from illnesses related to the exposure to toxic substances; relocation of those who want to be relocated; refraining from issuing additional permits to polluting facilities; reforming the regulatory system to address multiple and cumulative exposure to toxic chemicals; and requiring a safe distance between residences and polluting facilities. The Commission deemed the claims relating to equality and privacy admissible (Gonzalez, 2017). While the case seems to tilt in favor of the petitioners, a favorable decision may be purely symbolic if the United States resists compliance:

> The Mossville petition illustrates both the advantages and disadvantages of human rights-based approaches to environmental protection in a country like the United States that has generally resisted international human rights law and institutions. The enforcement mechanisms of the international human rights system are notoriously weak and reflect the unwillingness of states to relinquish sovereignty over domestic affairs. However, this does not mean that victories in human rights tribunals are necessarily hollow or pyrrhic. On the contrary, the reports and decisions of human rights bodies must be regarded as means rather than ends, as tools to be used by social movements in the struggle to expose and remedy human rights abuses.
>
> (Gonzalez, 2017: 310)

Furthermore, a decision in favor of the petitioners could: "name and shame" the United States; put pressure on US decision-makers to respect, protect, and fulfill human rights of Mossville residents; provide environmental justice advocates with an additional tool to harmonize US and international anti-discrimination law; and give an opportunity to educate regulators in the United States about the relationship between human rights and environmental protection (Gonzalez, 2017). Moreover, it provides an opportunity to clarify and develop international law regarding the right to a healthy environment and the obligation of states to regulate corporations and non-state actors. Although the United States responded to the petition by denying the existence of a right to a healthy environment as a stand-alone right under the American Declaration, in light of the recent Advisory Opinion of the Court, this denial may become redundant. Above all:

> The Mossville petition gives the IACHR the opportunity to address the intersectional and multi-dimensional nature of environmental injustice and the indivisibility of human rights. Mossville residents are living in an industrial sacrifice zone not just because they are black, but also because they are too poor to move. Racial discrimination and material deprivation are often inextricably intertwined.
>
> (Gonzalez, 2017: 314, footnotes omitted)

Whether or not the United States complies with an adverse decision against it – and in the current environment, it seems unlikely – this case certainly highlights the multi-faceted and intersectional nature of injustices suffered by the Mossville petitioners.

African system of human rights

The baby of the three regional systems – the African system – consisted of only the African Commission of Human Rights until 2005. It was established by the Banjul Charter on Human and Peoples' Rights adopted by the African Union in 1981. The Protocol establishing the African Court of Human Rights came into force in January 2005 (Protocol to the African Charter, 1998; Anton and Shelton, 2011). The African Charter explicitly recognizes a collective right to a healthy environment: "All peoples shall have the right to a general satisfactory environment favorable to their development" (African Charter, 1981: Article 24).

In the seminal case of *The Social and Economic Rights Action Center and the Center for Economic and Social Rights v. Nigeria* (*Ogoniland case*) (2001), the Commission elaborated on the scope of Article 24 and what the right to environment means:

> The right to a general satisfactory environment, as guaranteed under Article 24 of the African Charter or the right to a healthy environment, as it is widely known, therefore imposes clear obligations upon a government. It requires the State to take reasonable and other measures to prevent pollution and ecological degradation, to promote conservation, and to secure an ecologically sustainable development and use of natural resources. Article 12 of the International Covenant on Economic, Social and Cultural Rights (ICESCR), to which Nigeria is a party, requires governments to take necessary steps for the improvement of all aspects of environmental and industrial hygiene. The right to enjoy the best attainable state of physical and mental health enunciated in Article 16(1) of the African Charter and the right to a general satisfactory environment favourable to development (Article 16(3)) already noted obligate governments to desist from directly threatening the health and environment of their citizens.
>
> (Ogoniland case: para. 52)

The Commission further elaborated on the right to environment:

> Government compliance with the spirit of Articles 16 and 24 of the African Charter must also include ordering or at least permitting independent scientific monitoring of threatened environments, requiring and publicizing environmental and social impact studies prior to any major industrial development, undertaking appropriate monitoring and providing information to those communities exposed to hazardous materials and activities and providing meaningful opportunities for individuals to be heard and to participate in the development decisions affecting their communities.
>
> (Ogoniland case: para. 53)

As is evident from this decision, a distinct right to a healthy (or satisfactory) environment is far from being vague and indeterminate. It imposes clear obligations on states to prevent pollution, promote conservation, and secure sustainable development. To achieve those objectives, states have to undertake monitoring, require the preparation of environmental and social impact studies, and provide information and meaningful opportunities for individuals to participate in development decisions (in the *Pulp Mills* case (2010), the ICJ stated that transboundary environmental impact assessment is now part of customary international law).

Conclusion

This chapter surveyed the three regional systems of human rights which have generated an impressive body of jurisprudence on environmental rights and indigenous rights. The Advisory Opinion of the Inter-American Court in March 2018 is likely to be the dawn of a new era of litigation relating to environmental rights and rights of nature.

DISCUSSION QUESTIONS

1. Are the regional human rights systems the best institution to develop environmental rights?
2. Asia lacks a regional human rights system. How will they develop environmental rights if a substantive right to a healthy environment is not accepted at the international level?
3. Can the proliferation of regional systems and their jurisprudence lead to fragmentation of international law?
4. How can we ensure that there is coordination among regional human rights systems?
5. What is the relationship between regional human rights law and international human rights law?

Notes

1 This section draws from the author's book: Atapattu, S. (2016). *Human Rights Approaches to Climate Change: Challenges and Opportunities*. New York, NY: Routledge.

2 This section draws from Campbell-Duruflé, C. and Atapattu, S. (forthcoming). The Inter-American Court's Environment and Human Rights Advisory Opinion: Implications for International Climate Law. *Climate Law*. We are indebted to Campbell-Duruflé for providing an English translation of the summary of the Court's opinion.

References

Additional Protocol to the American Convention on Human Rights in the Area of Economic, Social and Cultural Rights (Protocol of San Salvador) (1988), OAS Treaty Series No 69 [Online]. Available at: www.oas.org/juridico/english/treaties/a-52.html [Accessed: 20 October 2018].

African Charter on Human and Peoples' Rights (1981) [Online]. Available at: www.achpr.org/instruments/achpr/ [Accessed: 20 October 2018].

American Declaration of the Rights and Duties of Man (1948). 30 April 1948, Hein's No. KAV 7225.

Anton, D.K. and Shelton, D. (2011). *Environmental Protection and Human Rights*. Cambridge: Cambridge University Press.

Arab Charter on Human Rights (1994). [Online]. Available at: www.humanrights.se/wp-content/uploads/2012/01/Arab-Charter-on-Human-Rights.pdf [Accessed: 1 October 2018].

ASEAN Declaration on Human Rights (2012) [Online]. Available at: https://asean.org/asean-human-rights-declaration/ [Accessed: 20 October 2018].

Asian Human Rights Charter (1998) [Online]. Available at: asianhumanrights-eng%20 (1).pdf [Accessed 20 October 2018].

Asia Pacific Forum (n.d.) [Online]. Available at: www.asiapacificforum.net/ [Accessed 20 October 2018].

Atapattu, S. (2016). *Human Rights Approaches to Climate Change: Challenges and Opportunities*. London: Routledge.

Atapattu, S. (2018). Climate Change under Regional Human Rights Systems. In: Duyck, S., Jodoin, S., and Johl, A. (eds.), *Routledge Handbook of Human Rights and Climate Governance*. London: Earthscan, pp. 128–31.

Atapattu, S. (2018). Extractive Industries and Inequality: Intersections of Environmental Law, Human Rights, and Environmental Justice. *Arizona State Law Journal* 50 (2): 431–54.

Boer, B. (2015). Environmental Law and Human Rights in the Asia-Pacific. In: Boer, B. (ed.), *Environmental Law Dimensions of Human Rights, The Collected Courses of the Academy of European Law*. Oxford: Oxford University Press, p.135.

Buergenthal, T. (2006). The Evolving International Human Rights System. *American Journal of International Law* 100: 783.

Cahill-Jackson, J. (2012). Mossville Environmental Action Now v. United States: Is a Solution to Environmental Injustice Unfolding? *Pace International Law Review Online Companion* 3: 173–4.

Chapman, M. (2010). Climate Change and the Regional Human Rights Systems. *Sustainable Development Law & Policy* 37–8: 60–1.

Council of Europe (2012). Manual on Human Rights and the Environment [Online]. Available at: www.echr.coe.int/LibraryDocs/DH_DEV_Manual_Environment_Eng.pdf [Accessed: 20 October 2018].

Earthjustice (2007a). Inter-American Commission on Human Rights to Hold Hearing on Global Warming [Online]. Available at: http://earthjustice.org/news/press/2007/inter-american-commission-on-human-rights-to-hold-hearing-on-global-warming [Accessed: 20 October 2018].

Earthjustice (2007b). Testimony of Martin Wagner before the Inter-American Commission on Human Rights [Online]. Available at: http://earthjustice.org/news/press/2007/global-warming-human-rights-gets-hearing-on-the-world-stage [Accessed: 20 October 2018].

Earthjustice (2013). Petition to the Inter-American Commission on Human Rights seeking Relief from the Violations of Rights of Arctic Athabaskan Peoples Resulting from Rapid Arctic Warming and Melting Caused by Emissions of Black Carbon from Canada [Online]. Available at: http://earthjustice.org/sites/default/files/AAC_PETITION_13-04-23a.pdf [Accessed: 20 October 2018].

Global Legal Action Network (GLAN) (2017). Crowdfunding Campaign for Climate Change Legal Action Launched [Online]. Available at: www.glanlaw.org/single-post/2017/09/24/Crowdfunding-campaign-for-climate-change-legal-action-launched [Accessed: 1 October 2018].

Gonzalez, C.G. (2017). Environmental Racism, American Exceptionalism, and Cold War Human Rights. *Transnational Law and Contemporary Problems* 26: 281.

Gordon, J. (2007). Inter-American Commission to hold Hearing after rejecting Inuit Climate Change Petition. *Sustainable Development Law & Policy* 7 [Online]. Available at: http://digitalcommons.wcl.american.edu/cgi/viewcontent.cgi?article=1239&context=sdlp [Accessed: 20 October 2018].

Hay, K. (2009). A Pacific Human Rights Mechanism: Specific Challenges and Requirements, 40 VUWLR 195.

International Law Commission (2001). Draft Articles on Responsibility of States for Internationally Wrongful Acts [Online]. Available at: http://legal.un.org/ilc/texts/instruments/english/draft_articles/9_6_2001.pdf [Accessed: 20 October 2018].

Magill, B. (2017). *Climate Liability News* [Online]. Available at: www.climateliabilitynews.org/2017/09/27/youth-climate-lawsuit-portugal-wildfires/ [Accessed: 20 October 2018].

Mathiesen, K. (2016). Kosovan Villagers to Take Coal Mine Woes to Human Rights Court, 19 December 2016 [Online]. Available at: www.climatechangenews.com/2016/12/19/kosovan-villagers-to-take-coal-mine-woes-to-echr/ [Accessed: 20 October 2018].

Kyoto Protocol (1997) [Online]. Available at: https://unfccc.int/process/the-kyoto-protocol [Accessed: 20 October 2018]

Kramer, L. (2015). The EU Courts and Access to Environmental Justice. In: Boer, B. (ed.), *Environmental Law Dimensions of Human Rights, The Collected Courses of the Academy of European Law*. Oxford: Oxford University Press, p.137.

Kravchenko, S. and Bonine, J. (2008). *Human Rights and the Environment: Cases, Law, and Policy*. Durham, NC: Carolina Academic Press.

Mossville Report No. 43/10 (2010) Petition 242-05.

Osofsky, H.M. (2009). The Inuit Petition as a Bridge? Beyond Dialectics of Climate Change and Indigenous Peoples' Rights. In: Burns, W.C.G. and Osofsky, H.M. (eds.), *Adjudicating Climate Change: State, National, and International Approaches*. Cambridge: Cambridge University Press, p.273.

Pavaoni, R. (2015). Environmental Jurisprudence of the European and Inter-American Courts of Human Rights: Comparative Insights. In: Boer, B. (ed.), *Environmental Law Dimensions of Human Rights, The Collected Courses of the Academy of European Law*. Oxford: Oxford University Press, p.69.

Pedersen, O. (2008). European Environmental Human Rights and Environmental Rights: A Long Time Coming? *Georgetown International Environmental Law Review* 21: 73.

Protocol to the African Charter on Human and Peoples' Rights on the Establishment of the African Court on Human and Peoples' Rights (1998) [Online]. Available at: www.achpr.org/instruments/court-establishment/ [Accessed: 20 October 2018].

Shelton, D. (2006). Human Rights and the Environment: What Specific Environmental Rights Have been Recognized? *Denver Journal of International Law & Policy* 35: 129.

Shelton, D. (2015). Whiplash and Backlash – Reflections on a Human Rights Approach to Environmental Protection. *Santa Clara Journal of International Law* 13: 11.

Spieler, P. (2010). *The La Oroya Case: The Relationship Between Environmental Degradation and Human Rights Violations*, HUM. RTS. BRIEF [Online]. Available at: www.corteidh.or.cr/tablas/r25885.pdf [Accessed: 1 September 2018]

UN Declaration on Rights of Indigenous Peoples (UNDRIP) (2007) [Online]. Available at: www.un.org/esa/socdev/unpfii/documents/DRIPS_en.pdf [Accessed: 20 October 2018]

United Nations Framework Convention on Climate Change (UNFCCC) (1992) [Online]. Available at: https://unfccc.int/resource/docs/convkp/conveng.pdf [Accessed: 20 October 2018].

Ward, T. (2011). The Right to Free, Prior and Informed Consent: Indigenous Peoples' Participation Rights within International Law. *Northwestern Journal of International Human Rights* 10: 54.

Cases

African Commission on Human and Peoples' Rights, The Social and Economic Rights Action Center and the Center for Economic and Social Rights v. Nigeria (2001). [Online]. Available at: www1.umn.edu/humanrts/africa/comcases/155-96.html

Aparicio Benito v. Spain (2006). (Application no. 36150/03), Judgment of 13 November 2006

Buckley v. United Kingdom (1996). ECHR 39

Community of La Oroya v. Peru (2009). Petition 1473-06, Inter-American Commission on Human Rights H.R., Report No. 76/09, OEA/Ser.L/V/II., doc. 51, petition 1473-06, 5 August 2009

Florea v. Romania (2010) (Application no. 37186/03), Judgment of 14 September 2010; This case is only available in French, however, a summary can be found on the page below, or through the press release on HUDOC. [Online]. Available at: http://sim.law.uu.nl/sim/caselaw/Hof.nsf/1d4d0dd240bfee7ec12568490035df05/83ff6f5c75014d78c1257797002d1151?OpenDocument

Hatton v. United Kingdom (2003). 2003-VIII Eur. Ct. H. R. 189, 228 All IACtHR decisions available at www.corteidh.or.cr/index.php/en/jurisprudencia

Kichwa Indigenous People of Sarayaku v Ecuador (2012). IACtHR

Lopez Ostra v. Spain (1994). Application No 16798/90 20 European Human Rights Reports 277 (Eur. Ct. H.R.)

Mary and Carrie Dann v. US (2002). IACtHR

Maya indigenous community of the Toledo District v. Belize (2004). IACtHR

Mayagna (Sumo) Awas Tingni Community v. Nicaragua (2011). IACtHR

Mossville Envtl. Action Now v. United States (2010). Case 242-05, Inter-Am. Comm'n H.R., Report No. 43/10, OEA/Ser.LV/II.138, doc. 47 2. [Online]. Available at: http://srenvironment.org/regional-decisions/inter-american-commission/

The Petition to the Inter American Commission on Human Rights Seeking Relief from Violation Resulting from Global Warming Caused by Acts and Omissions of the United States,

Summary of the Petition (2005). (the Inuit petition). [Online]. Available at: http://earthjustice.org/sites/default/files/library/legal_docs/petition-to-the-inter-american-commission-on-human-rights-on-behalf-of-the-inuit-circumpolar-conference.pdf
Powell and Rayner v. United Kingdom (1990). 172 Eur. Ct. H.R. (ser. A) at 18–20
Pulp Mills on the River Uruguay (Argentina v. Uruguay) (2010). ICJ Rep
Reparation for Injuries Suffered in the Service of the United Nations (Reparations Case) (1949). Advisory Opinion of 11 April 1949, ICJ Reports
Saramaka People v. Suriname (2007). IACtHR
Stoine Hristov v. Bulgaria (2009). 16 January 2009
Taşkin v. Turkey (2004). 2004-X Eur. Ct. H.R. 185
The Social and Economic Rights Action Center and the Center for Economic and Social Rights v. Nigeria (2001). African Commission on Human and Peoples' Rights, Comm. No. 155/96

Part II

Human rights of special relevance to environmental protection

5 Substantive rights

Introduction

International human rights law has evolved considerably over the last few decades. In fact, 2018 marks the 70th anniversary of the adoption of the Universal Declaration of Human Rights (UDHR). This watershed moment in history when the international community vowed to put the atrocities of the Second World War behind them, and the impact it has had on international relations cannot be underestimated. This is not to say that the international community has succeeded in eliminating all violations of human rights. Far from it. Recent history is replete with examples of many human rights violations around the world, some individual and others more widespread. However, states are more cautious, international and regional human rights institutions more vocal in their condemnation, and civil society organizations are more vigilant in monitoring violations. Citizens too are more aware of their rights.

These developments have also led to a more robust defense of environmental rights. In this chapter, we will discuss the substantive human rights that have been used by victims of environmental abuse. One of the interesting features of the environmental rights evolution is that victims have used both civil and political rights, as well as economic, social, and cultural rights to seek redress for environmental abuse. They have also used collective rights and procedural rights. Thus, environmental issues appear to transcend the full gamut of protected rights. We will discuss the substantive rights that have been the subject of claims in relation to environmental issues, both in litigation and complaints referred to UN treaty bodies under various human rights treaties. We will also discuss some of the concluding observations made by these treaty bodies on country reports submitted by states where relevant. Both successful and unsuccessful complaints will be discussed as they shed light on the link between the environmental issue in question and the protected right claimed to have been violated.

As noted, both civil and political rights, and economic, social, and cultural rights have been invoked in relation to environmental rights. The main rights under the International Covenant on Civil and Political Rights (ICCPR) (and other regional treaties) that have been invoked are the right to life, right to privacy and family life, and minority rights, especially in relation to the right to culture of indigenous peoples.

Similarly, several of the rights under the International Covenant on Economic, Social and Cultural Rights (ICESCR) have a direct link with environmental degradation. The rights to health, food, water, and an adequate standard of living are some examples.

Because the entry into force of the Optional Protocol to the ICESCR is rather recent, communications to the Committee on Economic, Social and Cultural Rights involving environmental issues is rather rare. However, the Committee has referred to environmental protection in its concluding observations.

In addition, the UN Human Rights Council has appointed different special procedures mandate-holders on various topics who clarify obligations relating to human rights covered by their mandate. Some of them have made a specific reference to environmental degradation and climate change in their work. These include: the Special Rapporteur on the human rights to safe drinking water and sanitation; the Special Rapporteur on the right to food; the Special Rapporteur on the right of everyone to the enjoyment of highest attainable standard of physical and mental health; the Special Rapporteur on adequate housing; the Special Rapporteur on the rights of indigenous peoples; and the Special Rapporteur on the human rights of migrants.

Substantive rights infringed by environmental degradation

Right to life

The right to life is protected under Article 6 of the ICCPR. It provides that "Every human being has the inherent right to life. This right shall be protected by law. No one shall be arbitrarily deprived of his life" (ICCPR: Article 6). This right was intended to protect people from arbitrary killings by states. While this is formulated in negative terms, in General Comment No. 6, the Human Rights Committee recognized that positive obligations emanated from it:

> The Committee noted that the right to life has been too often narrowly interpreted. The expression "inherent right to life" cannot properly be understood in a restrictive manner, and the protection of this right requires that States adopt positive measures. In this connection, the Committee considers that it would be desirable for States parties to take all possible measures to reduce infant mortality and to increase life expectancy, especially in adopting measures to eliminate malnutrition and epidemics.
>
> (1994)

General Comment No. 6 draws a clear link between the right to life and the right to health, with some scholars questioning whether malnutrition and epidemics are better suited to be included in relation to the right to health (Anton and Shelton, 2011). While this is arguable, the intention of the Committee in adopting the General Comment is clear – it wanted to expand the narrow interpretation of Article 6.

One of the early complaints involving Article 6 to be filed before the Human Rights Committee was *EHP et al. v. Canada*. Here, Port Hope Environmental Group submitted a communication against Canada on behalf of present and future generations of Port Hope, Ontario, Canada, including 129 Port Hope residents who had specifically authorized the environmental group (the "author") to act on their behalf. It alleged the violation of the right to life of the residents, including future generations, due to the radioactive waste remains in Port Hope. The author claimed that "the current state of affairs is a threat to the life of present and future generations of Port Hope, considering that excessive exposure to radio-activity is known to cause cancer and genetic defects" (Human Rights Committee, 1984).

The Committee dismissed the petition for non-exhaustion of local remedies, but recognized that the right to life could be impacted as a result of an environmental harm. It observed that the present communication raised serious issues with regard to the obligation of states to protect human life under Article 6 of the Covenant. It also considered that the author had standing to submit the communication on it's own behalf and on behalf of those who had expressly authorized it to do so. However, with regard to the reference to "future generations" the Committee treated it as "an expression of concern purporting to put into due perspective the importance of the matter raised in the communication" (Human Rights Committee, 1980).

General Comment No. 6 recognizes the right to life as the supreme right from which no derogation is permitted even in times of public emergencies. It further states that this right should not be interpreted narrowly and that it includes positive measures, such as those relating to reducing infant mortality and measures to eliminate malnutrition and epidemics (ICCPR: Article 4). Some scholars argue that "there is a strict duty upon States, as well as upon the international community as a whole, to take effective measures to prevent and safeguard against the occurrence of environmental hazards which threaten the lives of human beings" (Hunter, Durwood, and Zaelke, 2015: 1347–8 quoting Ramcharan, 1985: 310–1).

It is clear that the right to life can be infringed by environmental pollution or degradation. Admittedly, the issue must be of a serious nature in order to involve the right to life. In order to do so, victims need not have died. However, by the time an environmental issue poses a serious risk to the right to life, it might already be too late to provide relief to the victims.

Another complaint involved the proposed underground nuclear weapon testing by France and its potential impact on the right to life. The case of *Vaihere Bordes and John Temeharo v. France* (1995) involved the announcement by France that it intended to conduct a series of underground nuclear tests on the atolls of Mururoa and Fangataufa in the South Pacific. The authors of the communication – Bordes, Tauira (who later withdrew his communication), and Temeharo were all French citizens living in Papeete, Tahiti, French Polynesia. They claimed that the tests represented a threat to their right to life and the right not to be subjected to arbitrary interference with their privacy and family life. The authors relied on General Comment No. 6 on the right to life, as well as General Comment No. 14 on nuclear weapons and the right to life (OHCHR, 1984). General Comment No. 14 notes that "designing, testing, manufacture, possession and deployment of nuclear weapons are among the greatest threat to the right to life which confront mankind today" and urged the international community to prohibit the production of nuclear weapons and to recognize it as a crime against humanity. The complainants further alleged that the French authorities failed to take sufficient measures to protect their life and security and did not show that underground nuclear tests did not constitute a danger to the health of inhabitants of the South Pacific and the environment.

The authors relied on studies that showed the danger to life caused by the nuclear tests was due to their effects on the health of people living in the vicinity, as well as indirect risk through contamination of the food chain. France argued, *inter alia*, that the authors did not qualify as "victims" within the meaning of Article 1 of the Optional Protocol, because there had neither been a violation nor was there a real threat of such result. The Human Rights Committee noted that the issue was whether the conduct of underground nuclear tests by France on these atolls resulted in a violation of their right to life and the right to their family life. The Committee observed that:

> On the basis of information presented by the parties, the authors have not substantiated their claim that the conduct of nuclear tests between September 1995 and the beginning of 1996 did not place them in a position which they could justifiably claim to be victims whose right to life and to family life was then violated or was under a real threat of violation.
>
> (Human Rights Committee, 1995: para. 5.5)

The Committee noted, relying on its General Comment No. 14, that although the authors did not show that they were victims, it reiterated that "it is evident that the designing, *testing*, manufacture, possession and deployment of nuclear weapons are among the greatest threats to the right to life which confront mankind today" (Human Rights Committee, 1995: para. 5.9, emphasis in original).

The decision of the Committee that the authors did not substantiate their claim in this case is rather curious given that a few sentences later, the Committee reiterated its position that activities involving nuclear weapons *including testing* pose a great threat to the right to life of people. Here it was no secret that France engaged in nuclear testing. It was also clear that the complainants were living in the vicinity of these tests. The Committee seemed to have considered the damage in question to be rather remote and hence the authors did not fall within the definition of a victim. Nonetheless, this case is important because it recognized that there is a clear link between nuclear tests and the right to life.

In the submission in *Susila Malani Dahanayake and others v. Sri Lanka* (2004), the authors also invoked Article 6 claiming that:

> The right to life guaranteed by Article 6 of the Covenant has been interpreted by other treaty bodies, also by the Human Rights Committee, in a broad manner, and consequently claim a violation of their right to life, which includes a right to a healthy environment.

While the Committee concluded that the authors had not sufficiently substantiated their claim, both parties seemed to have accepted the link between the right to life and the right to a healthy environment. This is an important development.

Oneryildiz v. Turkey (2004) involved Articles 2, 8, and 13 of the European Convention of Human Rights. The applicants argued that the national authorities were responsible for the deaths of their close relatives and for the destruction of their property as a result of a methane explosion at a municipal rubbish tip near Istanbul. An expert group appointed by the District Court concluded in their report that the rubbish tip did not conform to the technical requirements and presented a number of dangers liable to give rise to major health risks for the inhabitants of the valley. They concluded that the tip exposed humans, animals, and the environment to all kinds of risks. It pointed out that the tip in question was not equipped with a system to prevent an explosion of the methane produced as a result of the decomposition of waste, and if an explosion occurs, the damage can be very substantial, in addition to the spread of disease even without an explosion. The Environment Office requested the Istanbul Governor's office to take action to prevent an explosion from occurring, and the mayor applied to the District Court to prevent the City Council and district councils from using the waste-collection site, and to get the tip closed, but this was opposed by the City Council. Unfortunately, in April 1993, a methane explosion occurred, causing a landslide that resulted in 39 deaths and destruction of property (Anton and Shelton, 2011: 445–56).

The European Court of Human Rights reiterated that its approach to the interpretation of Article 2 is guided by the object and purpose of the Convention as an instrument to protect individual human beings and "requires its provisions to be interpreted and applied in such a way as to make its safeguards practical and effective" (Anton and Shelton, 2011: 449). In this regard, the Court noted that:

> Article 2 does not solely concern deaths resulting from the use of force by agents of the State but also, in the first sentence of its first paragraph, lays down a positive obligation on States to take appropriate steps to safeguard the lives of those within their jurisdiction.
>
> (Anton and Shelton, 2011: 449)

The Court relied on *LCB v. the United Kingdom* and *Paul and Audrey Edwards v. the United Kingdom* (2002). It further noted that where the Court had to examine whether an infringement of Article 2 has taken place in situations where the act in question was public or not, and in the case of industrial activities which by their very nature are dangerous. Referring to cases involving toxic emissions from a fertilizer plant in *Guerra and Others* and nuclear tests in *LCB v. United Kingdom*, the Court noted:

> In this connection, contrary to what the Government appear to be suggesting, the harmfulness of the phenomena inherent in the activity in question, the contingency of the risk to which the applicant was exposed by reason of any life-endangering circumstances, the status of those involved in bringing about such circumstances, and whether the acts or omissions attributable to them were deliberate or are merely factors among others that must be taken into account in the examination of the merits of a particular case, with a view to determining the responsibility the State may bear under Article 2.
>
> (Anton and Shelton, 2011: 450)

The Court pointed out that the positive obligation to take all appropriate steps to safeguard life entails a primary duty to put in place legislative and administrative frameworks designed to provide effective deterrence against threats to the right to life. With regard to dangerous activities, and particularly with regard to the level of the potential risk to human lives, additional requirements apply, including licensing, regulation, and supervision of activity, making it compulsory to take practical measures to ensure the effective protection of citizens whose lives might be endangered by the inherent risk. The Court referred to the need to provide information to the public as a preventive measure. This right, recognized under Article 8 of the Convention (privacy and family life), could also be used in relation to the right to life.

This case is important for the detailed discussion of what constitutes a violation of Article 2 and the tests to be applied to determine whether such a violation has taken place. Most importantly, the Court laid down the principle that Article 2 imposes an *affirmative duty*, that states are under an obligation to prevent a violation of Article 2 by way of dangerous activities, whether operated by a public or private entity, and that the procedural right of access to information can be used to prevent a violation of Article 2.

Thus, it is important to note that it is not necessary to prove intent in order to establish a violation of the right to life in the environmental context. The cases under the right to life (and the *Yanomani case* involving the right to health, discussed later) do not involve allegations that the State acted with the purpose of threatening the plaintiff's lives: "Moreover, commentators observe that international legal liability for damaging the environment in ways that pose grave risks to life … should arise irrespective of whether the act or omission in question is deliberate, reckless or negligent" (Rick Herz, referred to in Hunter, Durwood, and Zaelke, 2015: 1349). Moreover, as the decisions under the right to life have shown, the victim need not have died to establish a violation:

> Under international law, environmental harms presenting grave risks to life also violate the right. For example, in [Port Hope], the Human Rights Committee found that the complainant made out a prima facie case based on the threat to life the waste dump presented.
>
> (Herz, referred to in Hunter, Durwood, and Zaelke, 2015: 1350)

Right to privacy and family life

The ICPPR includes a provision on the right to privacy and family life:

> 1 No one shall be subjected to arbitrary or unlawful interference with his privacy, family, home or correspondence, nor to unlawful attacks on his honour and reputation.
> 2 Everyone has the right to the protection of the law against such interference or attacks.
>
> (ICCPR: Article 17)

In addition, regional treaties recognize this right. Much of the jurisprudence on environmental rights brought before the European system revolves around the right to privacy and family life protected under Article 8 of the Convention ("Everyone has the right to respect for his private and family life, home and his correspondence"). Some of these cases were discussed in Chapter 4. Even cases involving the right to health have also invoked this provision, given that the European Convention does not include a right to health.

Lopez Ostra v. Spain (1990) involved noxious fumes from a waste treatment facility near the petitioner's home that did not comply with environmental standards and made the petitioner's daughter sick. The pediatrician recommended that they move from the area. Due to a malfunction in the facility, it released gas fumes, smells, and contamination which caused health problems and nuisance to many residents in the area, particularly those living in the applicant's district. The town council evacuated local residents and rehoused them free of charge for three months in 1988. Several medical and expert opinions were issued concerning the health effects on those living near the plant. A pediatrician stated that the applicant's daughter "presented a clinical picture of nausea, vomiting, allergic reactions, anorexia etc., which could only be explained by the fact that she was living in a highly polluted area" (Hunter, Durwood, and Zaelke, 2015: 1351). An expert report in 1993 indicated that gas concentrations in houses near the plant exceeded the permitted limit. It noted that the applicant's daughter and her nephew presented typical symptoms of chronic absorption of the gas in question and that there was a relationship of cause and effect between this clinical picture and the levels of gas.

The Court noted that "severe environmental pollution may affect individual's well-being and prevent them from enjoying their homes in such a way as to affect their private and family life adversely, without, however, seriously endangering their health" (Hunter, Durwood, and Zaelke, 2015: 1351). This pronouncement is rather strange in this case, because there was ample evidence that the applicant's daughter's health was seriously affected by the operations of the waste treatment plant, to the extent that they were forced to move out of their house. Had the European Convention included a provision on the right to health, this case would surely have come within its scope. The case also highlighted the close relationship between the right to health and the right to privacy and family life.

The Court pointed out that a fair balance has to be struck between the competing interests of the individual and the community as a whole, and in any case the State enjoys a certain margin of appreciation. While the Spanish authorities were not directly responsible for the emissions in question, the town allowed the plant to be built on its land and the state subsidized the plant's construction. Moreover, the family had to bear the nuisance caused by the plant for over three years before finally moving. They moved only when it was obvious that the situation could continue indefinitely:

> Having regard to the foregoing, and despite the margin of appreciation left to the respondent State, the Court considers that the State did not succeed in striking a fair balance between the interest of the town's economic well-being – that of having a waste-treatment plant – and the applicant's effective enjoyment of her right to respect for her home and her private and family life.
>
> (Hunter, Durwood, and Zaelke, 2015: 1352)

It is important to note that this involved a complaint by a private individual for harm caused by a privately-owned industrial plant. Under human rights law, cases must be brought against a government entity. The Court upheld the State's positive duty to take necessary measures to protect the individual's rights. Under principles of state responsibility states incur liability for activities of private individuals in two instances:

(a) if the state failed to apprehend the perpetrator and punish him/her for the violation; or
(b) if the state fails to take reasonable measures to control the activities of the private entity (individual or corporation).

(Draft Articles on State Responsibility, 2001)

Polluting activities are often carried out by private entities, and states incur responsibility for their own failure to control their activities, either by having proper legislation with relevant standards and regulations, licensing procedures, environmental assessment requirements, and/or by regularly monitoring their activities.

Minority rights and the right to culture

Interestingly, many of the communications under the ICCPR that relate to the environment have invoked Article 27 on minority rights, mostly involving indigenous groups and their right to culture. In the submission of *EP et al. v. Colombia*, the authors alleged that government policies and other development activities were causing havoc to the environment on the outlying islands on which the authors lived, "making farming impossible and depleting other traditional sources of sustenance such as fish and shellfish from mangrove swamps, and were also depriving them of fishing rights" (8). They invoked Article 27 on the basis that they had been dispossessed of their lands and that the Colombian government encouraged massive development of the islands causing severe environmental damage that affected the water table, making farming impossible which destroyed their traditional livelihoods. They also contended that the government permitted destruction of mangrove swamps by allowing electric power plants to freely dump hot, polluted water, and that the government had granted fishing rights to Honduras without regard to native interests, depriving them of their traditional means of survival. The committee found that the communication was inadmissible because local remedies had not been exhausted.

The communication of *Lubicon Lake Band v. Canada* (1994) also involved an indigenous group whose rights had been violated due to oil and gas development on lands traditionally used by the Band, leading to environmental destruction, and in the process violated a number of rights protected by the ICCPR. The Committee found a violation of Article 27: "Historical inequities, to which the State party refers, and certain more recent developments threaten the way of life and culture of the Lubicon Lake Band, and constitute a violation of article 27 so long as they continue."

While in several other cases, the Committee did not find a violation of Article 27, it nonetheless acknowledged the link between right to culture and environmental degradation. Examples include: *Ilmari Lansman et al. v. Finland* (1992), *Jouni Lasnman et al v. Finland* (1996) and *Mahuika v. New Zealand* (2000).

Right to health

Under Article 12 of the ICESCR, states have recognized "the right of everyone to the enjoyment of the highest attainable standard of physical and mental health." In order to fulfill this obligation, states have undertaken to take steps, *inter alia*, to improve all aspects of environmental and industrial hygiene.

That there is a close link between environmental degradation and the right to health is indisputable. The obligations relating to this right were further clarified by the Committee on Economic, Social and Cultural Rights in its General Comment No. 14, which was adopted in 2000. According to the General Comment:

> [t]he right to health embraces a wide range of socio-economic factors that promote conditions in which people can lead a healthy life, and extends to the underlying determinants of health, such as food and nutrition, housing, access to safe and potable water and adequate sanitation, safe and healthy working conditions, and *a healthy environment*. (Emphasis added)

It further noted that the right to health is an inclusive right that extends also to underlying determinants of health "such as access to safe and potable water and adequate sanitation, an adequate supply of safe food, nutrition and housing, healthy occupational and environmental conditions" (OHCHR, 1984). The right to health clearly depends on a healthy and clean environment.

In the petition filed by the Inuit of Canada and the United States against the United States in relation to climate change, the petitioners claimed, *inter alia*, that their right to health was violated because of the adverse impacts of climate change in the Arctic. Due to the warmer temperatures and increased sun intensity, health problems that were hitherto unknown in the Arctic, such as skin cancers and cataracts, sun burns, and immune system disorders have increased. Increased temperatures have also affected some animal species that the Inuit depend on as a source of protein, also leading to health problems. Climate change has forced Inuit peoples to rely on a store-bought diet instead of their traditional diet consisting of wild meat and wild plants. This has led to health problems. This case was discussed in detail in Chapter 4.

In its report on the Yanomani Indians of Brazil, the Inter-American Commission of Human Rights explored the link between environmental quality and the right to health. The Commission noted that the situation of indigenous communities with regard to healthcare, food, education, and land is extremely serious. They were experiencing problems of invasion by squatters, destruction of the environment by pollution from mercury, illegal exploitation of lumber and agriculture, and the inadequate size of their land being insufficient to sustain them. During 1974–76, construction work was carried out on the Northern Perimeter Highway, which was built through Yanomani territory. Because the construction workers were not vaccinated (nor vaccines provided to the indigenous communities), the Yanomani population of 13 villages caught epidemics that resulted in the death of one out of every four Indians. In addition, conflicts between immigrants and indigenous peoples escalated, also resulting in many deaths. The Commission concluded that:

> The situation of the Indian Peoples of Brazil in the areas of health, nutrition, and access to public services is worrying. Indicators demonstrate clearly discriminatory conditions as compared with standards and services for the general population …
>
> The Yanomani people have obtained full recognition of their right to ownership of their land. Their integrity as a people and as individuals is under constant attack by both invading prospectors and the environmental pollution they create, State protection against these constant pressures and invasions is irregular and feeble, so that they are constantly in danger and their environment is suffering constant deterioration.
>
> (Hunter, Durwood, and Zaelke, 2015: 1349)

Thus, it is important to note that, similarly to the right to life, it is not necessary to prove intent in order to establish a violation of the right to health. Shelton believes that:

> health is the bridge between guarantees of environmental quality and the full exercise of human rights. Those whose rights have been impacted by pollution and other environmental degradation have been able to pursue

> international litigation to seek a remedy for rights violations based on environmental harm.
>
> (2007: 59)

Marangopoulos Foundation for Human Rights (MFHR) v. Greece (2005) was the first complaint concerning environmental conditions under the European Social Charter adopted in 1996. The complaint invoked several provisions including Article 11 (right to health); Article 2(4) (right to reduced working hours or additional holidays for workers in dangerous or unhealthy occupations); Article 3(1) (safety and health regulations at work); and Article 3(2) (enforcement of safety and health regulations by measures of supervision). The complaint alleged that the state had not adequately presented negative environmental impacts of lignite mining or developed an appropriate strategy to prevent health hazards for the population. It also alleged that there was no legal framework to guarantee the safety of mine workers. The European Committee of Social Rights concluded that there was a violation of Articles 2(4), 3(2), and 11. With regard to the right to health, the Committee of Ministers stated:

> The Committee finds that Greek regulations satisfy all the requirements concerning information to the public about and their participation in the procedure for approving environmental criteria for projects and activities. However, the circumstances surrounding the granting and extension of several authorizations, and the publication on the Internet of such a complex document as the NAP1 for just four days, show that in practice the Greek authorities do not apply the relevant legislation satisfactorily.
>
> The Committee considers that the government does not provide sufficiently precise information to amount to a valid education policy aimed at persons living in lignite mining areas. Finally, very little has so far been done to organize systematic epidemiological monitoring of those concerned and no morbidity studies have been carried out.
>
> (Anton and Shelton, 2011: 443–4)

This conclusion refers to the importance of the procedural right to information in relation to fulfillment of the substantive right to health, drawing a clear link between procedural rights and substantive rights – both are significant in relation to the enjoyment of environmental rights. Procedural rights are discussed in Chapter 6.

Right to food and water

Article 11 of the ICESCR on the right to an adequate standard of living encompasses the right to food:

1 The States Parties to the present Covenant recognize the right of everyone to an adequate standard of living for himself and his family, including

> adequate food, clothing and housing, and to the continuous improvement of living conditions. The States Parties will take appropriate steps to ensure the realization of this right, recognizing to this effect the essential importance of international co-operation based on free consent.
>
> 2 The States Parties to the present Covenant, recognize the fundamental right of everyone to be free from hunger.
>
> (ICESCR, 1966: Article 11)

A notable omission here is a reference to right to water. This was rectified by the adoption of General Comment No. 15 on the Right to Water. The Committee on Economic, Social and Cultural Rights recognized that Articles 11 and 12 of the ICESCR include a human right to water:

> Water is a limited natural resource and a public good fundamental for life and health. The human right to water is indispensable for leading a life in human dignity. It is a prerequisite for the realization of other human rights.
>
> (General Comment No. 15, 2002)

The right to water and sanitation was subsequently endorsed by the UN General Assembly in 2010; it recognized that "the right to safe and clean drinking water and sanitation [is] a human right that is essential for the full enjoyment of life and all human rights" and it called upon

> States and international organizations to provide financial resources, capacity-building and technology transfer, through international assistance and cooperation, in particular to developing countries, in order to scale up efforts to provide safe, clean, accessible and affordable drinking water and sanitation for all.
>
> (GA resolution 64/292, 2010)

While General Assembly resolutions are not legally binding, the international community seems to have accepted a human right to water, although some states such as the United States and Canada have objected to such recognition. Some countries such as Bolivia have included the right to water in their constitutions.

In the Millennium Development Goals, adopted in 2000, the international community pledged to halve the number of people without access to water and sanitation by 2015. A similar pledge was made at the Johannesburg Summit in 2002 (United Nations). Sustainable Development Goal 6 provides: "Ensure availability and sustainable management of water and sanitation for all," with a target of ensuring universal and equitable access to safe and affordable drinking water for all by 2030 (United Nations: target 6.1, SDGs). Similarly, SDG 2 provides: "End hunger, achieve food security and improved nutrition and promote sustainable agriculture" with a target of ending hunger by 2030, and ensuring access by all people, in particular the poor and people in vulnerable situations, including infants, to safe, nutritious, and sufficient food all year round (United Nations: target 2.1, SDGs).

In its 7th report on the Situation of Human Rights in Cuba, the Inter-American Commission of Human Rights discussed the right to health extensively. It referred specifically to environmental hygiene and pointed out that "to preserve and care for a healthy population, an environment conducive to that goal is essential" (1983: para. 41).The report identified several factors that could have a significant impact on environmental hygiene: water supply, disposal of industrial and human waste, and garbage collection. The report noted that while many rural families relied on rivers for their water supply, many of these rivers are now polluted due to industrial waste. Pollution of rivers has led to the pollution of bays and coastal waters. Sanitation and sewerage are major issues as is pollution of air, water and soil which is increasing at a dangerous rate. The report pointed out that "unless methods are adopted to control it, it could undermine the successes that have been achieved in the health field" (1983: para. 61).

The African Commission on Human and Peoples' Rights has also pronounced on safe drinking water and the right to health. In the *Communications against Zaire* (1993), the Commission held that the failure by the government to provide basic services such as safe drinking water constituted a violation of Article 16 on the right to health (referred to Anton and Shelton, 2011: 475).

The *Ogoniland* case discussed below also dealt with the right to food. The Inuit petition, discussed above, also alleged that the petitioners' right to food was violated as a result of the consequences of climate change.

Right to use and enjoy property

The American Convention of Human Rights recognizes the right to enjoy property. Article 21 (OAS, 1969) provides:

1 Everyone has the right to the use and enjoyment of his property. The law may subordinate such use and enjoyment to the interest of society.
2 No one shall be deprived of his property except upon payment of just compensation, for reason of public utility or social interest, and in the cases and according to the forms established by law.

In the *Case of Mayagna (Sumo) Indigenous Community of Awas Tingni* (2001), the Inter-American Court of Human Rights affirmed for the first time the collective right of indigenous peoples to their land, resources, and environment. This case also highlighted the importance of consultations. The Court declared that Nicaragua had violated the Community's right to property when it granted concessions to a foreign company to log on the Community's traditional land without either consulting with the Community or obtaining its consent. Significantly, the Court defined property as those material goods capable of being acquired and all rights, all movable and immovable goods, and tangible and intangible goods, as well as any other intangible object to which a value can be assigned.

The Court pointed out that, in the context of this case, it was necessary to understand the concept of property in indigenous communities. In indigenous communities, ownership of land is not individual in nature, but rather collective. In other words, ownership is vested in the group and in the community. The Court then referred to the intimate relationship between land and indigenous culture:

> By virtue of the fact of their very existence, indigenous communities have the right to live freely on their own territories; the close relationship that the communities have with the land must be recognized and understood as a foundation for their cultures, spiritual life, cultural integrity and economic survival. For indigenous communities, the relationship with the land is not merely one of possession and production, but also a material and spiritual element that they should fully enjoy, as well as a means through which to preserve their cultural heritage and pass it on to future generations.
>
> (Hunter, Durwood, and Zaelke, 2015: 1353)

Furthermore, the Court stressed the need to take into account the customary law of indigenous peoples. It held that possession of land is sufficient to entitle indigenous communities without title to their land to obtain official recognition of their rights of ownership. This was the first case before the Inter-American Court to address the territorial rights of indigenous communities. Rights of indigenous peoples necessarily include land rights, since their culture and traditions depend upon access to their ancestral lands.

Similar sentiments were expressed in the *Case of Kichwa Indigenous Peoples of Sarayaku v. Ecuador, Reparations* (2002), referring to the communitarian tradition that recognizes a form of collective land tenure which, the Court pointed out, does not necessarily conform to the classic concept of property, but deserves equal treatment under Article 21 of the American Convention nonetheless. The Court went even further in this case and stated that protection of property rights is necessary to ensure their survival: "In other words, the right to use and enjoy the territory would be meaningless for indigenous and tribal communities if that right were not connected with the protection of natural resources in the territory" (Hunter, Durwood, and Zaelke, 2015: 1355).

In the case of *Saramaka People v. Suriname (Saramaka), Preliminary Objections, Merits, Reparations* (2007), the Inter-American Court concluded that the Saramaka people have a right to use and enjoy the natural resources that lie on and within their traditional territory and that are necessary for their survival. Restrictions can only be placed with their effective participation and they must receive reasonable benefits from any development. In addition, an independent environmental and social impact assessment must be carried out. We will discuss this case in more detail in Chapter 6 on procedural rights.

Right to a healthy environment

As discussed in Chapter 4, the African Charter on Human and Peoples' Rights is the only human rights treaty, albeit a regional one, to include a justiciable right to a healthy environment. According to Article 24, "All peoples shall have the right to a general satisfactory environment favorable to their development." This provision was interpreted by the Commission in the *The Social and Economic Rights Action Center and the Center for Economic and Social Rights v. Nigeria* (Ogoniland case) in 2001. This case arose out of activities of Shell Oil consortium in Nigeria which caused significant environmental damage on Ogoniland (an area inhabited by the Ogoni people, an indigenous group in Nigeria) as well as widespread human rights violations. The complaint alleged that the military government of Nigeria through this consortium was directly involved in oil production which resulted in massive oil spills which were avoidable, and serious contamination of waterways, soil, and air gave rise to short-term and long-term health problems in the community. Furthermore, no monitoring of activities was carried out, no consultations were held with the affected communities and the government did not require the corporation to carry out any environmental or health assessment impact studies. Moreover, members of the Nigerian army were at the disposal of the oil company taking ruthless measures against Ogoni people, destroying their property, threatening them, attacking the people, and setting fire to their houses and making hundreds homeless. The Nigerian government did not investigate these attacks and or hold the perpetrators accountable. In addition to the right to a satisfactory environment, the complaint alleged the violation of several other rights, including the right to food. The communication alleged violations of Articles 2 (non-discrimination), 4 (right to life), 14 (right to property), 16 (right to health), 18(1) (right to family), 21 (free disposal of wealth and natural resources), and 24 (right to a healthy environment) of the African Charter (1981).

The Commission elaborated on the right to a satisfactory environment as embodied in the African Charter on Human and Peoples' Rights as follows:

> The right to a general satisfactory environment, as guaranteed under Article 24 of the African Charter or the right to a healthy environment, as it is widely known, therefore imposes clear obligations upon a government. It requires the State to take reasonable and other measures to prevent pollution and ecological degradation, to promote conservation, and to secure an ecologically sustainable development and use of natural resources.
>
> Government compliance with the spirit of Articles 16 and 24 of the African Charter must also include ordering or at least permitting independent scientific monitoring of threatened environments, requiring and publicising environmental and social impact studies prior to any major industrial development, undertaking appropriate monitoring and providing information to those communities exposed to hazardous materials and activities and providing meaningful opportunities for individuals to be heard and to participate in the development decisions affecting their communities.
>
> (1981: paras. 52 and 53)

It is thus clear that specific obligations flow from the right to environment: the obligations to prevent pollution; require the preparation of environmental and social impact assessments; undertake appropriate monitoring and provide information to affected communities; and give affected communities meaningful opportunities to participate in development decisions affecting them. Thus, according to the African Commission, the right to a satisfactory environment as embodied in the African Charter comprises both substantive and procedural components, all of which are identified by the Independent Expert on Human Rights and the Environment in his mapping report which was discussed in Chapter 2.

Concluding observations of the Committee on Economic, Social and Cultural Rights

In addition to litigation before regional human rights fora and complaints before treaty bodies, the Committee on Economic, Social and Cultural Rights has discussed environmental issues in their concluding observations on various country reports.

In its concluding observations on the second periodic report of Ecuador, the Committee on Economic, Social and Cultural Rights "voiced concerns about the negative health and environmental impacts of natural resource extracting companies' activities on the exercise of land and cultural rights by the affected indigenous communities and the equilibrium of the ecosystem" (Anton and Shelton, 2011: 439). In response to Ecuador's 2012 report, the Committee again referred to the environmental impacts of mining and agribusiness projects and their effects on the peoples' ability to exercise their right to water (Knox, 2013a: 12). In its concluding observations on Honduras, the Committee referred to the State party's failure to address the problem of excessive deforestation which was negatively impacting the habitat of indigenous peoples (4). In its concluding observations on Paraguay's report, the Committee noted the impact of indiscriminate use of toxic agro-chemicals related to the expansion of soybean cultivation which was leading to deaths and illnesses among people, including children, the contamination of water supply, and the disappearance of ecosystems, and impacting the traditional food resources of the affected communities (12). Australia was told about the Committee's concern regarding the negative impact of climate change on the right to an adequate standard of living, including the rights to food and water affecting particular indigenous communities (14).

The Committee further discussed the effects on the right to food caused by the depletion of natural resources of the rainforest traditionally used by indigenous communities in Cameroon (15). Similarly, the Committee expressed its concern regarding the situation of indigenous peoples in the Russian Federation, many of whom live in poverty and "whose food supply is based on fishing and an adequate stock of reindeer and who are witnessing the destruction of their environment by widespread pollution" (16).

In its concluding observations on the report submitted by Colombia, the Committee pointed out that it wished to receive information on measures taken to address the ecological deterioration of the Amazon region and whether Colombia "had encountered any problems in reconciling development of its oil reserves with the maintenance of a balanced ecosystem" (Knox, 2013b: 15). Similar observations were made with respect to Ecuador in that it wished to receive information on how ecological deterioration of the Amazon region was affecting indigenous communities and the measures taken to address that situation. Similar observations have also been made with regard to Mexico, Guyana, Argentina, Belize, and Sweden (Knox, 2013b: 15).

The Committee has on several occasions expressed its concern about the availability of food and water for populations. For example, in its comments on Israel, the Committee expressed concern about the limited access to and distribution and availability of water for Palestinians in the occupied territories. The Committee also referred to the gap in the living conditions of Jews and Bedouins in the Negev: "the quality of living and housing conditions of the Bedouins continues to be significantly lower, with limited or no access to water, electricity and sanitation." Referring to General Comment No. 15 on water, the Committee strongly urged the state party to take immediate steps to ensure equitable access to water to all populations living in occupied territories.

In its comments on Yemen, the Committee expressed concerns about the living conditions of prisoners and detainees, especially women, with regard to access to healthcare facilities, adequate food, and safe drinking water. It also expressed concern about the persisting water crisis "which prevents access to safe and affordable drinking water, particularly for the disadvantaged and marginalized groups of society, and for rural areas" (Anton and Shelton, 2011: 473).

There are numerous instances where the Committee has linked the violation of rights protected by the Covenant with environmental degradation, pollution, and resource depletion in its concluding observations. In addition, the Committee has made statements and reports recognizing the impact of climate change on protected rights, rights of future generations, and the rights of indigenous peoples. The Committee has also recognized procedural rights, especially the right to participate in the decision-making process and the right to a remedy. Thus, while the jurisprudence of the Committee recognizing environmental rights is rather limited, there is no doubt that the Committee has explicitly recognized the link between environmental degradation and protected rights, especially rights to food, water, health, housing, and an adequate standard of living in its concluding observations.

Conclusion

In this chapter we discussed the substantive human rights that have been invoked in relation to environmental issues. It is significant that both civil and political rights, and economic, social, and cultural rights have been invoked by victims of environmental abuse. In addition, UN human rights treaty bodies are increasingly addressing the link between human rights and the environment, and in this chapter we surveyed some of the concluding observations by the UN Committee on Economic, Social and Cultural Rights where an explicit link has been made between human rights and resource extraction, mining, climate change, and various other forms of environmental degradation. Health is the obvious link between environmental degradation and human rights, but as we saw in this chapter, other substantive rights such as food, water, privacy and family life, property, and even the right to life have been invoked in relation to environmental issues.

In addition, these human rights bodies are increasingly expressing concern over the impact of climate change on the enjoyment of rights. Thus, for example, when the Inter-American Commission issued a press release prior to the climate conference in Paris in 2015, it noted that:

> The Inter-American Commission on Human Rights (IACHR) expresses its concern regarding the grave harm climate change poses to the universal enjoyment of human rights. Ahead of the upcoming Conference of Parties to the United Nations Framework Convention on Climate Change (COP21), to be held in Paris December 7–8, the IACHR urges the Member States of the Organization of American States (OAS) to work to ensure that any climate agreement reached there incorporates human rights in a holistic manner.
>
> (OAS, 2015)

Thus, human rights institutions are increasingly expressing their concern about environmental challenges, especially climate change. Moreover, they are applying environmental law principles in their decisions.

DISCUSSION QUESTIONS

1. What are the best substantive human rights to be used to articulate environmental rights?
2. What are the pros and cons of using substantive human rights to articulate environmental rights?
3. If a substantive right to a healthy environment is recognized, how should it be drafted?
4. What are the pros and cons of recognizing a substantive right to a healthy environment?
5. How do you think the courts should balance competing interests?

References

African Charter on Human and Peoples' Rights (1981) [Online]. Available at: www.achpr.org/files/instruments/achpr/banjul_charter.pdf [Accessed: 20 October 2018].

Anton, D.K. and Shelton, D. (2011). *Environmental Protection and Human Rights*. Cambridge: Cambridge University Press.

Boyle, A. (1996). The Role of International Human Rights Law in the Protection of the Environment. In: Boyle, A. and Anderson, M. (eds.), *Human Rights Approaches to Environmental Protection*. Oxford: Oxford University Press.

Committee on Economic, Social and Cultural Rights (CESCR) (2002). General Comment No 15.

Committee on Economic, Social and Cultural Rights (CESCR) (2004). Concluding Observations on Ecuador, UN Doc. E/C.12/1/Add.100, para. 12 (14 May 2004).

EHP v Canada (1984). CCPR/C/17/D/67/1980 [Online]. Available at: http://hrlibrary.umn.edu/undocs/html/67-1980.htm [Accessed: 20 October 2018].

Herz, R.L. (2000). Litigating Environmental Abuses under the Alien Tort Claims Act: A Practical Assessment, 40 *Virginia Journal of International Law* 40: 545.

Human Rights Committee (1994). General Comment 6 (1982): The Right to Life, adopted April 30, UN Doc HRI/GEN/1/Rev.7, at 128 [Online]. Available at: http://hrlibrary.umn.edu/peace/docs/hrcom6.htm [Accessed: 20 October 2018].

Hunter, D., Salzman, J., and Zaelke, D. (2015). *International Environmental Law and Policy*, 5th edn. St. Paul, MN: Foundation Press.

International Covenant on Civil and Political Rights (1966). [Online]. Available at: www.ohchr.org/Documents/ProfessionalInterest/ccpr.pdf [Accessed: 1 October 2018].

International Covenant on Economic, Social and Cultural Rights (1966). [Online]. Available at: www.ohchr.org/Documents/ProfessionalInterest/cescr.pdf [Accessed: 1 October 2018].

Inter-American Commission of Human Rights, The Situation of Human Rights in Cuba, 7th Report, OEA/Ser.L/V/II.61, Doc. 29 rev. 1, 4 October 1983, [Online]. Available at: http://www.cidh.org/countryrep/Cuba83eng/chap.13.htm [Accessed: 1 October 2018].

Knox, J.H. (2013a). Mapping Human Rights Obligations Relating to the Enjoyment of a Safe, Clean, Healthy and Sustainable Environment, Individual Report on the International Covenant on Economic, Social and Cultural Rights, Report No 1, prepared for the Independent Expert on the Issue of Human Rights Obligations relating to the Enjoyment of a Safe, Clean, Healthy and Sustainable Environment. Geneva: Office of the High Commissioner for Human Rights.

Knox, J.H. (2013b). Mapping Human Rights Obligations Relating to the Enjoyment of a Safe, Clean, Healthy and Sustainable Environment, Individual Report on the International Covenant on Civil and Political Rights, Report No 2, prepared for the Independent Expert on the Issue of Human Rights Obligations Relating to the Enjoyment of a Safe, Clean, Healthy and Sustainable Environment. Geneva: Office of the High Commissioner for Human Rights.

Office of the High Commissioner for Human Rights (OHCHR) (1984). CCPR General Comment No. 14: Nuclear Weapons and the Right to Life [Online]. Available at: www.facing-finance.org/en/database/norms-and-standards/ccpr-general-comment-no-14-nuclear-weapons-and-the-right-to-life/ [Accessed: 20 October 2018].

OAS (1969). American Convention of Human Rights [Online]. Available at: www.cidh.oas.org/basicos/english/basic3.american%20convention.htm [Accessed: 1 October 2018].

OAS (2015). IACHR Expresses Concern regarding Effects of Climate Change on Human Rights, 2 December 2015 [Online]. Available at: www.oas.org/en/iachr/media_center/PReleases/2015/140.asp. [Accessed: 20 October, 2018].

Ramcharan, B.G. (1985). The Right to Life, pp.1347–1348. Referred to in Hunter, D., Salzman, J., and Zaelke, D. (2015), *International Environmental Law and Policy*, 3rd edn. St. Paul, MN: Foundation Press, pp.310–11.

Shelton, D. (2007). Health and Environmental Protection: Linkages in Law and Practice. *Human Rights & Int'l Legal Discourse*, 1: 9.

UN General Assembly Resolution 64/292, The Human Right to Water and Sanitation, A/RES/64/292 (2010). [Online]. Available at: www.un.org/ga/search/view_doc.asp?symbol=A/RES/64/292 [Accessed: 1 October 2018].

United Nations (2002). Johannesburg Declaration on Sustainable Development, Johannesburg, World Summit on Sustainable Development.

Cases

Case of Guerra and Others v. Italy (116/1996/735/932)

Communication No 167/1984.

Communication 645/1995, UN Doc CCPR/C/57/D/645/1995

Communication 645/1995, UN Doc CCPR/C/57/D/645/1995

E.P. et al. v. Colombia (1988). Communication No. 318/1988, U.N. Doc. CCPR/C/39/D/318/1988 (1990). [Online]. Available at: http://hrlibrary.umn.edu/undocs/session39/318-1988.html [Accessed: 1 October 2018].

Ilmari Lansman et al. v. Finland (1994). 26 Oct 1994, UN Doc. CCPR/C/52/D/511/1992

Jouni Lasnman et al v. Finland (1996). 30 Oct 1996, UN Doc CCPR/C/587/D/671/1995 and 17 March 2005, UN Doc. CCPR/C/83/D/1023/2001

Kichwa Indigenous Peoples of Sarayaku v. Ecuador (2012). Reparations, Inter-American Court of Human Rights (2012). [Online]. Available at: http://corteidh.or.cr/docs/casos/articulos/seriec_245_ing.pdf [Accessed: 1 October 2018].

LCB v. the United Kingdom (14/1997/798/1001). [Online]. Available at: www.legal-tools.org/doc/8f2c19/pdf/ [Accessed: 25 September 2018]

Lopez Ostra v. Spain (1990). (Application no. 16798/90). [Online]. Available at: www.informea.org/sites/default/files/court-decisions/COU-157040.pdf [Accessed: 1 October 2018].

Lubicon Lake Band v. Canada (1990). Communication No. 167/1984 (26 March 1990), U.N. Doc. Supp. No. 40 (A/45/40) at 1 (1990). [Online]. Available at: http://hrlibrary.umn.edu/undocs/session45/167-1984.htm [Accessed: 1 October 2018].

Mahuika v. New Zealand (2000). 27 Oct 2000, UN Doc. CCPR/C/70/D/547/1993

Marangopoulos Foundation for Human Rights (MFHR) v. Greece (2005). No. 30/2005 [Online]. Available at: http://blogs2.law.columbia.edu/climate-change-litigation/wp-content/uploads/sites/16/non-us-case-documents/2007/20070607_Complaint-No.-302005_decision-1.pdf [Accessed: 1 October 2018].

Mayagna (Sumo) Indigenous Community of Awas Tingni v. Nicaragua (2001). Inter-American Court of Human Rights (2001). [Online]. Available at: www.corteidh.or.cr/docs/casos/articulos/seriec_79_ing.pdf [Accessed 1 October 2018].

Oneryildiz v. Turkey (2004). Application No 48939/99. [Online]. Available at: www.informea.org/sites/default/files/court-decisions/Oneryildiz_v_Turkey_2004.pdf [Accessed: 25 September 2018].

Paul and Audrey Edwards v. the United Kingdom (1996). (Application 46477/99)

Saramaka People v. Suriname (2007). Preliminary Objections, Merits, Reparations (2007), Inter-American Court of Human Rights (2012). [Online]. Available at: www.corteidh.or.cr/docs/casos/articulos/seriec_172_ing.pdf [Accessed: 1 October 2018].

Susila Malani Dahanayake and 41 other Sri Lankan citizens v. Sri Lanka (2006). 25 July 2006, UN Doc. CCPR/C/87/D/1331/2004

The Social and Economic Rights Action Center and the Center for Economic and Social Rights v. Nigeria, African Commission on Human and Peoples' Rights (Ogoniland case) (2001). Comm. No. 155/96 (2001). [Online]. Available at: www1.umn.edu/humanrts/africa/comcases/155-96.html [Accessed: 1 October 2018].

Vaihere Bordes and John Temeharo v. France (1996). Communication No. 645/1995, U.N. Doc. CCPR/C/57/D/645/1995 (1996). [Online]. Available at: http://hrlibrary.umn.edu/undocs/html/DEC64557.htm [Accessed: 1 October 2018].

Yanomani case (1985). Resolution N° 12/85, Case N° 7615. [Online]. Available at: www.cidh.org/annualrep/84.85eng/brazil7615.htm [Accessed: 25 September 2018].

6 Procedural rights

Introduction

The convergence between human rights and environmental protection is most prominent in relation to procedural rights. These rights, also called "access rights," are: access to information, participation in the decision-making process, and access to remedies. They crept into the environmental protection discourse through the environmental impact assessment process at the national level. These rights are sometimes referred to as "the three pillars of environmental democracy" and are established rights under international human rights law. The Rio Declaration on Environment and Development (1992) brought these rights within the environmental protection discourse:

> Environmental issues are best handled with the participation of all concerned citizens, at the relevant level. At the national level, each individual shall have appropriate access to information concerning the environment that is held by public authorities, including information on hazardous materials and activities in their communities, and the opportunity to participate in decision-making processes. States shall facilitate and encourage public awareness and participation by making information widely available. Effective access to judicial and administrative proceedings, including redress and remedy, shall be provided.
>
> (Principle 10)

These rights now form part of many environmental treaties. They also form the procedural components of sustainable development, thereby bringing human rights into the sustainable development discourse (Atapattu, 2006).

Principle 10 received a boost when it became the impetus for the adoption of the UNECE Convention on Access to Information, Public Participation in Decision-making and Access to Justice in Environmental Matters in 1998 (generally referred to simply as the "Aarhus Convention," named after the Danish city in which the Convention was adopted). Although a regional treaty, it was the first time that these procedural human rights were embodied in a binding *environmental* instrument. It is also open for accession by non-ECE states. Latin America and the Caribbean adopted their own instrument on this topic in March 2018 which is closely aligned with the Aarhus Convention. This is a significant development giving effect to the principle of transparency in relation to environmental matters. In this chapter, we will discuss the evolution of procedural rights under international law and how these rights have been applied in case law.

Procedural rights form part of the principles of good governance and comprise the procedural components of sustainable development (discussed in Chapter 1). Principles of good governance include: transparency, accountability, upholding the rule of law and respect for human rights, inclusivity, efficiency, and participation (Human Rights Council webpage). The linkages between human rights, environmental protection, and good governance are discussed in Chapter 16.

Evolution of participatory rights

Participatory rights emerged as part of the environmental impact assessment (EIA) procedures at the national level. The EIA is an important tool adopted in many countries to evaluate the impact of development projects on the environment. The United States was the first country in the world to mandate EIAs for major Federal projects, and now over 100 states incorporate some form of EIAs for development activities (Craik, 2008). The ICJ recognized this development in the *Pulp Mills case* where it said that carrying out environmental impact assessments, especially in the context of transboundary environmental damage, is now part of general international law (ICJ Reports, 2010).

One of the significant features of the EIA process is that they incorporate public participation. In many countries, EIA documents are public documents (or at least the members of the public have access to them), and members of the public can participate in the decision-making process. This was one of the early examples of the convergence between the human rights and environmental protection discourses.

It is generally accepted that decisions that are taken with the participation of those who are likely to be affected by a particular decision lead to better decisions and better projects in the long run. Many of the conflicts in relation to "development projects" have arisen because the affected public had not been provided with proper information or they had not been provided an opportunity to be heard. This is particularly seen in relation to indigenous peoples. In order to minimize such conflicts in the future, providing information in a timely manner to the potentially affected public and giving them an opportunity to be heard are important steps. The EIA process institutionalized this in many countries. However, participatory rights are not confined to the EIA process.

People's participation in governance is recognized in the ICCPR (1976). According to Article 25, every citizen shall have the right and the opportunity, without any distinctions mentioned in Article 2 and without unreasonable restrictions, among other things, to take part in the conduct of public affairs, directly or through freely chosen representatives.[1] Article 19 of the ICCPR (1976) includes freedom of information as part of the freedom of expression:

> Everyone shall have the right to freedom of expression; this right shall include freedom to seek, receive and impart information and ideas of all kinds, regardless of frontiers, either orally, in writing or in print, in the form of art, or through any other media of his choice.

The former Independent Expert on Human Rights and the Environment, John Knox, recognized in his mapping report that these procedural obligations are more or less established under international law:

> One of the most striking results of the mapping exercise is the agreement among the sources reviewed that human rights law imposes certain procedural obligations on States in relation to environmental protection. They include duties (a) to assess environmental impacts and make environmental information public; (b) to facilitate public participation in environmental decision-making – including by protecting the rights of expression and association; and (c) to provide access to remedies for harm. These obligations have bases in civil and political rights, but they have been clarified and extended in the environmental context on the basis of the entire range of human rights at risk from environmental harm.
>
> (Knox, 2013: para. 29)

These rights will be discussed separately in the next section.

Under Article 3 of the Aarhus Convention, the parties have undertaken to take the necessary legislative, regulatory, and other measures to implement the obligations under the Convention, namely information, public participation, and access to justice provisions in the Convention, including enforcement measures. In addition, the Convention embodies the principle of non-discrimination – the provisions are to be made available without discrimination as to citizenship, nationality, or domicile, or in the case of a legal person, without discrimination as to where it is registered. The Latin American Convention (P-10) (2018) includes a similar provision with regard to taking legislative, regulatory, administrative, or other measures within the framework of the domestic provisions to guarantee the implementation of the provisions of the Agreement (Article 4(3)).

It is important to note that the Aarhus Convention, a significant contribution to the law, does not frame the provisions in rights language, even though it does include a right to a healthy environment. Rather, they are framed as obligations of states. In a report prepared for the joint UNEP-OHCHR Expert Seminar on Human Rights and Environment held in 2002, Ebbesson points out that:

> Rather than using rights-oriented language, the Convention requires the parties to "ensure" that members of the public have access to information,

> are allowed to participate and have access to judicial review. Although the term "right" is generally avoided, the objective, structure and context of the Aarhus Convention are rights-oriented. In part, the Convention draws on notions of international human rights law. It is intended to provide for participatory, information and procedural rights in environmental matters, and a failure to do so implies a breach of the treaty.
>
> (2002, footnotes omitted)

Despite this critique, there is no doubt that the Aarhus Convention is a major step forward. It was the motivation for Latin American and Caribbean countries to adopt a similar document. Some have even called for a global convention on environmental democracy to be adopted along similar lines (Hunter, Salzman, and Zaelke, 2015).

The Regional Agreement on Access to Information, Public Participation and Justice in Environmental Matters in Latin America and the Caribbean ("ECLAC Agreement") is the latest treaty to include participatory rights. Closely aligned with the Aarhus Convention, this Agreement provides for "access rights," which are defined as "the right of access to environmental information, the right of public participation in the environmental decision-making process and the right of access to justice in environmental matters" (ECLAC Agreement, 2018: Article 2). Unlike the Aarhus Convention, the ECLAC Agreement is framed in rights language, reflecting the developments in this area. The Preamble specifically refers to Principle 10 of the Rio Declaration which was also the impetus for the Aarhus Convention. It also refers to the Rio+20 conference at which the Latin American countries declared their willingness to adopt a regional instrument on Principle 10 provisions. It emphasized that access rights are "interrelated and interdependent, and so each and every one of them should be promoted and implemented in an integrated and balanced manner," and that they contribute to the strengthening of democracy, sustainable development, and human rights. It is interesting that the Preamble refers to both human rights instruments and environmental instruments, including the UDHR, the Stockholm and Rio Declarations, Agenda 21, Agenda 2030, and the Sustainable Development Goals.

The objective of the ECLAC Agreement closely mirrors that of the Aarhus Convention:

> The objective of the present Agreement is to guarantee the full and effective implementation in Latin America and the Caribbean of the rights of access to environmental information, public participation in the environmental decision-making process and access to justice in environmental matters, and the creation and strengthening of capacities and cooperation, contributing to the protection of the right of every person of present and future generations to live in a healthy environment and to sustainable development.
>
> (2018: Article 1)

Like the Aarhus Convention, the ECLAC Agreement makes a clear link between procedural rights and the substantive right to a healthy environment. This formulation goes even further than that of the Aarhus Convention by according rights to future generations and including the inter-generational equity principle. This is a significant step forward. It is important to note that the subsequent treaties have built on Principle 10 of the Rio Declaration, itself a non-binding document. Thus, Principle 10 of the Rio Declaration has had a huge influence on developing procedural environmental rights.

Furthermore, the ECLAC Agreement includes a set of principles that guide the parties in implementing the Agreement. Reflecting the convergence of human rights and environmental protection, these are a mix of international human rights and environmental principles: equality and non-discrimination; transparency and accountability; non-regression and progressive realization; good faith; the preventive principle; the precautionary principle; inter-generational equity; maximum disclosure; permanent sovereignty over natural resources; sovereign equality; and the principle of *pro persona* (2018: Article 3). Even more importantly, the parties have agreed under the Agreement to explicitly provide for a substantive right to a healthy environment: "Each Party shall guarantee the right of every person to live in a healthy environment and any other universally-recognized human right related to the present Agreement" (2018: Article 4(1)). This is a significant development and endorses a right to a healthy environment in the Latin American and Caribbean region. This provision, coupled with the recent Advisory Opinion of the Inter-American Court of Human Rights (discussed in Chapter 4), has consolidated a substantive right to a healthy environment in the region.

Moreover, the parties have pledged to adopt the required legislative, regulatory, administrative, or other necessary measures within the framework of their domestic laws to guarantee the implementation of the provisions of the Agreement. They have also undertaken to provide guidance and assistance to the public, especially those in vulnerable situations (defined as persons or groups that face particular difficulties in fully exercising the rights in the Agreement). Thus, the parties have pledged to take proactive measures to *facilitate* the exercise of the rights in the Agreement, especially with regard to vulnerable groups. This goes much further than the mere provision of those rights. Overall, this Agreement breaks new ground in the field of environmental rights. It is heartening to see such a groundbreaking development in a region that has witnessed much repression and human rights violations.

Access to information

Freedom of information is generally accepted as a right of citizens to receive information held by government entities. Withholding crucial information is itself a form of repression. For example, in conflict situations, obtaining information about loved ones or those who have "disappeared" is impossible, and such information is withheld because it could expose the wrongdoings at the hands of the armed forces or the police or collusion by the government. On the other hand, having access to relevant information is a powerful tool. In the environmental context, relevant information is not often divulged to the public at the planning stage of a project. Often communities learn of mega-projects "when bulldozers arrive." By then all the decisions have been made, and it is too late to protest. Conflicts often take place because communities have not been informed of planned activities.

Moreover, in the financial world, various tools and legal principles have been developed to shield entities from scrutiny and liability. A good example is a new tool called "project finance": "the contractual framework that allows a project developer to pay for the cost of the project through its revenue" (Baker, 2015: 339). In other words, the project finances itself. However, in order to do so a "complex network of contracts and experts work together to offer the key benefit associated with project finance: risk mitigation" (Baker, 2015). Entities known as "special purpose entities" (SPE) have been devised to be the face of the project, contracting with banks, purchasers, suppliers, contractors, and a host of other entities essential to the success of the project. It also becomes the main risk-taker. Banks that provide overall funding accept project assets as security, thereby shielding the equity investors' assets from risk. The end result is that it becomes next to impossible for an affected community to seek redress from the responsible party, untangling this complex web of networks of entities and relationships: "In the event of extensive damage to the community or environment, the chances of recovery by the community are minimal, as they must compete with the project's creditors to recover from an insolvent SPE" (Baker, 2015). This illustrates the important role that information plays in society.

While many countries in the Global North have adopted freedom of information laws, many in the Global South are lagging behind. India, despite boasting of the largest democracy in the world, adopted its Right to Information Act only in 2005 (which replaced the former law of 2002). Some countries include it in their Bill of Rights (see Chapter 7). Some countries do both – the right to information is included as a fundamental right and is elaborated in a law that provides for the modalities, exceptions, fees, etc. For example, Sri Lanka amended its constitution to include a right to information in the bill of rights and also adopted a Right to Information Act in 2016.

Access to information is closely intertwined with public participation. Without one, the other doesn't exist. In order to participate in the decision-making process in a meaningful manner, access to information in a timely manner is necessary. However, the mere provision of information without having a way to participate in the decision-making process is also useless. Both are necessary if the public is to influence the decision being contemplated.

Access to environmental information is now considered as an established right. Many cases have also affirmed it. We will discuss these cases later in the chapter. The right to information can be construed narrowly or widely. At the one end of the spectrum is the narrowly construed right to seek information. Here the state's duty is not to interfere with the individual's effort to obtain information.

At the other end of the spectrum is the broader right to receive information, and the corresponding duty of the state to disseminate information about both public and private projects that might affect the environment. This expanded version imposes an obligation on states to actually educate the public about the possible adverse impacts of a project. Due to the complexity of projects and the technical nature of EIA reports, it is hard for the general public to understand the impact of projects fully. Disseminating information in a language that people understand plays an important role here. In most instances, the duty of states falls somewhere between these two extremes, and the law imposes an obligation on states to facilitate access to information by the public. In some countries, this right is limited to the potentially affected public, not to the general public at large.

In addition to being rights in themselves, rights to information and participation are essential tools for the exercise of other rights such as the right to health, food, water, sanitation, and adequate housing. In General Comment No. 15 on the right to water, the Committee on Economic, Social and Cultural Rights recognized that "individuals should be given full and equal access to information concerning water and the environment" (para. 48 reproduced in Knox, 2013: para. 32).

While many environmental treaties require states parties to facilitate and encourage public awareness and participation by making information widely available (examples include the Rotterdam Convention on the Prior Informed Consent Procedure for Certain Hazardous Chemicals and Pesticides in International Trade, the Stockholm Convention on Persistent Organic Pollutants, and the UN Framework Convention on Climate Change), the Aarhus Convention is the most elaborate application of these rights in the environmental context. It was also the first time that these procedural rights were brought explicitly within the environmental discourse.

The procedural rights included in the Aarhus Convention are clearly linked to the substantive right to a healthy environment. It recognizes that "every person has the right to live in an environment adequate to his or her health and well-being, and the duty, both individually and in association with others, to protect and improve the environment for the benefit of present and future generations" (UNECE, 1998: Preamble). The objective of the Convention is even more explicit:

> In order to contribute to the protection of the right of every person of present and future generations to live in an environment adequate to his or her health and well-being, each Party shall guarantee the rights of access to information, public participation in decision-making, and access to justice in environmental matters in accordance with the provisions of this Convention.
>
> (Article 2)

Thus, the Aarhus Convention links the substantive right to a healthy environment with procedural rights of access to information, public participation, and access to justice. This is a very significant development. Until the Aarhus Convention was adopted, no binding instrument specifically recognized procedural rights in relation to environmental issues. The Rio Declaration endorsed these rights and the Aarhus Convention took it one step further. In fact, the Preamble to the Convention specifically refers to Principle 10 of the Rio Declaration. Despite being a regional treaty (it applies to the ECE region), its significance cannot be over-emphasized (Atapattu, 2016).

According to the UNECE webpage, the Aarhus Convention is a new kind of environmental treaty. It:

- Links environmental rights and human rights
- Acknowledges that we owe an obligation to future generations
- Establishes that sustainable development can be achieved only through the involvement of all stakeholders
- Links government accountability and environmental protection
- Focuses on interactions between the public and public authorities in a democratic context

(UNECE, n.d.)

Moreover, it is not just an environmental treaty. It is about the relationship between people and their government and about governmental accountability, transparency, and responsiveness. Thus, the Aarhus Convention plays an important role combining human rights and good governance with sustainable development and environmental protection.

It is important to note that the mere availability of information is insufficient. Information should be *accessible* in a language that the public understands. If the information is too technical, then such information may not mean anything to the affected people. In situations where the information is too technical or complex, there may be an additional obligation on authorities to educate the public on the pros and cons of a particular project.

Article 4 of the Aarhus Convention (UNECE, 1998) elaborates on access to environmental information:

> Each Party shall ensure that … public authorities, in response to a request for environmental information, make such information available to the public, within the framework of national legislation, including, where requested and subject to subparagraph (b) below, copies of the actual documentation containing or comprising such information: (a) without an interest having to be stated; and (b) in the form requested.

Thus, the person requesting the information does not have to give a reason for requesting such information. This is a significant development. The Convention defines "the public concerned" as

> the public affected or likely to be affected by, or having an interest in, the environmental decision-making; for the purposes of this definition, non-governmental

> organizations promoting environmental protection and meeting any requirements under national law shall be deemed to have an interest.
>
> (UNECE, 1998: Article 2(5))

Article 4 includes the grounds on which information may be refused. Some of these reasons are: if the request is manifestly unreasonable; if the disclosure adversely affects the confidentiality of proceedings of public authorities; international relations; national defense or public security; confidentiality of commercial and industrial information, intellectual property rights; and confidentiality of personal data or interests of third parties. However, Article 4 makes it clear that these grounds must be interpreted in a restrictive way, "taking into account the public interest served by disclosure and taking into account whether the information requested relates to emissions into the environment" (UNECE, 1998: Article 4(4)).

Other obligations of parties with regard to provision of information include:

- updating environmental information regularly;
- establishing mandatory mechanisms ensure flow of information to public authorities regarding proposed projects that could have a significant impact on the environment;
- disseminating information held by a public authority immediately and without delay to members of the public who may be affected by any imminent threat to human health or the environment, whether caused by human activities or due to natural causes;
- ensuring that the information is accessible and transparent;
- including the information in electronic databases and ensuring access by the public;
- preparing and publishing national reports periodically;
- providing information to the public on relevant laws, international treaties, and other important documents pertaining to the environment.

(UNECE, 1998)

Article 5 of the ECLAC Agreement (2018) provides for access to environmental information: "Each Party shall ensure the public's right of access to environmental information in its possession, control or custody, in accordance with the principle of maximum disclosure." The definition of "environmental information" is also significant:

> "Environmental information" means any information that is written, visual, audio, and electronic, or recorded in any other format, regarding the environment and its elements and natural resources, including information related to environmental risks, and any possible adverse impacts affecting or likely to affect the environment and health, as well as to environmental protection and management"
>
> (2018: Article 2(e))

With regard to persons or groups in vulnerable situations, including indigenous peoples and ethnic groups, the Agreement imposes an obligation on the parties to guarantee that these groups or persons receive assistance in preparing their requests and obtain a response. This is a significant feature of the ECLAC Convention which seeks to ensure access to information by people in vulnerable situations under equal conditions.

Similar to the Aarhus Convention, access to environmental information can be refused in certain circumstances. One disconcerting provision that reflects the state sovereignty principle is that access to information can be refused in accordance with domestic legislation. By subjugating the groundbreaking provisions of the Agreement to domestic legislation, the Agreement has unfortunately undermined its own provisions.

Public participation

While access to information is at the core of principles of good governance, the mere provision of information or having information available and accessible by itself is meaningless if there is no avenue for the public to influence the decision-making process. Thus, access to information goes hand-in-hand with public participation. This right also evolved as a result of the EIA process at the national level. Like access to information, public participation can cover a wide array of rights. It ranges from comments limited to potential victims at one end of the spectrum, to public hearings at which any member of the public can participate, at the other end of the spectrum. In many countries, public participation is somewhere in the middle and is often limited to the EIA process.

Participatory rights have evolved from the baseline right of everyone to take part in the governance of the country and in the conduct of public affairs. As noted by Knox, "human rights bodies have built on this baseline in the environmental context, elaborating a duty to facilitate public participation in environmental decision-making in order to safeguard a wide spectrum of rights from environmental harm" (Knox, 2013: para. 36).

The need for public participation is recognized in many environmental treaties. Examples include the Espoo Convention on Environmental Impact Assessment in a Transboundary Context (1991) and the Convention on Biological Diversity (1992). Some treaties have taken this a step further and require prior informed consent. Examples include the Basel Convention on International Trade in Hazardous Waste; the Stockholm Convention on Persistent Organic Pollutants (2002); the Rotterdam Convention on Prior Informed Consent Procedure for Certain Hazardous Chemicals and Pesticides in International Trade (1998); the UNFCCC (1992); and the UN Convention to Combat Desertification (1994).

Knox points out in his Mapping Report that freedom of expression and association are important in relation to public participation in environmental decision-making. In addition, states are required to protect the life, liberty, and security of individuals exercising these rights. The growing threat to human rights/environmental rights defenders is a particular concern (Knox, 2014).

Article 6 of the Aarhus Convention provides for public participation. It provides that the public concerned shall be informed early in an environmental decision-making procedure, and in an adequate, timely and effective manner of: (a) the proposed activity; (b) the public authority responsible for making the decision; (c) the procedure envisaged, including the opportunities for public participation, and the time and venue of any public hearing; (d) where relevant information can be obtained from; (e) the authority to whom comments or questions can be submitted; (f) what relevant environmental information is available; and (g) if the activity is subject to a national or transboundary environmental impact assessment procedure.

The Convention requires the parties to adopt reasonable timeframes for the different phases of a project to allow sufficient time for the public to prepare and participate effectively during the decision-making process. Parties are only required to take "due account" of the outcome of the public participation, which may not be an adequate requirement.

Article 7 of the ECLAC provides for public participation in environmental decision-making and is framed as a right:

> Each Party shall ensure the public's right to participation and, for that purpose, commits to implement open and inclusive participation in environmental decision-making processes based on domestic and international normative frameworks.
>
> (ECLAC Agreement, 2018: Article 7(1))

In addition, parties are required to guarantee mechanisms for participation, not just in relation to decision-making processes, but also in relation to revisions, re-examinations, or updates with respect to projects and activities, as well as in relation to environmental permits that may have a significant impact on the environment or affect public health.

Parties are required to promote participation with respect to matters of public interest, such as land-use planning, policies, strategies, programs, plans, rules, and regulations which may have a significant impact on the environment. Parties must ensure that the public can participate in the decision-making process from an early stage so that "due consideration" can be given to their observations. Parties are required to provide the public with the necessary information in a clear, timely, and comprehensive manner. Reasonable timeframes should be provided to allow sufficient time to inform the public and for effective participation. The information to be provided and the procedure to be adopted are very similar to those in the Aarhus Convention. The Agreement also provides that:

> The public's right to participate in environmental decision-making processes shall include the opportunity to present observations through appropriate means available, according to the circumstances of the process. Before adopting the decision, the relevant public authority shall give "due consideration" to the outcome of the participation process. Once a decision has been made,

> the public must be informed in a timely manner of the decision, the reasons for the decision and how the observations of the public have been taken into consideration.
>
> (ECLAC, 2018: Article 7(7))

The Agreement includes three very important provisions – with regard to language, adapting to local conditions, groups in vulnerable situations, and observing laws relating to indigenous peoples and local communities:

> Each party shall establish conditions that are favorable to public participation in environmental decision-making processes and that are adapted to the social, economic, cultural, geographical and gender characteristics of the public.
>
> (Article 7(10))

> When the primary language of the directly affected public is different to the official languages, the public authority shall ensure that means are provided to facilitate their understanding and participation.
>
> (Article 7(11))

> The public authorities shall make efforts to identify and support persons or groups in vulnerable situations in order to engage them in an active, timely and effective manner in participation mechanisms.
>
> (Article 7(14))

> … Each party shall guarantee that domestic legislation and international obligations in relation to indigenous peoples and local communities are observed.
>
> (Article 7(15))

Clearly, the ECLAC Agreement incorporates novel concepts and principles that are not included in the Aarhus Convention. This is not surprising, given that during the 20 years since the Aarhus Convention was adopted, many developments have taken place, especially with regard to indigenous peoples and public participation.

Access to justice

Under international human rights law, states are under an obligation to provide an effective remedy when people's rights are violated. Under the ICCPR, states have undertaken:

(a) To ensure that any person whose rights or freedoms as herein recognized are violated shall have an effective remedy, notwithstanding that the violation has been committed by persons acting in an official capacity;

(b) To ensure that any person claiming such a remedy shall have his right thereto determined by competent judicial, administrative or legislative authorities, or by any other competent authority provided for by the legal system of the State, and to develop the possibilities of judicial remedy;
(c) To ensure that the competent authorities shall enforce such remedies when granted.

(ICCPR, 1976: Article 2(3))

This provision makes it clear that the remedies must be effective and that a competent authority must enforce such remedies when granted. It is immaterial whether the violation has been committed by a government official. Remedies that take an inordinately long time are not considered "effective."

Article 9 of the Aarhus Convention provides for access to justice with regard to the provisions included in the Convention. It provides that each party shall, within the framework of its national legislation, ensure that any person who considers that his/her request for information has been ignored or wrongfully refused, has access to a review procedure before a court of law or other independent body established by law. It also requires parties to provide access to a review procedure before a court of law or another independent and impartial body established by law to challenge the substantive and procedural legality of any decision, act, or omission subject to Article 6.

In addition, Article 9 requires each Party to ensure that members of the public have access to administrative or judicial procedures to challenge acts and omissions by private persons and public authorities which contravene provisions of its national law relating to the environment. Parties are also required to provide adequate and effective remedies, including injunctive relief, and be fair, equitable, timely, and not prohibitively expensive. Decisions shall be in writing. Decisions of courts, and, whenever possible, of other bodies, shall be publicly accessible.

It is important to note that access to justice pertains to two kinds of situations: (a) where an applicant's request for environmental information has been ignored, refused, or not dealt with in accordance with the law; and (b) where any member of the public having a sufficient interest challenges the substantive and procedural legality of any decision, act, or omission concerning specific activities which may affect the environment (Ebbesson, 2002).

Article 8 of the ECLAC Agreement provides for access to justice in environmental matters and requires each party to guarantee the right of access to justice in environmental matters in accordance with guarantees of due process. The ambit of Article 8 is wider than that of the Aarhus Convention. It requires the parties to ensure, in accordance with its domestic legislation, access to judicial and administrative mechanisms to challenge with respect to substance and procedure:

(a) any decision, action or omission relating to access to environmental information;
(b) any decision, action or omission relating to public participation process

(c) any other decision, action or omission that affects the environment adversely or violates laws or regulations relating to the environment.

(2018: Article 8(2))

Moreover, the ECLAC Agreement requires the remedies to be effective, timely, public, transparent, impartial, and not prohibitively expensive, and provides for broad active legal standing in accordance with domestic legislation; there is also the possibility of ordering precautionary and interim measures, *inter alia*, to prevent, stop, mitigate, or rehabilitate damage to the environment. It also provides for the reversal of the burden of proof, or a dynamic burden of proof when appropriate, and mechanisms for redress such as restitution, compensation, satisfaction, guarantees of non-repetition, assistance for affected persons, and financial instruments to support redress. With regard to persons or groups in vulnerable situations, the parties are required to establish support mechanisms, including free technical and legal assistance.

The case of *Velasquez Rodriguez* (1988) is important because the Court discussed the obligation of states to provide a remedy even when the violation is done by a private individual. It was not an environmental case, but the decision is equally applicable to all cases. The Inter-American Court held in this case that even if the attackers were private individuals, the failure of the government authorities to find the victim or to give any remedy to the family was itself a violation:

> An illegal act which violates human rights, and which is initially not directly imputable to a State (for example, because it is the act of a private person ...) can lead to international responsibility of the State, not because of the act itself, but because of the lack of due diligence to prevent the violations or to respond to it as required by the Convention .... Where acts of private parties that violate the Convention are not seriously investigated, those parties are aided in a sense by the government, thereby making the State responsible on the international plane.
>
> (paras. 172 and 177)

The scope of access to remedies has been widened by the adoption of Optional Protocols to several treaties by giving individuals the right to bring a petition to the treaty body established under that treaty against their state, provided local remedies have been exhausted. Not all treaties provide for the individual right of petition. Nine treaty bodies can hear individual petitions: the Committee on Civil and Political Rights, the Committee on the Elimination of Racial Discrimination, the Committee Against Torture, the Committee on the Elimination of all Forms of Discrimination Against Women, the Committee on the Rights of Persons with Disabilities, the Committee on Enforced Disappearances, the Committee on Migrant Workers, the Committee on Economic Social and Cultural Rights, and the Committee on the Rights of the Child (OHCHR, n.d.).

Giving individuals the right to petition an institution at the international level against its own state is a remarkable development as, under international law, typically only states can bring claims against each other. Indeed, even now, only states may be parties before the International Court of Justice in contentious cases (Statute of the International Court of Justice, 1945: Article 36). Although this right is limited to the human rights field, nonetheless, it is a significant departure from standard practice relating to standing before international fora. A similar example can be found in the investment field where corporations are allowed to bring arbitral proceedings against states under bilateral investment treaties signed between states.

Indigenous peoples

These participatory rights are the strictest in the context of indigenous peoples. Not only is participation to be facilitated, in some instances, the state is required to ensure the free, prior, and informed consent (FPIC) of the affected peoples. According to the UN Declaration on the Rights of Indigenous Peoples (2007), FPIC is required in several instances: (a) forcible removal from indigenous land (Article 10); (b) adopting and implementing legislation and administrative measures that may affect indigenous peoples (Article 19); (c) storage or disposal of hazardous materials on indigenous land (Article 29); and (d) development projects (Article 32). Some scholars consider FPIC to be part of the right to self-determination, rather than a derivative of participatory rights.

Article 18 of the Declaration provides for the right to participate in decision-making:

> Indigenous peoples have the right to participate in decision-making in matters which would affect their rights, through representatives chosen by themselves in accordance with their own procedures, as well as to maintain and develop their own indigenous decision-making institutions.

In addition, states are required to consult with indigenous peoples in good faith and obtain their free, prior and informed consent before adopting legislation or other administrative measures that may affect them (Article 19).

The Declaration also requires States to provide redress and where possible restitution. If restitution is not possible, fair and equitable compensation is to be provided. If military activities are going to take place on indigenous land, the state is required to consult with the indigenous people affected:

> States shall undertake effective consultations with the indigenous peoples concerned, through appropriate procedures and in particular through their representative institutions, prior to using their lands or territories for military activities.
>
> (Article 30(2))

The Inter-American Court of Human Rights confirmed the right of indigenous peoples to free, prior, and informed consultations in relation to oil concessions by Ecuador on indigenous lands in the *Sarayaku* case (2012). The Court stated that in order to ensure effective participation by members of an indigenous community in relation to development or investment plans within their territory, the State must consult the community in an active and informed manner and in accordance with their customs and traditions. The Court further said that the consultations must be undertaken in good faith, through culturally appropriate procedures, and must be aimed at reaching an agreement. Moreover, the state must ensure that the community is aware of the potential benefits and risks of the project so that they can decide whether to accept the proposed plan or not. Finally, the consultation must take into account traditional decision-making practices of the people or the community: "Failure to comply with this obligation or engaging in consultations without having regard to their essential characteristics compromises the State's international responsibility" (Hunter, Salzman, and Zaelke, 2015: 1362).

This decision makes it clear that the duties of states in the context of indigenous peoples extend beyond the mere duty to provide information – states are required to ensure that the indigenous community in question is aware of the potential risks and benefits of the proposed project. This requires the state to engage in raising awareness and providing information in a balanced and timely manner so that the community can decide for themselves whether the project is beneficial for them. The state must be mindful of the decision-making procedures of the community.

The principle of free, prior, and informed consent was affirmed in the *Case of Saramaka People v. Surinam* (2007). In this case, the Inter-American Court stated that where "large-scale development or investment projects" could have a major impact on indigenous peoples' territory, states must not only consult with the peoples, but also obtain their free, prior, and informed consent, according to their customs and traditions.

Case law

Several cases have affirmed procedural rights. In *Leander v. Sweden*, the European Court of Human Rights unanimously stated:

> The right to receive information basically prohibits a Government from restricting a person from receiving information that others wish or may be willing to impart to him. Article 10 [of the European Convention] does not, in circumstances such as those of the present case, confer on the individual a right of access to a register containing information on his personal position, nor does it embody an obligation on the Government to impart such information to the individual.
>
> (116 Eur. Ct. H.R. (ser. A) (1987): para. 74)

This case essentially endorsed a narrow formulation of the right to information, in that governments should not restrict this right. It does not require the government to provide information. This narrow formulation can be seen in relation to environmental cases as well. In *Guerra and others v. Italy*, which involved pollution resulting from the operation of a chemical factory near the applicants' town, the applicants invoked Article 10 on freedom of information, claiming that the government had failed to inform the public of the risks and the measures to be taken in the event of a serious accident. Endorsing a narrow formulation of Article 10, a grand chamber of the European Court of Human Rights affirmed that Article 10 generally only prohibited a government from interfering with a person's freedom to receive information. Interestingly, however, the Court recharacterized the case as one interfering with Article 8 of the Convention on the right to family, home, and private life and found a violation of that right. In this context, the Court noted that the people had waited a long time for essential information "that would have enabled them to assess the risks they and their families might run if they continued to live at Manfredonia, a town particularly exposed to danger in the event of an accident" (Anton and Shelton, 2011: 359). Similarly, in *Oneryildiz v. Turkey*, the Court found an obligation to provide information is implicit in the right to life.

This expansion of the right to information through other protected rights is significant. On the one hand, the Court has adopted a narrow formulation of Article 10 on freedom of information, whereas in the context of other rights, such as the right to privacy or the right to life, the Court has recognized an overt obligation on the part of the state to actually provide information. In this regard, it may be a good strategy to request information (or seek to establish a violation) in relation to another protected right – e.g., the right to privacy and family life – rather than rely on Article 10 to establish that the state failed to provide information, as its scope seems to be rather limited. However, in recent years, the court seems to have expanded the scope of Article 10, especially when the information is of interest to the public and sought by those who serve a "watchdog" function (Anton and Shelton, 2011: 359).

Thus, in *Tarsasag a Szabadsagjogokert v. Hungary* (2009), the court recognized that the public has a right to receive information that is of general interest. The case law has been developed mainly in relation to the freedom of the press. The function of the press includes the creation of fora for public debate. *Observer and Guardian v. United Kingdom* (1991) and *Thorgeir Thorgeirson v. Iceland* (1992) are relevant in this regard.

The case of *Claude Reyes et al v. Chile* (2006) came before the Inter-American Court of Human Rights. The rights alleged to be violated included Article 13 (freedom of thought and expression) and Article 25 (right to judicial protection) of the American Convention of Human Rights. The petitioners claimed that the State refused to provide all the information they requested from the Foreign Investment Committee on the forestry company Trillium and a deforestation project in Chile's Region XII that could be prejudicial to the environment and sustainable development of Chile. The Inter-American Commission that referred the case to Court (as no direct petition to Court by individuals is available under the Inter-American system) was of the view that the state did not provide a valid justification under Chilean law for the refusal, and that no effective judicial remedy was granted to contest the violation of the right to information. In addition, they "were not ensured the rights of access to information and to judicial protection, and there were no mechanisms guaranteeing the right of access to information" (Anton and Shelton, 2011: 369).

Elucidating on the scope of the right to freedom of thought and expression, the Court noted that it has established that this right includes "not only the right and freedom to express one's own thoughts, but also the right and freedom to *seek, receive and impart* information and ideas of all kinds" (emphasis in original, referring to Lopez Alvarez, 2006, Ricardo Canese, 2004, and Herrera Ulloa, 2004). The Court noted that, similar to the American Convention, other international human rights instruments such as the UDHR and ICCPR establish a positive right to seek and receive information. In this case, the Court noted that Article 13 of the Convention protects the right of all individuals to request access to State-held information, subject to the restrictions permitted by the Convention:

> Consequently, this article protects the right of the individual to receive such information and the positive obligation of the State to provide it, so that the individual may have access to such information or receive an answer that includes a justification when, for any reason permitted by the Convention, the State is allowed to restrict access to information in a specific case. The information should be provided without the need to prove direct interest or personal involvement in order to obtain it, except in cases in which a legitimate restriction is applied.
>
> (Anton and Shelton, 2011: 372–3)

The Court also pointed out that when information is delivered to an individual, such information can circulate in society, and therefore, Article 13 includes both an individual and social dimension. Moreover, in the Resolution adopted by the OAS General Assembly on June 3, 2006, the General Assembly urged states to respect and promote respect for everyone's access to public information and to adopt legislation or other provisions to give effect to this. Moreover, Article 4 of the Inter-American Democratic Charter emphasizes the importance of transparency in government activities, and Article 6 refers to the right and responsibility of all citizens to participate in decision-making relating to their own development and to promote diverse forms of citizen participation. Similarly, in the Declaration of Nueva Leon, adopted in 2004, states undertook to provide a legal and regulatory framework to guarantee the right of access to information, recognizing that "access to information held by the State, subject to constitutional and legal norms, including those on privacy and confidentiality, is an indispensable condition for citizen participation" (Anton and Shelton, 2011: 373). The Court also referred to the provisions on access to information in the UN Convention against Corruption, the Rio Declaration on Environment and Development, the Council of Europe Declaration (that declared that together with the right to freedom of expression, there should be a corresponding duty on public authorities to make information available on matters of public interest), and the seminal Aarhus Convention, as well as laws in many countries that provide for the freedom to obtain state-held information. In this regard, the Court noted that Chile had also made significant progress by adopting a constitutional amendment as well as a draft law on the right to information that was being processed.

The Court pointed out that access to public information is an essential component of democracy and greater transparency and that "in a representative and participative democratic system, the citizenry exercises its constitutional rights through a broad freedom of expression and free access to information" (Anton and Shelton, 2011: 373). It also referred to the close relationship between democracy and freedom of expression. The Court held that in order for individuals to exercise democratic control, the State must guarantee access to information of public interest that it holds. This leads to greater participation by individuals in the interests of society. The Court then went on to discuss the restrictions that can be imposed on the exercise of this right.

The importance of procedural rights was discussed by the Supreme Court of India in the case of *People's Union for Civil Liberties & Another v. UOI & Others* (2005). The Court noted that the right to information is a facet of "speech and expression" enshrined in Article 19 and thus, the right to information is indisputably a fundamental right.

The importance of environmental education was recognized by the Supreme Court of India in M.C. *Mehta* v. *Union of India and Others* (1991). The Petitioner in this case relied on Article 51A(g) of the Constitution of India which requires every citizen to protect and improve the natural environment, including forests, lakes, rivers, and wildlife, and to have compassion for living creatures. In order to fulfill this obligation, the petitioner argued that people needed to be better educated about the environment. The Court pointed out that although law is a regulator of human conduct, no law can effectively work unless it is accepted by people in society:

> No law works out smoothly unless the interaction is voluntary in order that human conduct may be in accordance with the prescription of law it is necessary that there should be appropriate awareness about what the law requires and there is an element of acceptance that the requirement of law is grounded upon philosophy which should be followed. This would be possible only when steps are taken in an adequate measure to make people aware of the indispensable necessity of their conduct being oriented in accordance with the requirements of law.
>
> (Atapattu, 2015: 111)

Noting that in a democratic polity dissemination of information is the foundation of the system, and that there is a general acceptance that protection of the environment is an indispensable necessity for life to survive on earth, the Court made the following directions: (a) require movie theaters, touring cinemas, and video parlors to exhibit free of charge at least two slides/messages on the environment in each show. The Ministry of Environment should prepare appropriate slide material for this purpose; (b) the Ministry of Information and Broadcasting should produce short information films on the environment; and (c) the University Grants Commission [the institution in charge of higher education] shall take steps to prescribe a course on the environment, and every school must enforce compulsory education on the environment. This case highlights the important role played by information and education in protecting the environment (Atapattu, 2015).

Human rights defenders

One of the biggest issues facing the human rights/environmental protection community is threats to the life and liberty of human rights defenders. Every year, human rights defenders are killed in many parts of the world and threats against their security continue. Latin America has seen many incidents of violence against human rights/environmental defenders, and it is no surprise that the ECLAC Agreement includes a provision on human rights defenders: "Each Party shall guarantee a safe and enabling environment for persons, groups and organizations that promote and defend human rights in environmental matters, so that they are able to act free from threat, restriction and insecurity" (2018: Article 9(1)).

In his mapping report, the Independent Expert on Human Rights and the Environment noted:

> The rights of freedom of expression and association are of special importance in relation to public participation in environmental decision-making. The Special Rapporteur on the situation of human rights defenders has said that those working on land rights and natural resources are the second-largest group of defenders at risk of being killed (A/HRC/4/37), and that their situation appears to have worsened since 2007.
>
> (UN, 2013: para. 18)

> Her last report described the extraordinary risks, including threats, harassment, and physical violence, faced by those defending the rights of local communities when they oppose projects that have a direct impact on natural resources, the land or the environment.
>
> (UN, 2013: para. 15)

States have obligations not only to refrain from violating the rights of free expression and association directly, but also to protect the life, liberty, and security of individuals exercising those rights (International Covenant on Civil and Political Rights (ICCPR), 1976: Article 2; Declaration on the Right and Responsibility of Individuals, Groups and Organs of Society to Promote and Protect Universally Recognized Human Rights and Fundamental Freedoms: Articles 2, 9, and 12). There can be no doubt that these obligations apply to those exercising their rights in connection with environmental concerns. The Special Rapporteur on the Situation of Human Rights Defenders has underlined these obligations in that context (UN, 2013: paras. 16 and 30), as has the Special Rapporteur on the Rights of Indigenous Peoples (OHCHR, 2013: para. 21), the Committee on Economic, Social and Cultural Rights (2016), the Inter-American Court of Human Rights (*Kawas Fernández v. Honduras*), and the Commission on Human Rights, which called upon States "to take all necessary measures to protect the legitimate exercise of everyone's human rights when promoting environmental protection and sustainable development" (resolution 2003/71).

In his 2017 report on human rights defenders, the Special Rapporteur on the Situation of Human Rights Defenders noted that those defenders working in the area of business and human rights are among the most vulnerable:

> The work of human rights defenders in the field of business and human rights is crucial to protecting the land and the environment, securing just and safe conditions of work, combating corruption, respecting indigenous cultures and rights and achieving sustainable development.
>
> (OHCHR, 2017: para. 1)

The report pointed out that these defenders are fighting against big businesses; however, defending and promoting human rights in the context of business is dangerous, and even deadly, work. For defending human rights over profit, privilege, and prejudice, ordinary people, communities, workers, and trade unionists face stigmatization, criminalization, physical attacks, and sometimes death. In many situations, these courageous people are being deprived of their most fundamental rights merely because they opposed powerful interests. The Special Rapporteur is deeply concerned that these defenders are suffering attacks by business actors overpowering and silencing them, which exerts a chilling effect on their work. This worrying trend is compounded by a lack of State action in response to such attacks. This applies to States on whose territory attacks occur, as well as to the home States of the business entities involved in the attacks (OHCHR, 2017).

The report points out that during 2015–2016 alone, there had been 450 incidents against human rights defenders documented. The actual number is likely to be significantly higher, as many of these incidents are not reported. Of these attacks, 25% were connected to companies with headquarters in just three countries – the United States, Canada, and China. The trend that we see around the world of intimidation, threats, reprisals, attacks, and even deaths of human rights defenders is very worrying and states should take urgent action to address this situation.

Conclusion

This chapter discussed the procedural rights that have played a critical role in relation to environmental issues. Protecting these rights leads to decisions that are sound that go a long way towards avoiding conflicts over natural resources. These three rights – access to information, participation in decision-making, and access to justice – are established rights under international law and have been brought within the environmental law framework by two regional treaties. With regard to indigenous peoples, their free, prior, and informed consent (FPIC) is required in certain circumstances.

DISCUSSION QUESTIONS

1. Are procedural rights more important than substantive rights?
2. What is the link between procedural rights and substantive rights?
3. How should states implement procedural rights?
4. What does the free, prior, and informed consent principle mean? Should we apply this higher standard to other vulnerable groups?
5. What is the link between procedural rights, sustainable development, and good governance?

Note

1 These are race, color, sex, language, religion, political or other opinion, national or social origin, property, birth, or other status (Article 2(1)).

References

Anton, D. and Shelton, D. (2011). *Environmental Protection and Human Rights*. Cambridge: Cambridge University Press.

Atapattu, S. (2006). *Emerging Principles of International Environmental Law*. New York, NY: Transnational.

Atapattu, S. (2015). The Role of Human Rights Law in Protecting Environmental Rights in South Asia. In: Hugland, L. and Stryker, R. (eds.), *Closing the Rights Gap: From Human Rights to Social Transformation*. Oakland, CA: University of California Press.

Atapattu, S. (2016). *Human Rights Approaches to Climate Change: Challenges and Opportunities*. London: Routledge

Baker, S. (2015). Project Finance and Sustainable Development in the Global South. In: Alam, S., Atapattu, S., Gonzalez, C., and Razzque, J. (eds.), *International Environmental Law and the Global South*. Cambridge: Cambridge University Press, p.338.

Basel Convention on the Control of Transboundary Movements of Hazardous Waste and their Disposal (1989). [Online]. Available at: www.basel.int/Portals/4/Basel%20 Convention/docs/text/BaselConventionText-e.pdf [Accessed: 1 October 2018].

Case Concerning Pulp Mills on the River Uruguay (Argentina v. Uruguay), Judgment of April 2010, ICJ Reports [Online]. Available at: www.icj-cij.org/files/case-related/135/135-20100420-JUD-01-00-EN.pdf [Accessed: 18 October 2018].

Committee on Economic, Social and Cultural Rights (2016). Human Rights Defenders and Economic, Social and Cultural Rights E/C.12/2016/2 7 October 2016.

Craik, N. (2008). *The International Law of Environmental Impact Assessment: Process, Substance and Integration*. Cambridge: Cambridge University Press.

Declaration of Nueva Leon (2004) [Online]. Available at: www.oas.org/juridico/english/summit_decl_nuevo_leon.pdf [Accessed: 20 October 2018].

Ebbesson, J. (2002). Information, Participation and Access to Justice: The Model of the Aarhus Convention, Joint UNEP-OHCHR Expert Seminar on Human Rights and the Environment, Background Paper No 5. Geneva, 14–16 January 2002.

Hunter, D., Salzman, J. , and Zaelke, D. (2015). *International Environmental Law and Policy*, 5th ed. St. Paul, MN: Foundation Press.

International Covenant on Civil and Political Rights (ICCPR) (1976) [Online]. Available at: www.ohchr.org/en/professionalinterest/pages/ccpr.aspx [Accessed: 18 October 2018].

Knox, J.H. (2013): Report of the Independent Expert on the issue of human rights obligations relating to the enjoyment of a safe, clean, healthy and sustainable environment. A/HRC/22/43, Human Rights Council (Mapping Report) [Online]. Available at: www.ohchr.org/en/issues/environment/srenvironment/pages/mappingreport.aspx [Accessed: 18 October 2018].

Kravchenko, S. (2009). Is Access to Environmental Information a Fundamental Human Right? *Oregon Review of International Law* 11: 227.

Latin America and Caribbean Treaty on Principle 10 (ECLAC Agreement)(2018) [Online]. Available at: https://repositorio.cepal.org/bitstream/handle/11362/39051/37/S1800064_en.pdf [Accessed: 18 October 2018].

Office of the High Commissioner for Human Rights (OHCHR) (n.d.). Human Rights Treaty Bodies – Individual Communications [Online]. Available at: www.ohchr.org/en/hrbodies/tbpetitions/Pages/IndividualCommunications.aspx [Accessed: 21 October 2018].

OHCHR (1998). Declaration on the Right and Responsibility of Individuals, Groups and Organs of Society to Promote and Protect Universally Recognized Human Rights and Fundamental Freedoms, UN GA resolution 53/144, 9 December 1998 [Online]. Available at: www.ohchr.org/en/professionalinterest/pages/rightandresponsibility.aspx [Accessed: 1 October 2018].

OHCHR (2000). Good Governance and Human Rights. [Online]. Available at: www.ohchr.org/en/issues/development/goodgovernance/pages/goodgovernanceindex.aspx [Accessed: 1 October 2018].

OHCHR (2003). Commission on Human Rights resolution 2003/71, Human rights and the environment as part of sustainable development [Online]. Available at: https://www.refworld.org/docid/43f3134dc.html [Accessed: 1 October 2018].

OHCHR (2003). Human rights and the environment as part of sustainable development, Commission on Human Rights resolution 2003/71, E/CN.4/RES/2003/71

OHCHR (2013). Report of the Special Rapporteur on the rights of indigenous peoples, Extractive industries and indigenous peoples A/HRC/24/41 (2013) [Online]. Available at: www.ohchr.org/EN/HRBodies/HRC/RegularSessions/Session24/Documents/A-HRC-24-41_en.pdf [Accessed: 24 September 2018].

OHCHR (2017). Report of the Special Rapporteur on the situation of human rights defenders (2017) A/72/170 [Online]. Available at: https://documents-dds-ny.un.org/doc/UNDOC/GEN/N17/220/75/PDF/N1722075.pdf?OpenElement [Accessed: 1 October 2018].

Rio Declaration on Environment and Development (1992). A/CONF.151/26 (Vol. I). [Online]. Available at: www.un.org/documents/ga/conf151/aconf15126-1annex1.htm [Accessed: 10 October 2018].

Rotterdam Convention on Prior Informed Consent Procedure for Certain Hazardous Chemicals and Pesticides in International Trade (1998). [Online]. Available at: www.jus.uio.no/lm/hazardous.chemicals.and.pesticides.prior.informed.consent.trade.rotterdam.convention.1998/portrait.a4.pdf [Accessed: 5 February 2019].

Statute of the International Court of Justice (1945). [Online]. Available at: www.icj-cij.org/en/statute [Accessed: 24 September 2018].

Stockholm Convention on Persistent Organic Pollutants (2002). [Online]. Available at: http://chm.pops.int/Portals/0/Repository/convention_text/UNEP-POPS-COP-CONVTEXT-FULL.English.PDF [Accessed: 1 October 2018].

UN (2013). Report of the Special Rapporteur on the situation of human rights defenders (2013), A/68/262 [Online]. Available at www.un.org/en/ga/search/view_doc.asp?symbol=A/68/262 [Accessed: 1 October 2018].

UN Convention on Biological Diversity (1992). [Online]. Available at: www.cbd.int/doc/legal/cbd-en.pdf. [Accessed: 5 February 2019].

UN Convention to Combat Desertification in Those Countries Experiencing Serious Drought and/or Desertification, Particularly in Africa (1994). [Online]. Available at: www.unccd.int/sites/default/files/relevant-links/2017-01/UNCCD_Convention_ENG_0.pdf [Accessed: 1 October 2018].

UNECE (1991). Convention on Environmental Impact Assessment in a Transboundary Context. [Online]. Available at: www.unece.org/fileadmin/DAM/env/eia/documents/legaltexts/Espoo_Convention_authentic_ENG.pdf [Accessed: 5 February 2019].

UNECE (1998). Convention on Access to Information, Public Participation in Decision-making and Access to Justice in Environmental Matters in 1998 (Aarhus Convention). [Online]. Available at: http://www.unece.org/fileadmin/DAM/env/pp/documents/cep43e.pdf [Accessed: 1 October 2018].

UNECE (n.d.). About the Convention – Introduction [Online]. Available at: www.unece.org/env/pp/introduction.html [Accessed: 18 October 2018].

UN Human Rights Council (2013). Report of the Special Rapporteur on the rights of indigenous peoples, James Anaya, Extractive industries and indigenous people, A/HRC/24/41, 1 July 2013 [Online]. Available at: www.ohchr.org/EN/HRBodies/HRC/RegularSessions/Session24/Documents/A-HRC-24-41_en.pdf [Accessed: 1 October 2018].

United Nations Declaration on the Rights of Indigenous Peoples, UN GA resolution 61/295, A/RES/61/295 [Online]. Available at: www.un.org/en/genocideprevention/documents/atrocity-crimes/Doc.18_declaration%20rights%20indigenous%20peoples.pdf [Accessed: 1 October 2018].

United Nations Framework Convention on Climate Change (1992). [Online]. Available at: https://unfccc.int/resource/docs/convkp/conveng.pdf [Accessed: 1 October 2018].

Cases

Claude Reyes et al v. Chile (2006)

Guerra and Others v. Italy (1996). 116/1996/735/932

Herrera-Ulloa v. Costa Rica (2004). Inter-American Court of Human Rights [Online]. Available at: www.corteidh.or.cr/docs/casos/articulos/seriec_107_ing.pdf [Accessed: 1 October 2018]

Kawas Fernández v. Honduras, Merits, Reparations and Costs, Judgment dated 3 April 2009 (Ser. C No. 196)

Kichwa People of Sarayaku v. Ecuador (2012)

Leander v. Sweden, App no 9248/81, A/116, (1987) 9 EHRR 433, IHRL 69 (ECHR 1987), 26th March 1987, European Court of Human Rights [ECtHR]

López-Álvarez v.Honduras (2006). Inter-American Court of Human Rights [Online]. Available at: www.corteidh.or.cr/docs/casos/articulos/seriec_141_ing.pdf [Accessed: 1 October 2018]

M.C. *Mehta v. Union of India and Others* (1991)

Observer and Guardian v. United Kingdom (ECHR 26 Nov, 1991)

Oneryildiz v. Turkey Application No. 48939/99; (2005) 41 EHRR 20; [2004] ECHR 657; 18 BHRC 145, (2005); [2004] Inquest LR 108

People's Union for Civil Liberties & Another v. UOI & Others (2005)

Pulp Mills on the River Uruguay (Argentina v. Uruguay) (2010): ICJ Reports

Ricardo Canese v. Paraguay (2004). Inter-American Court of Human Rights [Online]. Available at: http://corteidh.or.cr/docs/casos/articulos/seriec_111_ing.pdf [Accessed: 1 October 2018]

Saramaka People v. Suriname (2007)

Tarsasag a Szabadsagjogokert v. Hungary (ECHR, April 14, 2009)

Thorgeir Thorgeirson v. Iceland (1992)

Velasquez Rodriguez v. Honduras (1988) Judgment of 29 July 1988, Inter-Am. C.H.R. (Ser. C) No 4

7 Constitutional developments

Introduction

One of the features of modern constitutions is the inclusion of non-traditional rights in the bill of rights. This includes environmental rights. Older constitutions tend to include only civil and political rights, while newer constitutions include economic, social, and cultural rights, and even more progressive constitutions such as that of South Africa include a right to a healthy environment. As the supreme law of a country, constitutions play an important role in creating and protecting peoples' rights. Thus, the inclusion of rights, especially environmental rights, signifies the importance states attach to rights and the need to protect the rights of its people. This chapter will discuss these developments and what some scholars call the emergence of "global environmental constitutionalism" (May and Daly, 2015).

However, the absence of environmental rights in constitutions does not mean that they are irrelevant. Even in the absence of explicit environmental rights in the constitution, judiciaries across the world have developed environmental rights by interpreting existing rights creatively and expansively. The judiciaries have held that many of the rights in the constitution would be meaningless if people do not have a right to live in a clean environment. This chapter will discuss some of the salient cases from across the world.

Not all scholars, however, think that environmental rights should be entrenched or that they should be recognized at all. Similar to the debate we see at the international level (see Chapter 3), some argue that environmental rights are incapable of being defined or enforced. Some human rights scholars have argued that the recognition of new rights could lead to the dilution of existing core rights (Alston, 1982). Some environmentalists, on the other hand, feel that by recognizing environmental rights, we will be emphasizing the one species that has caused all the environmental problems in the first place: human beings. However, there is no denying that the rights framework has given a voice to marginalized communities and has enabled victims of environmental abuse to obtain relief. As some of the cases that we will discuss below show, these constitutional provisions can also be used in relation to *imminent* damage. Moreover, the language of rights is powerful and the entrenchment of rights in a constitution shows that these rights are special and important, and their infringement will lead to repercussions for the perpetrator and remedies for the victim.

A much newer development is the recognition of rights of nature in national constitutions. May and Daly believe that

> constitutionalization of rights of nature is part of a growing global movement highlighting the importance of the natural environment for its own sake and as a whole, rather than as an aggregation of resources to be harnessed by humans for various purposes.
>
> (2015: 255)

The importance of recognizing rights of nature was affirmed by the UN in the World Charter for Nature as far back as 1982. National judiciaries have given legal rights to inanimate objects such as rivers and forests. The constitutions of Bolivia and Ecuador include rights of nature.

Selected constitutional provisions on environmental rights

Overview

Over 100 constitutions in the world now recognize environmental rights (Boyd, 2012), ranging from policy statements to justiciable rights. In addition to national constitutions, state constitutions as well as indigenous constitutions also recognize environmental rights. In this section, we will discuss some of these provisions. According to Boyd, prior to the 1970s, only a handful of countries included constitutional provisions relating to the environment. These were rudimentary provisions on protecting natural beauty or conserving natural resources – examples include the Constitutions of Italy (1948), Malta (1964), Guatemala (1965), and San Marino (1974) (Boyd, 2012: 47). Similarly, the Constitutions of four Middle Eastern countries contained provisions relating to the protection and preservation of natural resources: Kuwait (1962), the United Arab Emirates (1971), Bahrain (1973), and the Syrian Arab Republic (1973) (Boyd, 2012: 47).

The first substantial provisions relating to environmental protection appear in the constitutions of Switzerland (1971), Panama (1972), Greece (1975), Papua New Guinea (1975), India (1976), and Portugal (1976). According to Boyd, every year since the adoption of the 1972 Stockholm Declaration, "at least one nation has written or amended its constitution to include or strengthen provisions relating to environmental protection" (2012: 47). During the 1980s, the leaders in "greening" constitutions were Latin American and Western European states. This accelerated in the 1990s, with African and Eastern European countries following suit. This trend mirrors observations relating to patterns of constitutionalization. Scholars have described five scenarios involving political events that led to constitutional change: (a) post-Second World War reconstruction; (b) decolonization; (c) transition to democracy; (d) dual transition to democracy and market economy; and (e) incorporation of international law into domestic law (Boyd, 2012: 49).

According to Boyd, the peak year for adopting environmental rights was 1992, with 18 constitutions incorporating new environmental provisions in that year alone. The Earth Summit in Rio and the attention paid to environmental issues at the international level possibly led to this development. Broken down by decade, the number of constitutions with environmental provisions is given in Table 7.1.

It is not surprising that the decade that saw the highest number of constitutional provisions was also the decade during which the highest number of treaties were adopted at the international level (Hunter, Salzman, and Zaelke, 2015: 168). In the environmental protection field, action at the national level tends to mirror developments at the international level.

Despite these developments, 46 members of the UN do not have any constitutional provisions on environmental protection. Not surprisingly, the UK is one of them (as it has no written constitution), and 29 of these nations are former British colonies. It is interesting that almost all of the English-speaking countries in the Americas lack constitutional environmental provisions, while all 22 non-English speaking states have constitutional environmental provisions. The nation that is considered as the birthplace of environmental law – the United States – ironically lacks any constitutional environmental provisions. This could be due to the division that Boyd observes between common law countries and civil law countries. Common law countries, especially those with older constitutions, seldom include economic, social, and cultural rights in their Bill of Rights. Their focus is almost exclusively on civil and political rights. In contrast, of the 77 states with exclusively civil law systems, 72 have constitutional environmental provisions. Of the 25 nations with mixed legal systems (combining both common law and civil law), 23 have constitutional environmental provisions.

While some tend to think that most constitutions adopted or amended since the 1970s have constitutional environmental provisions, this is not the case. Despite pleas from environmentalists, Canada did not include an environmental provision in its 1982 constitution. Similarly, an experts' committee in Ireland elected to recommend the inclusion of the government's duty to protect the environment, rather than the inclusion of a right to a healthy environment in

Table 7.1 Number of constitutions with environmental rights

Decade	*Number of constitutions with environmental rights*
Pre-1972	6
1973–1979	17
1980–1989	22
1990–1999	69
2000–2009	27
2010–2011	6

Source: Boyd, 2012: 49.

its 1996 Report reviewing the Constitution (Constitution Review Group, 1996). Although it has been amended several times since then, it still lacks any environmental protection provisions. In total, 147 countries have added environmental provisions since the early 1970s, while 40 countries chose not to do so, despite adopting other constitutional reforms. In Sri Lanka, where several attempts at constitutional reform have taken place, the drafters have played a game of hide-and-seek – including an environment right, then taking it out and then including it again, so it is not clear what its ultimate fate will be in the final draft that is submitted to Parliament (if that ever materializes). Currently, the constitutional reform process is on hold for political reasons.

An analysis of environmental constitutional provisions

Not all the constitutions that include environmental provisions have a substantive, justiciable right to a healthy environment. In this section, we will discuss some of the constitutional provisions that embody environmental provisions, including environmental rights. Constitutional environmental provisions can be divided into several categories: (a) those that include substantive environmental rights including related rights, such as the right to water; (b) those with procedural environmental rights; (c) constitutions that include individual and/or state environmental duties; (d) policy directives or principles of state policy; (e) sustainable development, public trust, and future generations; and (f) other miscellaneous environmental provisions (May and Daly, 2015). Boyd uses a similar categorization, but does not distinguish between (e) and (f), instead putting them all together under a "catch all" category (2012).

As of 2012, 92 constitutions include substantive environmental rights, 30 constitutions include procedural rights (with some overlapping), 83 constitutions provide for individual responsibility, while 140 provide for government duty (this presumably includes provisions on state policy or directive principles) (Boyd, 2012). While some overlap is inevitable, this gives a good picture of environmental provisions in world constitutions.

Substantive environmental rights

The inclusion of substantive environmental rights is usually a feature of newer constitutions, although it would be wrong to generalize. While Boyd counts 92 constitutions that include substantive environmental rights, May and Daly's count is lower, at 75. Below are a few selected examples.

The Constitution of Bolivia adopted in 2009 provides that:

> Everyone has the right to a healthy, protected, and balanced environment. The exercise of this right must be granted to individuals and collectives of present and future generations, as well as to other living things, so they may develop in a normal and permanent way.
>
> (Constitution of Bolivia: Article 33)

With regard to justiciability, the Constitution provides:

> Any person, in his own right or on behalf of a collective, is authorized to take legal action in defense of environmental rights, without prejudice to the obligation of public institutions to act on their own in the face of attacks on the environment.
>
> (Constitution of Bolivia: Article 34)

It is thus clear that the substantive right protected under Article 33 has a remedy included in the Constitution in the event that right is violated.

The Constitution of Angola provides that "all citizens shall have the right to live in a healthy and unpolluted environment" (Article 24(1)), while the Constitution of Chile provides: "all have the right to live in an environment free from contamination." The environmental provisions in the Constitution of South Africa are perhaps the best-known:

> Everyone has the right (a) to an environment that is not harmful to their health or well-being; and (b) to have the environment protected, for the benefit of present and future generations, through reasonable legislative and other measures that
>
> (i) prevent pollution and ecological degradation;
> (ii) promote conservation; and
> (iii) secure ecologically sustainable development and use of natural resources while promoting justifiable economic and social development.
>
> (Article 24)

While most constitutional provisions refer to the right to live in a healthy ("balanced," "favorable," "ecologically balanced," or "clean" are some of the adjectives used) environment, some constitutions link environmental provisions with development. Thus, the Constitution of the Democratic Republic of Congo provides: "All persons have the right to a healthy environment that is favorable to their development" (Article 53). The Constitution of the Dominican Republic is much more detailed:

> Every person has the right, both individually and collectively, to the sustainable use and enjoyment of the natural resources; to live in a healthy, ecologically balanced and sustainable environment for the development and preservation of the various forms of life, of the landscape and of nature.
>
> (Article 67(1))

The Constitution of El Salvador frames this right in relation to children: "Every child has the right to live in familial and environmental conditions that permit his integral development, for which he shall have the protection of the State" (Article 34). A few constitutions refer to the right to obtain compensation for violations. Thus, for example, the Constitution of Ukraine provides: "Everyone has the right to an environment that is safe for life and health, and to compensation for damages inflicted through the violation of this right" (Article 50). The Constitution of Belarus is similar: "Everyone is entitled to a wholesome environment and to compensation for loss or damage caused by violation of this right" (Article 46).

A few constitutions refer to future generations. The Constitution of Argentina refers to future generations as well as human development: "All residents enjoy the right to a healthy, balanced environment which is fit for human development and by which productive activities satisfy current necessities without compromising those of future generations" (Article 41). Similarly, the Bolivian Constitution refers to both present and future generations, and takes it one step further and accords rights to other living things as well (Article 33).

It is also important to point out that some constitutions confine these rights to citizens while most constitutions refer to "person, individual or everyone." This is significant, as most rights are accorded to citizens as it is the link of nationality that binds the state to its people. Most states accord certain rights to citizens only, such as the right to vote. However, rights of a more fundamental nature, such as the right to life, freedom against torture, etc., are typically accorded to everyone irrespective of the nationality. It is significant that the right to a healthy environment is not confined to citizens in most countries where this right is recognized. It is not clear, however, just looking at the environmental provisions, whether non-citizen victims can obtain any relief under the Constitution, or whether remedies are confined to citizens.

The North–South divide with regard to substantive environmental rights is also worth pointing out. Of the 75+ countries that include a substantive environmental right in their constitutions, only two are from the Global North (three if you add France with its Charter of the Environment); these are Belgium and Norway. While Belgium's provision refers to human dignity ("Everyone has the right to lead a life worthy of human dignity … [including] the right to enjoy the protection of a healthy environment" (Article 23(4)), the Norwegian provision is linked to health and biodiversity ("every person has a right to an environment that is conducive to health and to natural surroundings whose productivity and diversity are preserved" (Section E)). This North–South divide requires more analysis, but it is noteworthy that almost all the countries with substantive constitutional environmental rights are from the Global South. There could be several reasons for this – these constitutions tend to be newer; these countries are also experiencing negative consequences of environmental degradation, often the result of wanton action by their former colonial masters; most countries are also seeing the negative impact of investment by multinational companies; and finally, the rights language is powerful, and it is possible that many countries were taking cognizance of the developments internationally.

Although the US Constitution does not include an environmental right, five state constitutions do – Hawaii, Illinois, Massachusetts, Montana, and Pennsylvania (Kravchenko and Bonine, 2008). The Montana Constitution provides in Article II, Section 3: "All persons are born free and have certain inalienable rights. They include the right to a clean and healthful environment." Article IX, Section 1 provides:

1 The State and each person shall maintain and improve a clean and healthful environment in Montana for present and future generations …
2 The legislature shall provide adequate remedies for the protection of the environmental life support system from degradation and provide adequate remedies to prevent unreasonable depletion and degradation of natural resources.

The reference to future generations is significant. It is, however, not clear whether the section on remedies actually provides a judiciable right to human beings. The wording is not very clear.

Procedural environmental rights

According to Boyd, 30 national constitutions provide procedural rights specifically relating to the environment. Procedural rights include the right to information relating to the environment, right to participate in the decision-making process and the right to remedies, and the right to challenge particular decisions or unconstitutional laws. In this section, we will provide some examples. The Constitution of the Czech Republic provides: "Everybody is entitled to timely and complete information about the state of the environment and natural resources" (Article 35(2)), while the Chechnyan constitution is more elaborate, combining several things: "Everyone has the right to favorable environmental surroundings, reliable information about its condition and to compensation for damage caused to his/her health or property through ecological violations of the law" (Article 39). The Russian Federation provides: "Everyone shall have the right to a favorable environment, reliable information about its condition, and to compensation for the damage caused to his or her health or property by ecological violations" (Article 42). The Constitution of Serbia provides: "Everyone shall have the right to healthy environment and the right to timely and full information about the state of environment" (Article 74).

Thus, some of these constitutions combine the substantive right with procedural rights, and Boyd believes that "this suggests that procedural rights are viewed as a complement to rather than a substitute for, substantive environmental rights" (Boyd, 2012: 66). Indeed, the Aarhus Convention, discussed in Chapter 6, makes this link clear when it provides that the objective of adopting procedural rights is to facilitate the realization of a substantive right to a healthy environment.

Countries with past authoritarian regimes that withheld information from their citizens seem more likely to include access to environmental information in the bill of rights. Thus, procedural environmental rights are most commonly found in Eastern Europe and Latin America (Boyd, 2012: 67). Similarly, countries that had a traumatic experience with lack of information relating to an environmental disaster are also more likely to incorporate environmental procedural rights. This is the case with Ukraine, where vital information regarding high levels of radiation was withheld from the public after the Chernobyl disaster (Boyd, 2012: 67). The Constitution of Ukraine provides: "Everyone is guaranteed the right of free access to information about the environmental situation … and also the right to disseminate such information" which is considered to be "the leading example of a constitutionally recognized right to information" (May and Daly, 2015: 244).

Some countries employ a hybrid approach, with legislation supplementing the constitutional provisions. For example, Sweden's Constitution incorporates its Freedom of Press Act, while the freedom of information laws in New Zealand and Canada are considered quasi-constitutional (May and Daly, 2015: 242). Similarly, a constitutional amendment to Sri Lanka's constitution added a right to information as a fundamental right and its Right to Information law adopted in 2016 made this right operational (SHR, 2016).

Many constitutions incorporate generic procedural rights, and these are applicable in relation to environmental issues. Likewise, many countries, especially industrialized countries, have right to information laws, although not framed as a constitutional right, that are applicable in relation to environmental issues. Moreover, with regard to the environmental impact assessment process, many countries consider them as public documents subject to public participation. As discussed in Chapter 6, procedural rights are now considered to be part of customary international law binding on states.

Procedural rights are closely related to principles of good governance. Without access to timely and reliable information, it is hard for individuals to enjoy their rights. Many authoritarian regimes withhold information as a way of repressing the public. This is prevalent in countries where "disappearances" are a common practice. Withholding information is another way to ensure that civil society groups cannot challenge government decisions. It also leads to a lack of transparency which in turn can lead to a lack of accountability. Thus, May and Daly recognize that "The ability of the public to receive information from the government in a timely fashion is a cornerstone of good governance, especially in democratic societies" and that access to timely information is possibly the most central among the three pillars of access rights (2015: 239).

Environmental duties

Many constitutions frame environmental issues not as a right, but as a duty to protect the environment either by individuals, the state, or both. This seems to be the approach of older constitutions. Individual responsibility for protecting the constitution is provided for in 83 constitutions, which is marginally less than the total number of constitutions with a substantive right to a healthy environment, while 140 constitutions provide for the government's duty to protect the environment (Boyd, 2012). Some countries combine the substantive right with environmental duties. The Constitution of the Democratic Republic of the Congo, for example, frames it as a duty to defend: "All persons have the right to a healthy environment. They have the duty to defend it" (2005: Article 53). Similar wording is found in the Constitutions of Guinea and Mozambique. Ethiopia, on the other hand, imposes a duty on everyone – government and citizens alike: "Government and citizens shall have the duty to protect the environment" (Article 92(4)). Some Constitutions use softer language, such as the Constitution of Kyrgyzstan, which states that "everyone should care for the environment, flora and fauna" (Article 48(3)). The Constitution of Madagascar frames it as a duty to respect the environment: "Everyone shall have the duty to respect the environment" (Article 39).

By far, the commonest form of constitutional provisions relating to the environment is framed as governmental duties. While most constitutions, such as that of Sweden, provide for a general duty to protect the environment ("The public institutions shall promote sustainable development leading to a good environment for present and future generations" (Article 2)), Portugal's constitution has very elaborate provisions on this, which looks more like legislation than a constitutional provision:

1 In order to ensure enjoyment of the right to the environment within an overall framework of sustainable development acting via appropriate bodies and with the involvement and participation of citizens, the state shall be charged with:

(a) preventing and controlling pollution and its effects and the harmful forms of erosion;
(b) conducting and promoting town and country planning with a view to a correct location of activities, balanced social and economic development and the enhancement of the landscape;
(c) creating and developing natural and recreational reserves and parks and classifying and protecting landscapes and places, in such a way as to guarantee the conservation of nature and the preservation of cultural values and assets that are of historic or artistic interest;
(d) promoting the rational use of natural resources, while safeguarding their ability to renew themselves and maintain ecological stability, with respect for the principle of inter-generational solidarity;
(e) acting in cooperation with local authorities, promoting the environmental quality of rural settlements and urban life, particularly on the architectural level and as regards the protection of historic zones;
(f) promoting the integration of environmental objectives into the various policies of a sectorial nature;
(g) promoting environmental education and respect for environmental values;
(h) ensuring that tax policy renders development compatible with the protection of the environment and quality of life.

Bhutan's constitution is unique in this regard. Not only does it impose a general obligation on the state to protect the environment, it also enacts a specific duty with regard to preserving the forest cover. Moreover, the substantive right to a healthy environment is also framed as a duty of the government:

2 The Royal Government shall:

(a) protect, conserve and improve the pristine environment and safeguard the biodiversity of the country;
(b) prevent pollution and ecological degradation;
(c) secure ecologically balanced sustainable development while promoting justifiable economic and social development; and
(d) ensure a safe and healthy environment.

3 The Government shall ensure that, in order to conserve the country's natural resources and to prevent degradation of the eco-system, a minimum of sixty percent of Bhutan's total land shall be maintained under forest cover for all time.

(Boyd, 2012: 59)

Other constitutions include a general duty to protect the environment although there is range of such duties. The Bangladesh Constitution adopts soft language – "The State shall *endeavor* to protect and improve the environment and to preserve and safeguard the natural resources, biodiversity, wetlands, forests and wild life for the present and future citizens" (Article 18A, emphasis added), while other constitutions impose a stricter duty. Thus, for example, the Constitution of Croatia provides "The State shall ensure conditions for a healthy environment" (Article 69), while the Constitution of Costa Rica provides a supplementary role for environmental legislation:

> Every person has the right to a healthy and ecologically balanced environment. The State shall guarantee, defend and preserve this right. The law will determine the corresponding responsibilities and sanctions.
>
> (Article 50)

The East Timor Constitution links the duty to sustainable development: "The State shall promote actions aimed at protecting the environment and safeguarding the sustainable development of the economy" (Section 61(3)), while the Constitution of Egypt links it with other legislation and a substantive right: "Safeguarding the environment is a national duty, and the law shall regulate the right to a good environment and the measures necessary to safeguard it" (Article 59). Some constitutions, similar to the one of Bhutan, impose an obligation to not only protect the environment, but also to improve it. Thus, the Constitution of Sri Lanka provides: "The State shall protect, preserve and improve the environment for the benefit of the community" (Article 27(14)).

Principles of state policy/policy directives

In several countries, environmental provisions are included in the section on directive principles of state policy or as a policy directive or, as we saw in the previous section, as a state duty. This is usually the case in older constitutions. The Constitution of Armenia, for example, provides: "The Government … shall ensure the implementation of state policies in the area ... of ... environmental protection" (May and Daly, 2015: 73), whereas the Constitution of Italy provides "The state has exclusive legislative power with respect to ... the protection of the environment, [and] the ecosystem" (Article 117(s)). The Constitution of Malawi includes a very elaborate provision:

> The State shall actively promote the welfare and development of the people of Malawi by progressively adopting and implementing policies and legislation aimed at ... manage[ing] the environment responsibly in order to (i) prevent the degradation of the environment, (ii) provide a healthy living and working environment for the people of Malawi, (iii) accord full recognition to the rights of future generations by means of environmental protection and the sustainable development of natural resources, and (iv) conserve and enhance the biological diversity of Malawi.
>
> (Article 13(d))

The Mozambique Constitution imposes an obligation on the state and local authorities to adopt policies to protect the environment and promote the rational use of all natural resources (Article 90(2)), while the Constitution of Peru provides "The State determines national environmental policy. It promotes the sustainable use of its natural resources" (Article 67). Many of the Constitutions in South Asia include environmental provisions in the directive principles chapter, as opposed to the Bill of Rights. The situation in the Philippines is similar (see the *Minors Opposa* case discussed in Chapter 8). However, as the cases discussed in Chapter 8 show, this has not stopped the judiciary in the region from interpreting these provisions broadly and articulating environmental rights.

Provisions relating to sustainable development, public trust, and climate change

As environmental constitutionalism has matured, newer constitutions are adopting even more novel provisions. These include references to sustainable development, climate change, and the public trust doctrine, and, even more innovatively, rights of nature.

As discussed in Chapter 1, sustainable development and sustainability have developed phenomenally over the years, especially since the Rio Conference in 1992. They have also crept into constitutions as well as judicial decisions. Inherent in the concept of sustainable development is the need to protect the rights of future generations. Many of the substantive constitutional environmental rights refer to future generations as well. Nearly 20 constitutions refer to a goal of sustainability or sustainable development, although in most cases these provisions do not seem to be justiciable (May and Daly, 2015: 262). However, as the judicial decisions in Chapter 8 show, these non-justiciable provisions can be a great tool in the hands of an innovative judiciary. About a dozen constitutions refer to both sustainable development and future generations. Mozambique's constitution requires the state to adopt policies with a view to guaranteeing the right to the environment within the framework of sustainable development aimed at guaranteeing the rational utilization of natural resources and safeguarding the capacity to regenerate, ecological stability, and the rights of future generations (May and Daly, 2015: 339).

May and Daly (2015) recognize the significance of incorporating these new concepts in constitutions:

> The incorporation of sustainability into domestic constitutions has great potential to advance both sustainability and constitutionalism. Laxman and Ansari, for example, have examined the relationship among sustainable management, the utilization of the environment, and the constitutional safeguards of environmental rights. They note how constitutional provisions help bridge the gap left by international and domestic laws, even given the array of sustainability provisions already in existence. Even though the vast majority of these provisions create no judicially enforceable rights, they nonetheless affirm national values of environmental sustainability to which courts and others may advert.
>
> (264, footnotes omitted)

The Constitution of Colombia refers to sustainable development in the context of natural resources: "The State shall plan the handling and use of the natural resources in order to guarantee their sustainable development, their conservation, restoration, or substitution" (Article 80). Some constitutions refer to sustainability in the context of substantive environmental rights. Thus, for example, the Constitution of Benin provides: "Every person has the right to a healthy, satisfactory and sustainable environment and has the duty to defend it. The State sees to the protection of the environment" (Article 27). The Constitution of Belgium refers specifically to sustainable development: "In the exercise of their respective competences, the Federal State, the Communities and the Regions pursue the objective of sustainable development in its social, economic and environmental aspects, taking into account the solidarity between the generations" (Article 7bis).

Another "new" principle included in constitutions is public trust, although it is not a new principle by any means. In fact, it is an ancient principle that can be traced to Roman law. Six constitutions refer to public trust, or some variation of it, in relation to natural resources, requiring policy-makers, rulers, or citizens to hold resources in trust for current or future generations (May and Daly, 2015: 267). The Ugandan Constitution provides, *inter alia*:

> The Government or a local government as determined by Parliament by law, shall hold in trust for the people and protect, natural lakes, rivers, wetlands, forest reserves, game reserves, national parks and any land to be reserved for ecological and touristic purposes for the common good of all citizens.
>
> (May and Daly, 2015: 342)

The other constitutions that refer to this principle are those of Ethiopia and Papua New Guinea. Swaziland and Ghana vest this trust in their leaders, while Tanzania's constitution holds citizens accountable to hold resources in trust for future generations. Similarly, the Bhutanese Constitution provides: "[e]very Bhutanese is a trustee of the Kingdom's natural resources and environment for the benefit of the present and future generations and it is the fundamental duty of every citizen to contribute to the protection of the natural environment" (May and Daly, 2015: 267).

In several cases from South Asia, the courts have applied the public trust doctrine, despite the absence of it in the constitution. Examples are M.C. *Mehta v. Kamal Nath and others* (Supreme Court of India) and *Bulankulame and others v. Minister of Industrial Development and Others* and *Sugathapala Mendis and another v. Chandrika Kumaratunga and others* from the Supreme Court of Sri Lanka. In the *Bulankulame case*, the Supreme Court of Sri Lanka went one step further and adopted the principle of shared responsibility, stating that the public trust doctrine is too restrictive.

Climate change is another issue that seems to be making an appearance in constitutions, albeit marginally. Only the Dominican Republic's constitution explicitly refers to climate change. It requires the adoption of a plan "that assures the efficient and sustainable use of the natural resources of the Nation, in accordance with the need of adaptation to climate change" (May and Daly, 2015: 271). Considered as one of the most challenging issues to confront humanity, climate change is a global issue, and with the adoption of the Paris Agreement, almost the entire international community (with the notable exception of the United States) have made emission reduction commitments called Nationally Determined Contributions (NDCs) – the United States also did so under President Obama, although President Trump withdrew from the agreement – which require mitigation measures to be taken at the domestic level. Incidentally, one of the cases that is currently pending in the United States (which has so far survived challenges of various kinds) is based on the atmospheric trust doctrine. This case, as well as others, will be discussed in Chapter 8.

Another right that is increasingly being included in constitutions is the right to water. Some of the impetus for this inclusion was the wave of water privatization that took place in the 1990s in many parts of the world, especially Latin America, at the behest of the World Bank. In many parts of the world, water is becoming a critical issue, with some areas experiencing water stress. Climate change will exacerbate the situation. Security experts are worried that water scarcity could also lead to conflicts. Water wars have already broken out in Bolivia and other places. Constitutions of at least 14 states include a right to water. The South African Constitution is possibly the best known for its recognition of a human right to water: "Everyone has the right to have access to … (b) sufficient food and water" enforced by regular reports on progressive realization (Article 27.1).

Some constitutions frame it as part of environmental rights. For example, the Constitution of Cambodia provides: "The State shall protect the environment and balance of abundant natural resources and establish a precise plan of management of land, water, air, wind, geology, ecological systems, mines, energy, petrol and gas, rocks and sand, gems, forests and forestrial products, wildlife, fish and aquatic resources" (Article 59). The Nigerian Constitution imposes an obligation on the state to "protect and improve the environment and safeguard the water, air and land, forest and wild life of Nigeria" (Article 20), while the Laos Constitution imposes an obligation on everybody to "protect the environment and natural resources: land, underground, forests, fauna, water sources and atmosphere" (Article 17). Uganda includes water in the provision on sustainable development and requires the State to "take all possible measures to prevent or minimize damage and destruction to land, air and water resources resulting from pollution or other causes" (Article 27(i)).

Lessons from global environmental constitutionalism

In their seminal work on global environmental constitutionalism, May and Daly note that "environmental constitutionalism is a relatively recent phenomenon at the confluence of constitutional law, international law, human rights, and environmental law" (May and Daly, 2015: 1). They recognize that over 100 constitutions worldwide have incorporated some kind of environmental right in their constitutions and that certain constitutional principles relating to environmental protection have emerged. Unlike previous works on the link between human rights and environmental protection and similar topics, these scholars use comparative constitutionalism – "the practice by constitutional courts of comparing and contrasting texts, contexts, and outcomes elsewhere" – as their framework (May and Daly, 2015: 4).

Although judicial decisions of other countries have no binding effect, judges often look at other jurisdictions to see what lessons they can draw from their counterparts, especially in new and emerging areas of law, such as environmental rights. This is also true of areas of law that emerged at the international level, and where states strive to apply a set of global principles at the national level. In such situations, how other judiciaries apply, mold, and interpret these international principles hold important lessons for other judiciaries as well, with the caveat that environmental issues and ecosystems are unique to each country. One region where judiciaries have consistently consulted each other's decisions is South Asia where the judiciary have played an important role in developing environmental rights. Their jurisprudence is discussed in Chapter 8. Although there are no constitutional courts in South Asia and the Supreme Court in each country plays this role, they have consistently relied on decisions of their counterparts, thereby generating something akin to a regional environmental constitutionalism.

One of the significant features of environmental constitutionalism is that it mirrors the convergence of human rights and environmental protection at the international level (Anton and Shelton, 2011). Despite this, international law lags behind national developments, as the former does not yet recognize a distinct right to a healthy environment, although "greening" of international human rights law has occurred (Boyle and Anderson, 1996). National judiciaries have interpreted existing rights in an expansive and creative manner to include a right to a healthy environment and the right to be free from pollution. In addition, these judiciaries have applied international environmental law principles directly, thereby incorporating these principles into national law.

Hayward identifies several reasons as the rationale for a constitutional approach to environmental protection: first, it acknowledges the importance that a society places on environmental protection. Constitution is the supreme law of the land and its provisions supersede all other laws. Second, constitutional provisions can promote the coordination of environmental measures within the country, as in many countries these laws and measures have developed in an *ad hoc* manner. Third, it can promote the coordination of environmental protection measures between states, as many environmental problems are transboundary and increasingly, global, in nature.

Harmonization of principles and laws is necessary, and there is an increasing convergence between constitutional provisions and international laws, especially in the areas of environmental law and human rights law. Fourth, environmental protection need not depend on narrow majorities in legislative bodies. Finally, it can spur citizen involvement in environmental protection measures (Hayward, 2005: 5–7):

> In sum, providing for environmental protection at the constitutional level has a number of potential advantages: it entrenches a recognition of the importance of environmental protection; it offers the possibility of unifying principles for legislation and regulation; it secures these principles against the vicissitudes of routine politics, while at the same time enhancing possibilities of democratic participation in environmental decision-making processes.

Constitutional entrenchment, of course, does not necessarily imply that environmental protection should be articulated as constitutional rights. Many countries, especially those with older constitutions, incorporate them as guiding principles of state policy or do not refer to them at all in the constitution. On the other hand, the non-inclusion of a reference to environmental protection or rights in the constitution does not, by itself, necessarily mean that the state does not consider it sufficiently important. On the contrary, as the Supreme Court of the Philippines noted in the celebrated *Minors Oposa Case* (1993):

> While the right to a balanced and healthful ecology is to be found under the Declaration of Principles and State Policies and not under the Bill of Rights, it does not follow that it is less important than any of the civil and political rights enumerated in the latter. Such a right belongs to a different category of rights altogether for it concerns nothing less than

> self-preservation and self-perpetuation – aptly and fittingly stressed by the petitioners – the advancement of which may even be said to predate all governments and constitutions. ***As a matter of fact, these basic rights need not even be written in the Constitution for they are assumed to exist from the inception of humankind.*** If they are now explicitly mentioned in the fundamental charter, it is because of the well-founded fear of its framers that unless the rights to a balanced and healthful ecology and to health are mandated as state policies by the Constitution itself, thereby highlighting their continuing importance and imposing upon the state a solemn obligation to preserve the first and protect and advance the second, the day would not be too far when all else would be lost not only for the present generation, but also for those to come – generations which stand to inherit nothing but parched earth incapable of sustaining life.
>
> (187, emphasis added)

As noted by May and Daly (2015), as well as Boyd, over 100 countries now include "environmental rights" (interpreted broadly) in their constitutions, whether in the Bill of Rights (usually justiciable), or in the Directive Principles Chapter (not typically justiciable) (Boyd, 2012: 76). Moreover, 147 out of 193 national constitutions incorporate some form of environmental protection provisions, and the right to live in a healthy environment enjoys explicit recognition in 92 constitutions. Boyd believes that "No other human right has achieved such a broad level of constitutional recognition in such a short period of time" (2012: 76). The significance of these developments cannot be under-estimated:

> Overall, these constitutional developments appear to reflect a rapid evolution of human values, with environmental protection gaining the requisite moral importance to merit inclusion in written documents that express a society's most cherished and deeply held values.
>
> (Boyd, 2012: 77)

Boyd identifies the convergence of "three striking and related developments" in the world that led to the recognition of a constitutional right to a healthy environment: (a) the adoption of a wave of new constitutions, as well as amending existing ones; (b) the human rights revolution; and (c) the global environmental crisis (2012: 3). However, it would be a mistake to assume that the main objective of constitutional guarantees of environmental rights is to promote litigation. Rather, these guarantees aim to "reduce the level of harm being inflicted on humans and the Earth and to redistribute inequitable allocations of environmental harms … and that litigation is merely a means to an end" (Boyd, 2012: 28).

The constitutional guarantee to a healthy environment could contribute to a broad range of outcomes, including: stronger environmental regulation and implementation of existing environmental laws; increasing accountability; strengthening democracy by promoting greater citizen participation and access to information; protecting vulnerable communities, including future generations; and enabling a better balance with competing norms, such as economic and social rights (Boyd, 2012: 28). If the right to a healthy environment is also recognized at the international level, then states will be required to report on the status of the right in their country reports to the relevant human rights treaty body. Even in the absence of a specific right, states report on their environmental obligations in their country reports, and treaty bodies have started referring to environmental issues and environmental obligations in their concluding observations. The concluding observations made by the Committee on Economic Social and Cultural Rights on the Fifth Period Report of Australia is a good example (2017).

Of course, some of these outcomes can be achieved without constitutional guarantees, but being enshrined in the Constitution ensures a higher level of protection that is lacking in normal legislation. However, not everyone is convinced that constitutional environmental rights are a good idea. Similar arguments have been made against the recognition of a right to a healthy environment at the international level (Alston, 1982). Opponents have argued that such a right is vague, cannot be enforced, will open floodgates to litigation, is anthropocentric, is likely to be ineffective, is not justiciable, etc. (Boyd, 2012: 33). However, as our experience with regional human rights institutions and with national judiciaries has shown, many of these concerns have proven to be unfounded. In fact, many judiciaries have shown that even in the absence of constitutional environmental rights, they are open to interpreting existing rights expansively to *include* environmental rights. Indeed, most human rights have achieved coherence and maturity because judiciaries around the world had the opportunity to interpret them. They have been able to balance them against other rights as well as evaluate them vis-à-vis society at large. There is no logical reason as to why a right to a healthy environment should not be given the same opportunity. As Boyd points out:

> Many of the arguments marshaled against the right to a healthy environment have been leveled at other human rights and found wanting. All human rights are phrased in brief, general terms whose meaning evolves over time. Rights are almost always balanced against competing rights, so absolutism is a non-issue.
>
> (2012: 44)

The absence of a specific right to a healthy environment under international human rights law may even have been the impetus for such recognition at the domestic level in many countries either in their constitutions or through judicial interpretation. Global environmental constitutionalism offers many lessons for the global community and the environmental rights revolution is likely to continue into the foreseeable future.

DISCUSSION QUESTIONS

1. What are the pros and cons of including environmental rights in constitutions?
2. If you were drafting a new constitution for your country, would you include environmental rights? If so, how would you formulate them?
3. Why do you think constitutions adopted after 1990s include environmental rights?
4. Do you think there is now a "global environmental constitutionalism" with specific principles?
5. What are the most important environmental principles that have contributed to recognizing constitutional environmental rights?

References

Alston, P. (1982). A Third Generation Solidarity Rights: Progressive Development or Obfuscation of International Human Rights Law? *Netherlands International Law Rev* 29: 307–22.

Anton, D. and Shelton, D. (2011). *Environmental Protection and Human Rights*. Cambridge: Cambridge University Press.

Atapattu, S. (2015). The Role of Human Rights Law in Protecting Environmental Rights in South Asia. In: Haglund, L. and Stryker, R. (eds.), *Closing the Rights Gap: From Human Rights to Social Transformation*. Oakland, CA: University of California Press.

Atapattu, S. (2002). The Right to a Healthy Life or the Right to Die Polluted?: The Emergence of a Human Right to a Healthy Environment under International Law. *Tulane Environmental Law Journal* 16: 65–126.

Belgian Constitution [Online]. Available at:www.dekamer.be/kvvcr/pdf_sections/publications/constitution/GrondwetUK.pdf [Accessed: 10 October 2018].

Bosselmann, K. (2015). Global Environmental Constitutionalism: Mapping the Terrain. *Widener Law Review* 21: 171.

Boyd, D. (2012). *The Environmental Rights Revolution: A Global Study of Constitutions, Human Rights, and the Environment*. Vancouver: UBC Press.

Boyle, A. and Anderson, M. (eds.) (1996). *Human Rights Approaches to Environmental Protection*. Oxford: Oxford University Press.

Burger, M. (2003). Bipolar and Polycentric Approaches to Human Rights and the Environment. *Columbia Journal of Environmental Law* 28: 371–93.

Committee on Economic Social and Cultural Rights (2017). *Concluding Observations on the Fifth Period Report of Australia*, E/C.12/AUS/CO/5, 23 June 2017. Geneva: United Nations.

Constitution of Belarus (1994) [Online]. Available at: www.constituteproject.org/constitution/Belarus_2004.pdf?lang=en [Accessed: 10 October 2018].

Constitution of Benin (1990) [Online]. Available at: www.constituteproject.org/constitution/Benin_1990.pdf?lang=en [Accessed: 10 October 2018].

Constitution of Bolivia (2009) [Online]. Available at: www.constituteproject.org/constitution/Bolivia_2009.pdf [Accessed: 1 October 2018].

Constitution of Cambodia (1993) [Online]. Available at: www.constitution.org/cons/cambodia.htm [Accessed: 10 October 2018].

Constitution of Chechen Republic (1992) [Online]. Available at: www.refworld.org/docid/3ae6b5fb8.html [Accessed: 10 October 2018].

Constitution of Chile (1980) [Online]. Available at: http://confinder.richmond.edu/admin/docs/Chile.pdf [Accessed: 1 October 2018].

Constitution of Colombia (1991) [Online]. Available at: http://confinder.richmond.edu/admin/docs/colombia_const2.pdf [Accessed: 10 October 2018].

Constitution of Costa Rica (1949) [Online]. Available at: www.constituteproject.org/constitution/Costa_Rica_2011.pdf [Accessed: 10 October 2018].

Constitution of Egypt (2014) [Online]. Available at: www.constituteproject.org/constitution/Egypt_2014.pdf [Accessed: 10 October 2018].

Constitution of El Salvador (1983) [Online]. Available at: http://confinder.richmond.edu/admin/docs/ElSalvador1983English.pdf [Accessed: 10 October 2018].

Constitution of Ghana (1992) [Online]. Available at: www.constituteproject.org/constitution/Ghana_1996.pdf?lang=en [Accessed: 10 October 2018].

Constitution of Guinea (2010) [Online]. Available at: www.constituteproject.org/constitution/Guinea_2010.pdf?lang=en [Accessed: 10 October 2018].

Constitution of Madagascar (2010) [Online]. Available at: www.constituteproject.org/constitution/Madagascar_2010.pdf [Accessed: 10 October 2018].

Constitution of Norway (1814) [Online]. Available at: www.constituteproject.org/constitution/Norway_2014.pdf [Accessed: 10 October 2018].

Constitution of Peru (1993) [Online]. Available at: www.constituteproject.org/constitution/Peru_2009.pdf?lang=en [Accessed: 10 October 2018].

Constitution of Portugal (1976) [Online]. Available at: www.constituteproject.org/constitution/Portugal_2005.pdf [Accessed: 10 October 2018].

Constitution of South Africa (1996) [Online]. Available at: www.justice.gov.za/legislation/constitution/SAConstitution-web-eng.pdf [Accessed: 1 October 2018].

Constitution of Sweden (1974) [Online]. Available at: www.constituteproject.org/constitution/Sweden_2012.pdf?lang=en [Accessed: 10 October 2018].

Constitution of the Argentine Nation (1994) [Online]. Available at: www.biblioteca.jus.gov.ar/argentina-constitution.pdf [Accessed: 10 October 2018].

Constitution of the Czech Republic (1992) [Online]. Available at: www.usoud.cz/fileadmin/user_upload/Tiskova_mluvci/Ustava_EN_ve_zneni_zak_c._98-2013.pdf [Accessed: 10 October 2018].

Constitution of the Democratic Republic of Timor-Leste (2002) [Online]. Available at: www.fd.uc.pt/g7+/pdfs/Timor.pdf [Accessed: 10 October 2018].

Constitution of the Democratic Socialist Republic of Sri Lanka (1978) [Online]. Available at: www.parliament.lk/files/pdf/constitution.pdf [Accessed: 10 October 2018].

Constitution of the Federal Democratic Republic of Ethiopia (1994) [Online]. Available at: www.wipo.int/edocs/lexdocs/laws/en/et/et007en.pdf [Accessed: 10 October 2018].

Constitution of the Independent State of Papua New Guinea (1975) [Online]. Available at: www.unesco.org/education/edurights/media/docs/600e78096209b63b86f0135f52694b257b4b0c0e.pdf [Accessed: 10 October 2018].

Constitution of the Italian Republic (1947) [Online]. Available at: www.senato.it/documenti/repository/istituzione/costituzione_inglese.pdf [Accessed: 10 October 2018].

Constitution of the Kingdom of Bhutan (2008) [Online]. Available at: www.nationalcouncil.bt/assets/uploads/docs/acts/2017/Constitution_of_Bhutan_2008.pdf [Accessed: 10 October 2018].

Constitution of the Kingdom of Swaziland (2005) [Online]. Available at: www.wipo.int/edocs/lexdocs/laws/en/sz/sz010en.pdf [Accessed: 10 October 2018].

Constitution of the Kyrgyz Republic (2010) [Online]. Available at: www.refworld.org/pdfid/3ae6b5ae0.pdf [Accessed: 10 October 2018].

Constitution of the People's Republic of Bangladesh (1972) [Online]. Available at: http://bdlaws.minlaw.gov.bd/print_sections_all.php?id=367 [Accessed: 10 October 2018].

Constitution of the Republic of Angola (2010) [Online]. Available at: http://extwprlegs1.fao.org/docs/pdf/ang72591ENG.pdf [Accessed: 1 October 2018].

Constitution of the Republic of Armenia (1995) [Online]. Available at: www.constituteproject.org/constitution/Armenia_2015.pdf?lang=en [Accessed: 10 October 2018].

Constitution of the Republic of Croatia (2010) [Online]. Available at: www.constituteproject.org/constitution/Croatia_2010.pdf?lang=en [Accessed: 10 October 2018].

Constitution of the Republic of Malawi (1994) [Online]. Available at: www.sdnp.org.mw/constitut/dtlindx.html [Accessed: 10 October 2018].

Constitution of the Republic of Mozambique (1990) [Online]. Available at: http://confinder.richmond.edu/admin/docs/moz.pdf [Accessed: 10 October 2018].

Constitution of the Republic of Serbia (2006) [Online]. Available at: https://www.ilo.org/dyn/natlex/docs/ELECTRONIC/74694/119555/F838981147/SRB74694%20Eng.pdf [Accessed: 10 October 2018].

Constitution of the Republic of the Philippines (1987) [Online]. Available at: www.officialgazette.gov.ph/constitutions/1987-constitution/ [Accessed: 10 October 2018].

Constitution of the Republic of Uganda (1995) [Online]. Available at: https://ulii.org/ug/legislation/consolidated-act/0 [Accessed: 10 October 2018].

Constitution of the Russian Federation (1993) [Online]. Available at: www.constitution.ru/en/10003000-01.htm [Accessed: 10 October 2018].

Constitution of the State of Montana (1972) [Online]. Available at: https://leg.mt.gov/bills/mca_toc/Constitution.htm [Accessed: 10 October 2018].

Constitution of the United Republic of Tanzania (1977) [Online]. Available at: https://rsf.org/sites/default/files/constitution.pdf [Accessed: 10 October 2018].

Constitution of Ukraine (1996) [Online]. Available at: www.constituteproject.org/constitution/Ukraine_2004.pdf [Accessed: 10 October 2018].

Democratic Republic of Congo (2005) [Online]. Available at: www.constitutionnet.org/sites/default/files/DRC%20-%20Congo%20Constitution.pdf [Accessed: 1 October 2018].

Gellers, J.C. (2017). *The Global Emergence of Constitutional Environmental Rights*. New York, NY: Routledge.

Hayward, T. (2005). *Constitutional Environmental Rights*. Oxford: Oxford University Press.

Hunter, D., Salzman, J., and Zaelke, D. (2015). *International Environmental Law and Policy*, 5th ed. St. Paul, MN: Foundation Press.

Kotze, L. (2016). *Global Environmental Constitutionalism in the Anthropocene*. Oxford: Oxford University Press.

Kravchenco, S. and John B. (2008). *Human Rights and the Environment: Cases, Law and Policy*. Durham, NC: Carolina Academic Press.

Lao People's Democratic Republic's Constitution (1991) [Online]. Available at: www.constituteproject.org/constitution/Laos_2003.pdf?lang=en [Accessed: 10 October 2018].

May, J.R. and Daly, E. (2015). *Global Environmental Constitutionalism*. Cambridge: Cambridge University Press.

Parvez, H. and Azfar, A. (2004). Securing Environmental Rights Through Public Interest Litigation in South Asia. *Virginia Environmental Law Journal* 22: 215–47.

Puvimanasinghe, S.F. (2009). Sustainable Development in the Courts: Towards a Jurisprudence of Sustainable Development in South Asia: Litigation in the Public Interest. *Sustainable Development Law & Policy* 10: 41.

Razzaque, J. (2002). *Human Rights and the Environment: The National Experience in South Asia and Africa*, Joint UNEP-OHCHR Expert Seminar on Human Rights and the Environment, Background Paper No 4. Geneva, January 2002. [Online] Available at: www2.ohchr.org/english/issues/environment/environ/bp4.htm [Accessed: 1 September 2018].

Report of the Constitution Review Group (1996). Dublin: Stationery Office [Online]. Available at: http://archive.constitution.ie/reports/crg.pdf [Accessed: 1 October 2018].

Republic of Nigeria (1989) [Online]. Available at: www.concourt.am/armenian/legal_resources/world_constitutions/constit/nigeria/nigeri-e.htm [Accessed: 10 October 2018].

Shelton, D. (1991). Human Rights, Environmental Rights and Right to Environment. *Stanford Journal of International Law* 28: 103–38.

Smith, D. (1999). Human Rights and the Environment: Clean Air and a Clean Environment as Fundamental Human Rights. *Colorado Journal of International Environmental Law & Policy Year Book* 149.

Taylor, P.E. (1998). *An Ecological Approach to International Law: Responding to Challenges of Climate Change*. London: Routledge.

UNECE (1998). Convention on Access to Information, Public Participation in Decision-making and Access to Justice in Environmental Matters in 1998 (Aarhus Convention). [Online]. Available at: www.unece.org/fileadmin/DAM/env/pp/documents/cep43e.pdf [Accessed: 1 October 2018].

UNFCCC (2015). Paris Agreement [Online]. Available at: https://unfccc.int/sites/default/files/english_paris_agreement.pdf [Accessed: 1 October 2018].

United Nations Conference on Environment and Development (Rio Conference) (1992) [Online]. Available at: www.un.org/geninfo/bp/enviro.html [Accessed: 1 October 2018].

Cases

Bulankulame and others v. Minister of Industrial Development and Others (1999), S.C. Application No. 884/99 (F/R) (Supreme Court of Sri Lanka). [Online]. Available at: www.elaw.org/content/sri-lanka-bulankulama-v-min-industrial-development-eppawala-case-sc-application-no-88499-fr [Accessed: 20 October 2018].

M.C. *Mehta v. Kamal Nath and others* (2000) (Supreme Court of India). [Online]. Available at: www.informea.org/en/court-decision/mc-mehta-v-kamal-nath-and-ors [Accessed: 1 October 2018].

Minors Oposa v. Factoran, Supreme Court of the Philippines (1993), 33 ILM 173. [Online]. Available at: http://hrlibrary.umn.edu/research/Philippines/Oposa%20v%20Factoran,%20GR%20No.%20101083,%20July%2030,%201993,%20on%20the%20State's%20Responsibility%20To%20Protect%20the%20Right%20To%20Live%20-in%20a%20Healthy%20Environment.pdf [Accessed: 1 October 2018].

Sugathapala Mendis and another v. Chandrika Kumaratunga and others (2008) (Supreme Court of Sri Lanka). [Online]. Available at: www.lawnet.gov.lk/2008/12/31/sugathapala-mendis-and-another-v-chandrika-kumaratunga-and-others-waters/ [Accessed: 20 October 2018].

8 Selected national cases

Introduction

Judiciaries around the world have articulated environmental rights in several ways. They have interpreted existing rights in an expansive and creative manner. They have also expanded standing (*locus standi*) so that NGOs could bring cases on behalf of groups of victims which has paved the way for public interest litigation. Judiciaries have also used international environmental law principles in cases, effectively incorporating these principles into national law. In this chapter, we will take a virtual tour around the world, discussing some of the seminal cases on environmental rights.

In the previous chapter, we discussed different types of constitutional provisions from around the world that have been used as the foundation for environmental rights, some justiciable, some not. Generally, whether constitutional environmental rights are only statements of aspirations or are enforceable depends on whether they will be protected by the judiciary and whether a remedy exists. If they cannot be enforced or if the judiciary is not willing to interpret them broadly, then constitutional entrenchment becomes meaningless. They are merely aspirational or are intended to guide policy direction. However, in many countries, judiciaries have articulated environmental rights even in the absence of specific provisions: "National courts have given a range of answers to the question whether environmental rights in national constitutions actually give rise to a judicial remedy" (Kravchenko and Bonine, 2008: 67). These range from: (a) mere statement of aspiration; (b) as a basis for opening access to the courts; (c) as policy guidance to the legislature and executive; (d) as a judicially enforceable check on the legislature; (e) as a command for specific actions by the legislature; (f) as a displacement for executive authority; and (g) as the basis for private remedies for damages or injunctions (part of tort law) (Kravchenko and Bonine, 2008: 67–8). In addition, where states have ratified optional protocols giving victims access to regional or international human rights bodies, they act as an additional scrutiny of state action itself. In this chapter, we will discuss some of the salient national cases that have discussed the link between human rights and the environment.

Selected cases from around the world

Asia

South Asia has been one of the most vibrant regions that have articulated environmental rights. Judiciaries there have developed environmental rights by interpreting existing rights broadly to encompass environmental rights. One of the earliest cases of the Supreme Court of India was *Subhash Kumar v. State of Bihar* (1991), where the Court observed that the right to life enshrined in Article 21 of the Indian Constitution included the right to enjoyment of pollution-free water and air for the full enjoyment of life:

> The right to life is a fundamental right under article 21 of the Constitution and it includes the right of enjoyment of pollution-free water and air for full enjoyment of life. If anything endangers or impairs that quality of life in derogation of laws, a citizen has the right to have recourse to article 32 of the Constitution for removing the pollution of water or air which may be detrimental to the quality of life.

In many of these cases, the courts had to strike a balance between economic development, on the one hand, and the right of people to live in a clean environment, on the other. The *Vellore Citizens Welfare Forum v. Union of India* (1996) involved environmental damage and the public health hazard caused by tanneries in the State of Tamil Nadu. The Supreme Court of India noted that while these tanneries brought in a considerable amount of revenue and generated employment, the industry had no right to destroy the ecology and create a health hazard. It noted that: "Sustainable development, and in particular the polluter pays principle and the precautionary principle, have become part of customary international law." In *AP Pollution Control Board II v. MV Nayudu* (2001), the Supreme Court of India observed that the right to have access to drinking water is fundamental to life and it is the duty of the State under Article 21 of the Constitution to provide clean drinking water to its citizens. The Court further noted that there is a need to balance the right to a healthy environment and the right to sustainable development. Similarly, in *Narmada Bachao Andolan v. Union of India* (2000), the Court noted that water is a basic human need and is part of the right to life and human rights enshrined in Article 21 of the Indian Constitution. The Court reiterated the importance of the precautionary principle, sustainable development, and the "polluter pays" principle, as well as the public trust doctrine.

Like in India, the Constitution of Bangladesh does not include a right to environment. But its judiciary has interpreted the right to life clause in an expansive manner to articulate environmental rights. In *Dr Mohiuddin Farooque v. Bangladesh, represented by the Secretary, Ministry of Irrigation, Water Resources and Flood Control* (1996), the Supreme Court of Bangladesh considered the issue whether the "fundamental right to life included the protection and preservation of the environment, the ecological balance and an environment free from pollution essential for the enjoyment of the right to life." While the Court noted that the Bangladeshi Constitution does not contain a provision akin to Article 48A of the Indian Constitution (which imposes a duty on citizens to protect the environment), it pointed out that Articles 31 and 32 on the right to life encompass within their ambit the protection and preservation of the environment, ecological balance free from pollution of air and water, and sanitation, without which life can hardly be enjoyed:

> Articles 31 and 32 of our constitution protect the right to life as a fundamental right. It encompasses within its ambit, the protection and preservation of environment, ecology free from pollution of air and water, sanitation without which life can hardly be enjoyed and omission contrary thereto will be violative of the said right to life.

This trend has been followed in subsequent cases. The space given by the Court in this case has opened the door to similar litigation in Bangladesh, and the courts have developed a robust body of law on environmental issues.

A similar trend can be seen in Pakistan. Pakistan's Constitution does not contain any reference to the environment either in the Bill of Rights (which is rarely seen in Constitutions adopted prior to the 1990s) or in the directive principles chapter (which is more common). The only reference to the environment is in a schedule to the Constitution which says that ecology may be the subject of legislation by both the provinces as well as the Federation (Hassan and Azfar, 2004).

It is noteworthy that the lawyers for the petitioners in *Shehla Zia and others v. WAPDA* relied on the extensive jurisprudence in India that recognized that the right to life included the right to a quality of life. They also relied on the constitutionally guaranteed right to dignity under Article 14 of the Pakistani Constitution. The Pakistani Supreme Court rose to the occasion and held that a wide meaning should be given to the word "life" to enable a man not only to sustain life but also to enjoy it:

> The word "life" has not been defined in the Constitution, but it does not mean, nor can it be restricted only to the vegetative or animal life or mere existence from conception to death. Life includes all such amenities and facilities which a person in a free country is entitled to enjoy with dignity, legally and constitutionally.

It further noted that where life of citizens is degraded, the quality of life is adversely affected, and health hazards are created affecting a large number of people. With regard to scientific uncertainty, the Court noted that the precautionary principle should be applied, and that a balance should be struck between economic progress and prosperity and minimizing possible hazards. Thus, the Court confirmed that a policy of sustainable development should be adopted. With regard to the Rio Declaration, the Court noted that it "would serve as a great binding force and to create discipline among the nations while dealing with environmental problems." In *Salt Miners Labor Union v. Industries and Mineral Development*, the Supreme Court, citing *Shehla Zia*, articulated that "the right to have unpolluted water is the right of every person wherever he lives."

Bokhari v. Federation of Pakistan (2003) concerned the grounding and collapse of a ship in the port of Karachi in 2003, leading to a major oil-spill, which caused massive environmental damage. In the Supreme Court, the ability of the legal system to respond was found to be totally lacking due to many reasons, including lack of preparedness and failure to ratify relevant international conventions. The Court considered the concept of public interest litigation (PIL) as it had evolved in India and Pakistan, where it was said to be particularly useful because of the realities of poverty, illiteracy, and institutional fragility. In Pakistan, PIL had been used in a wide range of social issues, from environmental pollution to the prevention of exploitation of children. The public interest suit of *Irfan v. Lahore Development Authority* (2003) involved air and noise pollution from rickshaws, mini-buses, and other vehicles and the non-performance of statutory duties by the relevant authorities, charged with ensuring a pollution-free environment for citizens.

In a landmark case in neighboring Nepal, *Suray Prasad Sharma Dhungel v. Godavari Marble Industries and Others*, the Supreme Court of Nepal held that a clean and healthy environment is part of the right to life clause under Article 12(1) of the Nepal Constitution. The Court noted in this case:

> Since clean and healthy environment is an indispensable part of a human life, the right to a clean, healthy environment is undoubtedly embedded within the Right to Life. It is clear that the constitutional perimeter in which the applicant had filed the writ petition, has been substantively changed form (sic) the commencement of Article 26(4) of the Constitution of the Kingdom of Nepal 1990, because this Article has taken environmental conservation as one of the basic Directive Principles of the State.

The Court also acknowledged the importance of sustainable development, stating that while industry is the foundation of development of a country, it is essential to maintain environmental balance along with industry:

> It is essential to establish balance between the need to provide continuity to developmental activities and priority to the protection of the environment … Development is for the interest, prosperity of human being. Therefore, life of human being is the end. Development is the means to live happily; human being cannot live (sic) clean and healthy life without clean and healthy environment. Therefore, safety of the environment is the means. Environment protection measures should be initiated taking into account this fact.

In a more recent case, the Supreme Court of Nepal elaborated on the significance of drinking water for the purpose of maintaining public health. In *Prakash*

Mani Sharma and others v. Nepal Drinking Water Corporation and Others (2001), the Court, while not explicitly recognizing that water is a fundamental right, nonetheless held that providing pure drinking water is the responsibility of the welfare state.

In *Prakash Mani Sharma v. His Majesty's Government Cabinet Secretariat and Others* (2003), the Supreme Court of Nepal held that a healthy environment is a prerequisite for the protection of personal freedoms under the Constitution. Relying on the concept of sustainable development, the Court acknowledged that environment and development should proceed harmoniously, and environmental protection cannot be ignored for the sake of development. Moreover, environmental destruction cannot be encouraged in the name of development.

The Supreme Court of Sri Lanka has also handed down several significant judgments articulating environmental rights using the equality provision under Article 12(1) of its Constitution. The Court has also been receptive to the use of international environmental law principles in adjudicating national cases. Thus, in the *Gunaratne v. Homagama Pradeshiya Sabha* case (1998), the Supreme Court noted that public participation and transparency are essential if sustainable development is to be achieved. These principles have since gained ground internationally.

In the *Bulankulama v. Minister of Industrial Development and others (Eppawala case)* (1999), the Supreme Court of Sri Lanka noted that "the organs of the state are guardians to whom the people have committed the care and preservation of the resources of the people." The Court articulated the principle of shared responsibility applied to natural resources and the environment. The Supreme Court also relied on several principles of the Stockholm Declaration and Rio Declaration, and pointed out that as a member of the UN, Sri Lanka could hardly ignore those provisions. It also relied extensively on international environmental law principles in elucidating the meaning of sustainable development. Drawing inspiration from Judge Weeramantry's separate opinion in the *Case Concerning the Gabcikovo Nagymaros Project (Hungary v. Slovakia)*, the Court referred to the need to balance the needs of the present generation with those of posterity. Sustainable development ***does not*** mean that further development must be halted.

The Supreme Court noted that in this case the salutary provisions of the law relating to notification and publication of EIAs had not been observed, and it called upon the relevant parties to comply with the law, stressing that nobody is above the law. The Court essentially rejected the confidentiality provisions in the proposed agreement as being contrary to the goal of sustainable development. It stressed the importance of procedural principles, and referred to Principle 10 of the Rio Declaration which includes the principles of access to information, public participation, and access to remedies, noting that it calls for better citizen participation in environmental decision-making and access to environmental information "for they can help to ensure greater compliance by States of international environmental standards through the accountability of their governments."

Mundy v. Central Environmental Authority and others concerned several appeals relating to the building of Sri Lanka's first modern highway, the Southern

Expressway linking its capital city of Colombo with the Southern city of Matara, an important step in terms of infrastructure development towards enhancing industry, trade, and investment. Protracted litigation opposing the project and its different alternative routes involved allegations of potential damage to human rights, including large-scale displacement and injury to the environment, including sensitive ecosystems. The Court of Appeal had upheld the developmental interest, holding that when balancing the competing interests, the conclusion necessarily had to be made in favour of the greater interests of the community, which would benefit immensely from the project. The Court gave highest priority to the public interest in development, then to the environmental damage to wetland ecosystems, and lastly, to the human interests of affected persons. Several persons appealed to the Supreme Court with regard to particular sections of the route which resulted in the taking of their lands with no arrangements for compensation. The Supreme Court varied the order and ordered compensation under the *audi alteram partem* principle of natural justice and Constitutional Article 12(1) on equality and equal protection.

The case of *Centre for Environmental Justice v. Ministry of Agriculture, Environment, Irrigation and Mahaweli Development et al.* concerned the protection of a major national park in Sri Lanka, forming a wetland of international importance under the Ramsar Convention on Wetlands, and alteration of the boundaries of this park by the government. It was argued that this alteration would pose a further threat to the ecosystem, already endangered by landfills, aquaculture farms, fisheries, pollution, mining of minerals, and clearing of mangroves. The petition argued that the authorities were acting in breach of several international conventions, including the Wetlands, Cultural and Natural Heritage and Biodiversity Conventions and the Bonn Convention on Migratory Species of Wild Animals, and other instruments, including the Johannesburg Declaration, and relevant articles of the Sri Lankan Constitution.

The seminal case of *Minors Oposa v. Secretary of the Department of Environment and Natural Resources* (1994) from the Philippines, also discussed in Chapter 7, is possibly the best-known case that relates to environmental rights. It involved a challenge by a group of minors of timber license agreements on the ground that deforestation was causing environmental damage to themselves as well as future generations. The Supreme Court of the Philippines noted:

> While the right to a balanced and healthful ecology is to be found under the Declaration of Principles and State Policies and not under the Bill of Rights, it does not follow that it is less important than any of the civil and political rights enumerated in the latter. Such a right belongs to a different category of rights altogether for it concerns nothing less than self-preservation and self-perpetuation – aptly and fittingly stressed by the petitioners – the advancement of which may even be said to predate all governments and constitutions. *As a matter of fact, these basic rights need not even be written in the Constitution for they are assumed to exist from the inception of humankind.*
>
> (Emphasis added)

In *Metropolitan Manila Development Authority v. Concerned Residents of Manila Bay* (2008), a group of residents in Manila filed a complaint against several government agencies for the cleanup, rehabilitation, and protection of Manila Bay. They based their claims on constitutional rights, international law, and various national statutes. Tests conducted during trial court proceedings revealed fecal coliform bacteria at various beaches greatly in excess of what was considered permissible by the government. The Court, referring to its earlier decision in the *Oposa* case (1994) reaffirmed the constitutional right to a healthy environment:

> [t]he right to a balanced and healthful ecology need not even be written in the Constitution for it is assumed, like other civil and political rights guaranteed in the Bill of Rights, to exist from the inception of mankind and it is an issue of transcendental importance with intergenerational implications. Even assuming the absence of a categorical legal provision specifically prodding petitioners to clean up the bay, they and the men and women representing them cannot escape their obligation to future generations of Filipinos to keep the waters of the Manila Bay clean and clear as humanly as possible. Anything less would be a betrayal of the trust reposed in them.

The Malaysian High Court decision concerning the *Bakun Hydro-Electric Dam project* (1996) (BHEP) is significant. The project was approved by the Malaysian Cabinet in September 1993 at a cost of US $5.4 billion. In April 1995, three natives who would be affected by the project filed action against the company, the Director General of Environment, and the Malaysian Government. The main contention revolved around the issue of EIA and public participation. In delivering the judgment, the High Court of Malaysia remarked:

> Under the guidelines prescribed by the Director-General, as provided for under the EQA [Environmental Quality Act] itself, a valid assessment of an EIA prepared by the project proponent of the prescribed activities cannot be made without some form of public participation. This is essential, for interaction between people and their environment is fundamental to the concept of environmental impact. For this, a right is vested on the plaintiffs to obtain and be supplied with a copy of the EIA coupled with the right to make representations and be heard. While waiting to exercise their rights and being assured by the executive through their leaders – including those directly in charge that the relevant procedures of the EQA will be adhered to – the Minister suddenly strikes a mortal blow ... [in gazetting amendments to the Federal EIA law to exclude certain activities in Sarawak]. Though it is claimed by the defendants that this amendment order only alters the procedure in the evaluation of the EIA on the BHEP, in substance and in fact and visible to all, it tantamounts *(sic)* to the removal of the entire rights of the plaintiffs to participate and give their views before the EIA is approved. *This Court shall not stand idly by to witness such injustice especially when the plaintiffs have turned to this institution to seek redress.*
>
> (Raman, 1997: 95, emphasis added)

The Court pointed out that "the interaction between the public and their environment is fundamental to the concept of impact" and that it is mandatory for the authorities to hear the views of the public first, before granting approval: "It makes a mockery of the whole issue to say the EIA can be approved first and if the public has any constructive ideas, they can submit later. This certainly is illogical, deprived of good sense and sound reasoning" (High Court Judgment, 1996 in Raman, 1997: 96). While the case did not address the issue of a right to environment *per se*, it explicitly endorsed the importance of participatory rights.

Although the Federal Constitution of Malaysia does not include a provision on the environment, in *Tan Tek Seng v. Suruhanjaya Perkhidmatan Pendididan* (1996), the Court interpreted the right to life and liberty expansively to include a right to a healthy environment:

> the expression "life" appearing in art (5) does not refer to mere existence. It incorporates all those facets that are an integral part of life itself and those matters which go to form the quality of life. Of these are right to seek and be engaged in lawful and gainful employment and to receive those benefits that our society has to offer to its members. *It includes the right to live in a reasonably healthy pollution free environment.*
>
> (Rahman, 2010: 61, emphasis in original)

This is remarkably similar to the jurisprudence generated from South Asia. In the case of *Adong bin Kuwau and others v Kerajaan Negeri Johor and Anor* (1997), the right of indigenous peoples to an adequate standard of living, deprivation of the produce of the forest for themselves as well as descendants were brought in through the right to property, even though the judge made no explicit reference to a right to a healthy environment. However, in a later case, the Court of Appeal interpreted the right to life provision in a more restrictive manner (Rahman, 2010).

North America

Paradoxically, the United States, considered as one of the most litigious countries in the world, is lagging behind when it comes to advancing environmental rights. As noted in the previous chapter, the US Constitution makes no mention of environmental rights, and no attempt has been made to articulate them through other rights. This may, however, be changing in light of the adverse consequences of climate change.

Five state constitutions, on the other hand, include a right to a healthy environment. Montana is one of them. The case of *Montana Environmental Information Center v. Department of Environmental Quality* (USA) (1999) involved a challenge to overturn legislation. The Plaintiffs alleged that the Department of Environmental Quality (DEQ) illegally amended the mineral exploration license of Seven-Up Pete Joint Venture (SPJV) for a massive open-pit gold mine to allow it to discharge groundwater containing high levels of arsenic and zinc into shallow aquifers without requiring a non-degradation review under the MCA. The plaintiffs alleged that the MCA violated the right to a clean and healthy environment. On appeal, in overturning the decision of the District Court, the Supreme Court of Montana concluded that

> the right to a clean and healthful environment guaranteed by Article II, Section 3 and those rights provided for in Article IX, Section 1 were intended by the Constitution's framers to be interrelated and interdependent and that state or private action which implicates either, must be scrutinized consistently.
>
> (Kravchenko and Bonine, 2008: 77)

The Court further stated:

> We conclude, therefore, that the District Court erred when it held that Montana's constitutional right to a clean and healthy environment was not implicated, absent a demonstration that public health is threatened or that current water quality standards are affected to such an extent that a significant impact has been had on either the Landers Fork or Blackfoot River.
>
> We conclude that the constitutional right to a clean and healthy environment and to be free from unreasonable degradation of the environment is implicated based on the Plaintiffs' demonstration that the pumping tests proposed by SPJV would have added a known carcinogen such as arsenic to the environment in concentrations greater than the concentrations present in the receiving water.
>
> (Kravchenko and Bonine, 2008: 77)

Juliana et al. v. USA et al.,[1] filed in the US District Court for the District of Oregon, broke new ground in litigation on climate change. A group of children, ably assisted by Professor Mary Wood of University of Oregon Law School, filed action against the US government, including then-President Obama and various government officials and agencies, because they asserted that the government had known for decades that CO2 pollution has been causing catastrophic climate change and failed to take action to cut down fossil fuel emissions. Not only have they failed to take such action, they took action to increase carbon pollution through fossil fuel extraction, production, consumption, transportation, and exportation. The plaintiffs sought immediate action to reduce atmospheric CO2 concentrations to 350 ppm by 2100 which they asserted was doable, but action must be taken immediately.

The plaintiffs claimed the violation of their rights as follows:

1. Their right to life and liberty, in violation of their substantive due process rights
2. Their equal protection rights under the Fifth Amendment by denying them protections afforded to previous generations
3. The implicit right, via the Ninth Amendment, to a stable climate and an ocean and atmosphere free from dangerous levels of CO2
4. The public trust doctrine secured by the Ninth Amendment, by denying future generations essential natural resources.

The State for its part moved to dismiss all claims. They argued that (a) the plaintiffs lacked standing; (b) the case raised non-justiciable political questions; (c) the case failed to state a constitutional claim; and (d) the public trust doctrine does not provide a cognizable federal cause of action. The District Court discussed the claim under three headings: (a) standing to sue and justiciability; (b) causation; and (c) redressability of the injury. Regarding standing and justiciability, the Magistrate Judge Coffin noted:

> But the intractability of the debates before Congress and states legislatures and the alleged valuing of short-term economic interest despite the cost to human life, necessitates a need for the courts to evaluate the constitutional parameters of the action or inaction taken by the government. This is especially true when such harms have an alleged disparate impact on a discrete class of society.
>
> (*Juliana* case, April 2016: 8)

With regard to causation, the Judge noted, relying on *National Audubon Society Inc. v. Davis* (2002: 9), that a causal chain does not fail "simply because it has several links, provided those links are not hypothetical or tenuous and remain plausible." In cases where there are numerous third parties whose independent decisions collectively have a significant effect on plaintiffs' injuries, the causal chain may be too weak to support standing. However, in this case, the Judge pointed out that there is an alleged strong link between all the third party decisions, given the government's regulation of CO2 emissions. If the allegations in the complaint are true, the failure to regulate CO2 emissions has resulted in "a danger of constitutional proportions" to public health.

With regard to redressability, the Judge was of the view that at this stage of the proceedings the court must determine whether it can fashion a remedy to address the alleged harm should plaintiffs prove it. Relying on *Wolfson v. Brammer* (2010: 11), the Judge pointed out: "Redressability does not require certainty, but it does require a substantial likelihood that the injury will be redressed by a favorable judicial decision."

It is noteworthy that Judge Coffin referred to the decision of the Dutch court in *Urgenda Foundation v. The State of The Netherlands* (2015), where the Hague District Court ordered the reduction of greenhouse gas emissions by at least 25% by 2020, rejecting the government's argument that a reduction of Netherland's emissions would be ineffectual compared to global emissions. The Judge pointed out that regulation by the United States, in combination with reduction efforts of other countries, "may very well have sufficient impact to redress the alleged harms" (12). Thus, the Judge declined to dismiss the case for lack of standing.

With regard to the argument advanced by the defendants that there is no constitutional right to be free from CO2 emissions, the Judge pointed out that the plaintiffs do not assert such a right. What they do assert is that the defendants' action/inaction with regard to their obligations concerning regulating environmental pollution has violated their substantive due process rights and has done so in favor of older generations. With regard to the arguments made against the public trust doctrine, the court pointed out that when combined with EPA's duty to protect public health from airborne pollutants and the government's public trust duties, the allegations in the complaints provide a substantive due process claim. For these reasons, the court refused to dismiss the claims relating to the public trust doctrine. U.S. District Court Judge Ann Aiken, upholding Judge Coffin's above recommendation, made this groundbreaking statement: "Exercising my 'reasoned judgment,' I have no doubt that the right to a climate system capable of sustaining human life is fundamental to a free and ordered society" (Juliana case, 2016: 32).

The Trump administration fought tooth and nail to get the case dismissed, taking the case all the way to the US Supreme Court. In a marked departure from recent cases, the US Supreme Court unanimously ruled in favor of the 21 youth plaintiffs in the constitutional climate lawsuit filed against the federal government. The Court denied the Trump administration's application for stay, preserving the US District Court's trial start date of October 29, 2018. The Court also denied the government's "premature" request to review the case before the District Court heard all of the facts that support the youths' claims at trial. Legal, human rights, and environmental communities are awaiting the outcome of this case with bated breath. Whichever way the decision goes, the youth plaintiffs and their lawyers deserve credit for the hard fight they are giving and for persevering against all odds. They certainly brought attention to the issue of climate change and advanced the novel concept of "atmospheric trust doctrine" (Wood, forthcoming).

The latest in a long line of cases brought against the Trump administration was filed in the US District Court for the Eastern District of Pennsylvania in 2017. In *Clean Air Council, SB through his Guardian and BB through his Guardians v. USA, Donald Trump as President and others* (2017), the plaintiffs argued that the US government's acts to roll back regulations previously directed at reducing the United States' contribution to climate change, and its reliance on junk science to wage a war on facts, data, and reliable principles increases the United States' contribution to climate change, intensifying its effects, and thereby endangering the lives and welfare of US citizens in violation of the due process clause of the Constitution. In addition to causing adverse health impacts on the two minor plaintiffs (aged 7 and 11 years respectively), they also argued that there could be irreversible impacts of climate change on the planet and future generations.

Relying on the 2009 report of the UN Office of the High Commissioner for Human Rights, the plaintiffs pointed out that climate change "impacts, directly and indirectly, an array of internationally guaranteed human rights" (Petition, *Clean Air Council, SB through his Guardian and BB through his Guardians v. USA, Donald Trump as President and others* (2017)). They pointed out that the US Government had known for decades that climate change presents a clear danger to American people, as a result of which the United States has experienced severe droughts, heatwaves, floods, superstorms, and catastrophic weather events, including wildfires in California, causing loss of lives and billions of dollars in property damage. The plaintiffs relied on *Massachusetts v. EPA* (2007), where the Supreme Court held that greenhouse gases (GHG), including CO2, are pollutants and must be regulated by the EPA if found to endanger public health or welfare under the Clean Air Act pursuant to which the EPA issued its "endangerment findings" that GHG emissions endangered public health and the welfare of current and future generations.

The plaintiffs also relied on the reports of the Intergovernmental Panel on Climate Change (IPCC), especially its 2014 report which attributed the warming of the planet to anthropogenic activities. They pointed out that "climate change not only threatens life, but the most basic life-sustaining necessities including clean air, pure water, sufficient food and adequate shelter." Moreover, the fourth National Climate Assessment, echoing the words of the IPCC's 5th assessment report, found it "extremely likely" that anthropogenic forces are the primary cause of climate change. Thus:

> By acting to prevent the devastating and life-endangering consequences of climate change, the United States Government has acknowledged its obligation as the trustee of our country's natural resources not to take affirmative acts to enhance, increase, or intensify those dangers.
>
> (*Clean Air Council, SB through his Guardian and BB through his Guardians v. USA, Donald Trump as President and others*, 2017)

In addition to these scientific reports, the plaintiffs relied extensively on international and regional instruments on environmental protection as well as human rights: the Stockholm Declaration on the Human Environment, especially Principle 1 referring to a healthy environment; the Rio Declaration on Environment and Development, which again declared in Principle 1 that "human beings are entitled to a healthy and productive life in harmony with nature"; the African Charter on Human and Peoples' Rights that recognizes a human right to a satisfactory environment; the American Convention on Human Rights that recognized a right to a healthy environment; the Hague Declaration on the Environment that recognized the "right to live in dignity in a viable global environment"; and the UN General Assembly resolution that recognized that "all individuals are entitled to live in an environment adequate for their health and well-being." The plaintiffs also referred to the many resolutions of the UN Human Rights Council that have repeatedly recognized that a stable climate system is necessary for the realization of human rights. Similarly, national and international courts have also upheld the responsibility of states to protect the environment, including the decision by the International Criminal Court to prosecute environmental crimes. The plaintiffs concluded that

> a consensus of nations has recognized a governmental responsibility not to endanger life, including the fundamental right to a life-sustaining climate system. Rolling back environmental protections is contrary to this consensus, as it recklessly exposes Plaintiffs to the harms caused by manmade contributions to climate change based on junk science.
>
> (*Clean Air Council, SB through his Guardian and BB through his Guardians v. USA, Donald Trump as President and others*, 2017)

Arguing that courts should be guided by customary international law and international human rights law when determining fundamental rights, the plaintiffs pointed out that nearly half of all US State Constitutions include environmental provisions "with at least sixteen recognizing substantive constitutional rights to a healthy, stable, or quality environment."[2] Moreover,

> increased CO2 concentrations jeopardize Plaintiffs' welfare by diminishing their capacity to provide for their basic human needs, safely raise families, practice their religious and spiritual beliefs, maintain their bodily integrity, and lead lives with access to clean air, water, shelter, and food.
>
> (*Clean Air Council, SB through his Guardian and BB through his Guardians v. USA, Donald Trump as President and others*, 2017)

The conduct will have even worse consequences for future generations.

In addition to the cause of action based on due process, the plaintiffs relied on the public trust doctrine which, they contended, has been part of US common law since the country was founded and exists to prevent sovereign governments from alienating natural resources that the government holds in trust for the common benefit of all citizens. The United States has taken the position that it holds in trust public lands and the natural resources on them, and it has the right and duty to protect and preserve those resources. Thus, as sovereign trustees, the defendants have a duty to refrain from "substantial impairment of the public trust." Moreover:

> Defendants have implemented Rollbacks that largely eliminate the protections intended to combat the devastating effects of climate change. In doing so, Defendants have engaged in affirmative acts in contravention of their duty of case as trustees to manage the life-sustaining resources of the public trust in the best interests of the present and future beneficiaries of the public trust, including Plaintiffs.
>
> (*Clean Air Council, SB through his Guardian and BB through his Guardians v. USA, Donald Trump as President and others*, 2017)

This is possibly the first time that international human rights law is relied upon so extensively in a national environmental case in the United States. It will be interesting to see how the courts will handle this case.

In July 2018, the Attorney General of Rhode Island filed an action against major oil companies, including BP and Exxon, for their role in causing climate change and causing damage to the coast and communities. While this case was filed on the basis of creating a public nuisance, not on the right to a healthy environment, it is nonetheless significant as it was the first time that a US state brought a lawsuit against oil companies. A similar lawsuit that was filed in 2012 by the Village of Kivalina, Alaska, against several oil and coal companies was dismissed by the US Court of Appeals 9th Circuit. In *Village of Kivalina and City of Kivalina v. Exxon Mobil and others*, the Court held:

> In sum, the Supreme Court has held that federal common law addressing domestic greenhouse gas emissions has been displaced by Congressional action. That determination displaces federal common law public nuisance actions seeking damages, as well as those actions seeking injunctive relief. The civil conspiracy claim falls with the substantive claim. Therefore, we affirm the judgment of the district court. We need not, and do not, reach any other issue urged by the parties.
>
> Our conclusion obviously does not aid Kivalina, which itself is being displaced by the rising sea. But the solution to Kivalina's dire circumstance must rest in the hands of the legislative and executive branches of our government, not the federal common law.
>
> (*Kivalina v. Exxon Mobil*, 2012: 16)

This case led a scholar to conclude that it threw "federal public nuisance claims out the window for relief from pollution caused by GHG emissions, leaving plaintiffs such as Kivalina with what seems like no judicial avenue to obtain relief" (Johnson, 2013: 563).

In the neighboring Canada, Ecojustice's case on the right to a healthy environment is currently pending. Ecojustice is requesting the government to include a right to a healthy environment in the Canadian Charter:

> By enshrining the right to a healthy environment in the Charter, we can align our highest law with our most deeply-held values. More concretely, a Charter right to a healthy environment could prevent harmful rollbacks of environmental laws – like the sweeping changes buried in the 2012 federal omnibus budget bills. It would also make it easier for people living in toxic hotspots like Sarnia's Chemical Valley to seek justice. Overall, countries that recognize environmental rights in their constitutions have smaller per capita ecological footprints, which is a major win for people and the planet.
>
> (Ecojustice webpage)

Latin America

The constitutions of many countries in Latin America include environmental rights as well as rights of nature. Thus, it is not surprising that several important decisions on environmental rights have come from that region. *Mendoza Beatriz Silva and others v. National Government of Argentina and others in regard to damages suffered* (2008) was a landmark decision from Argentina. Article 41 of the Argentine Constitution recognizes a right to a healthy and balanced environment fit for human development. It also recognizes the need to ensure that rights of future generations are not endangered. The plaintiffs in this case filed action against the national government and several private companies seeking compensation for damage resulting from pollution of the basin, the cessation of contaminating activities, and a remedy for collective environmental damage. The Court's initial ruling in 2006 required the government to conduct an EIA. A directive in 2007 ordered the government to establish a comprehensive cleanup and restoration plan for the river. Prior to its final ruling, issued in 2008, the Court took the innovative step of requesting the intervention of the University of Buenos Aires to advise it on the adequacy of the cleanup plan presented by the government. The objectives set by the court for the cleanup program were mainly two-fold: improving the quality of life of the people of the basin area and the environmental restoration of the river basin's components. The Court also issued a comprehensive cleanup program:

- to organise within 30 days a system of public information on the internet for the general public that includes all information related to the clean-up programme;
- to eliminate all industrial pollution by relocating businesses or requiring them to undertake a treatment plan to stop pollution;
- to improve drinking water, and sewage and storm drainage systems;
- to put in place specific health programmes to meet the needs of the river basin population;
- to form a committee of interested non-governmental organisations in collaboration with the national ombudsman to monitor compliance with the cleanup programme.

(*Mendoza Beatriz Silva and others v. National Government of Argentina and others in regard to damages suffered*, 2008)

The courts in Colombia have found an enforceable right to a healthy environment under the right to life provision (Article 11 of the Constitution). In a case involving noxious contamination from an asphalt facility, the trial court held:

> Everyone has the right to enjoy and live in a healthy environment. This should be regarded as a fundamental human right, which is a prerequisite and basis for the exercise of other human, economic and political rights. It should be recognized that a healthy environment is a *sine qua non* condition for life itself and that no right could be exercised in a deeply altered environment.
>
> (Kravchenko and Bonine, 2008: 70)

Similarly, in *Victor Ramon Castrillon Vega v. Federación Nacional de Algodoneros y Corporation Autonoma Regional del Cesar*, the Constitutional Court of Colombia held that the toxic fumes from an open pit violated the right to life of those who were living in close proximity.

A recent landmark case from Colombia, *Demanda Generaciones Futuras v. Minambiente*, filed by a group of 25 young plaintiffs assisted by DeJusticia, an influential think-tank in Colombia, involved deforestation of the Amazon. The

Colombian Supreme Court handed down a *tutela*, a rights-based injunction, on climate change. The plaintiffs, who were between 7 and 25 years of age and living in cities that are most at risk due to climate change, argued that deforestation of the Amazon is contributing to global warming and affecting their rights, as well as those of future generations. They argued that their rights to life, health, and a healthy environment are being violated by the government. They pointed out that under the Paris Agreement and national law, the government has undertaken commitments to reduce deforestation and the emission of greenhouse gases and the obligation to "reduce the net rate of deforestation to zero in the Colombian Amazon by 2020" (English translation available on the DeJusticia website). Despite this commitment, deforestation has increased by 44% from the figure reported in 2015. The plaintiffs argued that deforestation in the Amazon has dire consequences for other parts of the country, including alteration of the water cycle, alteration of the soils to absorb water, changes in the water supplies, and increasing global warming. These consequences affect their enjoyment of a healthy environment.

An amicus brief filed on behalf of the Director of Climate Science, Awareness and Solutions at the Earth Institute at Columbia University, New York, United States, stated that the precautionary principle requires us to act now to avert calamitous consequences of climate change, and that sea level rise and ocean acidification can result in the infringement of fundamental rights of the present generation, and the rights of future generations will be more severely affected:

> Accordingly, the principle of intergenerational equity compels action without further delay so as not to burden disproportionately young persons and future generations. As well, the principles of solidarity, participation, and the best interest of children counsel consideration of interests retained by persons beyond those wielding present political authority. Considered interests, as well, must not be limited to those within the specific region of this Court's usual jurisdiction. Neither should they be limited to those of the present generation.
>
> (*Demanda Generaciones Futuras v. Minambiente*)

The Supreme Court pointed out that the fundamental rights of life, health, the minimum subsistence, freedom, and human dignity are substantially linked and determined by the environment and the ecosystem. Without a healthy environment, subjects of law and sentient beings in general will not be able to survive, much less protect those rights, for our children or for future generations. Neither can the existence of the family, society, or the state itself be guaranteed.

The increasing deterioration of the environment is a serious attack on current and future life and on other fundamental rights; it gradually depletes life and all its related rights. The inability to exercise the fundamental rights to water, to breathe pure air, and to enjoy a healthy environment is making Colombians sick. It also increases the lack of fresh water and decreases the ability to enjoy a dignified life. Therefore, in this case, the exceptional proceeding of the *tutela* is sufficiently demonstrated to resolve in depth the problems raised, because the jurisprudential assumptions for this purpose are met, given the connectedness of the environment with fundamental rights.

The Supreme Court ordered the Presidency of the Republic, the Ministry of Environment and Sustainable Development, and the Ministry of Agriculture and Rural Development, in coordination with other relevant actors and with the participation of the plaintiffs, the affected communities, and the interested population in general to formulate a short-, medium-, and long-term action plan within four months to counteract the rate of deforestation in the Amazon, tackling the impacts of climate change. They were also ordered to prepare an inter-generational pact for the life of the Colombia Amazon to adopt measures aimed at reducing deforestation to zero. Moreover, in order to protect this ecosystem that is vital for our global future, the Colombian Amazon is recognized as a "subject of rights" entitled to protection, conservation, maintenance, and restoration.

The 1980 Constitution of Chile contains several environmental rights, including a right to live in an environment free of contamination. It also includes a right to enforce these rights under the Constitution. In *Comunidad de Chanaral v. Codelco Division el Salvador*, the Supreme Court held that the right to a clean environment is owed not only to individuals and communities, but also to future generations. In *Petro Flores v. Corporacion Del Cobre, Codelco*, which involved contamination from a government-run copper mine, the Court stated, after a site visit, that "the preservation of nature and the conservation of the environmental heritage is an obligation of the State, according to our Fundamental Constitution, and the polluting act which the respondent voluntarily executes is arbitrary from any point of view" (Kravchenko and Bonine, 2008: 71).

Europe

While many of the environmental rights cases have come before the European Court of Human Rights, as discussed in Chapter 4, individual countries have also dealt with such cases. The case of *Senih Ozay v. Ministry of the Environment and Eurogold Madencilik* (1997) came before the Supreme Administrative Court of Turkey. This lawsuit involved a request to cancel the license granted to Eurogold Mining Corporation by the Ministry of Environment to operate gold mines in Izmir and the surrounding villages of Bergama, Ovacik. Article 56 of the Constitution of the Republic of Turkey provides that: "Each individual has the right to live in a healthy and balanced environment. It is the duty of the State and citizens to develop the environment, protect environmental health and prevent environmental pollution." Taking into account the environmental impact assessment report and several expert reports on the gold mine, the Supreme Court of Turkey held that the method of operation of the gold mine was unsafe and had a direct and indirect effect on human health.

The case of *Urgenda Foundation v. The State of The Netherlands* (2015) did not involve human rights *per se*, but it is important from the point of view of climate change litigation and the court's directive to the State. Rejecting the argument by the Netherlands that a reduction in its emissions would be ineffectual in light of the emissions by other countries, the Court ordered the government to reduce greenhouse gas emissions nationwide by at least 25% by 2020:

> The state should not hide behind the argument that the solution to the global climate problem does not depend solely on Dutch efforts. Any reduction of emissions contributes to the prevention of dangerous climate change and as a developed country the Netherlands should take the lead in this.

This case survived an appeal by the government recently. Citing the jurisprudence of the European Court of Human Rights, IPCC reports, UNEP reports, international climate agreements (UNFCCC, the Kyoto Protocol, the Paris Agreement) and COP decisions, EU directives, and the consequences of climate change, The Hague Court of Appeal stated:

> To summarise, from the foregoing it follows that up till now the State has done too little to prevent a dangerous climate change and is doing too little to catch up, or at least in the short term (up to end-2020). Targets for 2030 and beyond do not take away from the fact that a dangerous situation is imminent, which requires interventions being taken now. In addition to the risks in that context, the social costs also come into play. The later actions are taken to reduce, the quicker the available carbon budget will diminish, which in turn would require taking considerably more ambitious measures at a later stage, as is acknowledged by the State (Statement of Appeal 5.28), to eventually achieve the desired level of 95% reduction by 2050. In this context, the following excerpt from AR5 (cited in legal ground 2.19 of the judgment) is also worth noting: "*(…) Delaying mitigation efforts beyond those in place today through 2030 is estimated to substantially increase the difficulty of transition to low-longer-term emissions levels and narrow the range of options consistent with maintaining temperature change below 2°C relative to pre-industrial levels.*"
>
> (*Netherlands v Urgenda Foundation*, 2018, English translation: para. 71)

This is a great victory, not just for the Netherlands, but also for climate advocates in the entire international community. Indeed, the array of documents cited by the Court, ranging from environmental to human rights, and from legal to scientific, is striking. As John Knox noted:

> As a matter of international human rights law, the decision is a very big deal …. It applies the European Court's jurisprudence to climate change in a clear

and concrete way, and provides a precedent for other such suits in Europe and beyond.

(Knox, 2018)

A similar case was filed in Belgium by a group of concerned citizens. In *VZW Klimatzaak v. Kingdom of Belgium* (2016), a Belgian court of the first instance issued a declaratory judgment against the Belgian government. Specifically, the court determined that the federal and regional governments of Belgium have failed to reduce greenhouse gas emissions to the extent required by law, namely 40% below 1990 levels by 2020 and 87.5% below 1990 levels by 2050. The appeal by the government is pending.

The *Urgenda* case was also the impetus for a case currently pending in the United Kingdom. In Fall 2017, Plan B, a charitable group supporting "strategic legal action" against climate change, launched judicial review proceedings that challenged the failure of the UK government to revise its 2050 carbon target under the UK Climate Change Act 2008: "Plan B alleges that the Secretary of State for Business, Energy, and Industrial Strategy's failure to make more ambitious targets for 2050 is beyond the scope of its authority, irrational, and *a breach of fundamental human rights*" (Day, 2017, emphasis added). This is the first case to challenge the United Kingdom's climate targets.

In *Greenpeace Nordic Association v. Norway Ministry of Petroleum and Energy* (2017), the petitioners argued that the Ministry's lease of blocks of the Barents Sea to oil and gas developers violated the constitutional guarantee of a "right to an environment that is conducive to health and to a natural environment whose productivity and diversity are maintained" (Article 112). Petitioners alleged that maintenance of that environment requires staying within a global emissions budget consistent with a warming of not more than 1.5°C or 2°C recognized by the Paris Agreement, and that leasing of Barents Sea to oil and gas exploitation will be contrary to that goal (UN Environment, 2017).

In *Lliuya v. RWE AG*, a Peruvian farmer filed an action in a German court against a German utility. He sought damages to offset the costs of protecting his town from melting glaciers for which he alleged RWE is partly responsible. The Court, however, held that there was no "linear causal chain" linking the alleged injury and RWE's emissions – "rather many emitters had created the risk of flood confronting the plaintiff's town such that the root cause of the risk could not be ascribed to RWE in particular" (UN Environment, 2017: 21). This case highlighted the legal hurdles that complainants have to jump over to establish damage caused by climate change.

Africa

The Constitution of South Africa is celebrated as one of the most progressive Constitutions in the world. As discussed in Chapter 7, it embodies economic, social, and cultural rights, including the right to a healthy environment. The case of *Earthlife Africa Johannesburg v. Minister of Environmental Affairs et al.* (2017) before the High Court of South Africa, Gauteng Division in Pretoria involved the environmental impacts associated with the decision to build a 1200 MW coal power plant which requires, among other things, an environmental authorization. Earthlife argued that the Chief Director was obliged to consider the climate change impacts of the proposed power plant before granting authorization, which he failed to do. The Court pointed out that "the environmental impact assessment process is a key means of promoting sustainable development, by ensuring that the need for development is sufficiently balanced with full consideration of the environmental impacts of a project with environmental impacts." Moreover, courts when interpreting legislation are required to promote the purpose, spirit, and objects of the Bill of Rights. This includes the fundamental justiciable environmental right in Section 24 of the Constitution:

> Section 24 recognizes the interrelationship between the environment and development. Environmental considerations are balanced with socio-economic considerations through the ideal of sustainable development. This is apparent from section 24(b)(iii) which provides that the environment will be protected by securing ecologically sustainable development and use of natural resources while promoting justifiable economic and social development. Climate change poses a substantial risk to sustainable development in South Africa … Sustainable development is at the same time integrally linked with the principle of intergenerational justice requiring the state to take reasonable measures to protect the environment "for the benefit of present and future generations" and hence adequate consideration of climate change. Short-term needs must be evaluated and weighed against long-term consequences.

The Court pointed out, relying on the decision of the Court of Appeal of the State of California in *Communities for a Better Environment v. City of Richmond* (2010), that mitigation measures in relation to climate change should be identified and formulated during the environmental impact report process and before final approval was sought. It is important to point out here that the court in this case considered that climate change posed a substantial risk to sustainable development of the country, and that the relevant officials were under an obligation to ensure that a climate impact assessment was prepared, although the law may not specifically require it. The Court also recognized that sustainable development and the inter-generational equity principle are intrinsically linked.

In *Gbemre v Shell Petroleum Development Company Nigeria Limited and Others* (2005), a Nigerian federal court ruled that oil companies must stop flaring gas in the Niger Delta. Jonah Gbemre, a representative of the Iwherekan community in the Niger Delta filed suit against the Nigerian government and Shell. The court held that the practice of gas flaring is unconstitutional, as it violates the guaranteed fundamental rights of life and dignity of human persons provided in the Constitution of the Federal Republic of Nigeria and the African Charter on Human and Peoples' Rights.

Australia and the Pacific

The *McArthur River Mine* case from Australia raised a number of human rights issues. In 2003, owners of the mine proposed to expand the operation of the mine which involved the rerouting of the MacArthur River. Despite concerns over serious environmental damage, public health, and cultural factors, the Minister for Mines approved the expansion due to pressure from the Federal Government. After challenges and changes to the law, fresh approval was issued. A discussion paper prepared by the Environment Defenders office in New South Wales and Victoria points out that:

> A number of human rights issues are raised by this case. The mine expansion had the potential to cause serious environmental harm, affecting the health and well-being of those Aboriginal communities living near the mine. Human rights that had the potential to be breached as a result included the rights to health, life and to culture, as well as the "right to a clean and healthy environment". There was also a failure to respect participatory rights, such as the right to participate in public affairs, in the way that the government ensured the ongoing operation of the mine by passing special legislation to defeat the successful challenge to the Government's decision-making processes.
>
> (Environment Defenders Office, 2009: 15)

In a case from New Zealand, *Thomson v. Minister for Climate Change Issues*, the petitioner challenged New Zealand's INDC, alleging that it fell short of the emissions reductions required by its Climate Change Response Act of 2002. One of its purposes is to enable New Zealand meet its international obligations under the UNFCCC. This case is pending.

Two cases from New Zealand related to "climate refugees." In *Ioane Teitiota v. The Chief Executive of the Ministry of Business, Innovation and Employment*, a Kiribati citizen appealed the denial of refugee status arguing that the effects of climate change, including rising sea levels and environmental degradation are forcing citizens to leave the island. The Supreme Court of New Zealand concluded that climate-induced displacement did not qualify for refugee status under the 1951 Geneva Convention. The Court noted, however, that "environmental degradation resulting from climate change or other natural disasters could … create a pathway into the Refugee Convention or protection person jurisdiction" (UN Environment, 2017: 32). In *In re: AD*, a family from Tuvalu appealed the decision denying them resident visas, arguing that they would be at risk of suffering adverse impacts of climate change if they were deported to Tuvalu. New Zealand found that the family had established "exceptional circumstances of a humanitarian character" but declined to do so on the basis of climate change, despite acknowledging that climate change impacts could affect the enjoyment of human rights.

National human rights commissions

While this chapter discussed selected cases at the national level, innovative approaches to environmental rights are not confined to the judiciary. National human rights institutions are starting to play an important role in relation to environmental rights. We will discuss two examples here. In Mexico, a constitutional reform in 1999 gave its National Human Rights Commission full autonomy as an agency. It can receive and investigate complaints on human rights violations and make recommendations to the government, including outlining corrective actions. Prior to the inclusion of a right to a healthy environment in its Constitution, Mexico's National Human Rights Commission, CNDH, had issued several recommendations with regard to environmental protection. In one case, the CNDH found that untreated wastewater being released into a river violated, among other things, the human right to an adequate environment and the right to water.

Perhaps the most celebrated case that is before a National Human Rights Commission is the complaint filed by Greenpeace Southeast Asia and others against "Carbon Majors" in the Philippines (Greenpeace, 2017). It is a petition requesting an investigation of the responsibility of the Carbon Majors for human rights violations or threats of violations resulting from the impacts of climate change. In the petition, the complainants argued:

> The Carbon Majors should be held accountable for violations or threats of violations of Filipinos' rights (a) to life; (b) to the highest attainable standard of physical and mental health; (c) to food; (d) to water; (e) to sanitation; (f) to adequate housing; and (g) to self-determination resulting from the adverse impacts of climate change. Special attention should be paid to marginalized and disadvantaged people and communities particularly vulnerable to the effects of climate change, including women, children, persons with disabilities, those living in extreme poverty, indigenous peoples, and displaced persons; as well as the right of Filipinos to development.

The National Human Rights Commission of the Philippines accepted the petition and it is currently pending before the Commission.

Conclusion

This chapter discussed seminal cases from around the world on environmental rights, including several cases that dealt with climate change. Litigation relating to climate change is increasing around the world and the decision in the *Urgenda* appeal gives us reason for optimism. In addition, national human rights commissions are also making their mark on environmental rights and climate change. While litigation is often the last resort in a long struggle to obtain justice, it is likely that the trend we saw in this chapter will continue. It would be hard to reverse the principles that have developed over the years in relation to environmental rights.

DISCUSSION QUESTIONS

1. What are the innovative approaches that judiciaries around the world have adopted to articulate environmental rights?
2. What are the relevant environmental law principles that judiciaries have applied?
3. What are the competing claims that courts have to balance with environmental rights?
4. What are the substantive and procedural rights that have been used by courts to articulate environmental rights?
5. If a substantive right to a healthy environment is adopted under international law, will it make a difference to these cases?

Notes

1 Similar cases that were filed across the United States were all dismissed. Only this case survived.

2 Contrast this with the discussion in Chapter 7 which states that five US state constitutions recognize a right to a healthy environment.

References

Day, J. (2017). UK Government Threatened with Legal Action on Climate Change (21 November 2017) [Online]. Available at: www.lexology.com/library/detail.aspx?g=390b4c7e-99b2-4851- 84ba-4fc74250af36 [Accessed: 1 September 2018].

DeJusticia. (2018). In Historic Ruling, Colombian Court Protects Youth Suing the National Government for Failing to Curb Deforestation. (5 April 2018) [Online]. Available at: www.dejusticia.org/en/en-fallo-historico-corte-suprema-concede-tutela-de-cambio-climatico-y-generaciones-futuras/. [Accessed: 5 February 2019].

EarthJustice (2005). *Environmental Rights Report: Human Rights and the Environment.* Oakland, CA: EarthJustice.

Ecojustice. Right to a Healthy Environment [Online]. Available at: www.ecojustice.ca/case/right-to-a-healthy-environment/ [Accessed: 1 October 2018].

Environment Defenders Office (NSW) and Environment Defenders Office (Victoria) (2009). Protection of Human Rights and Environmental Rights in Australia, Discussion Paper, 21 May 2009. Sydney: Environment Defenders Office.

Environmental Law Alliance Worldwide (ELAW) (n.d.). South Asian cases [Online] Available at: www.elaw.org/resources.

Greenpeace (2017). The Climate Change and Human Rights Petition [Online]. Available at: www.greenpeace.org/seasia/ph/press/releases/Worlds-largest-carbonproducers-ordered-to-respond-to-allegations-of-human-rights--abuses-from-climatechange/The-Climate-Change-and-Human-Rights-Petition/ [Accessed: 28 February 2017].

Intergovernmental Panel on Climate Change (IPCC) (2014). Climate Change 2014, Synthesis Report, Summary for Policymakers [Online]. Available at: www.ipcc.ch/pdf/assessment-report/ar5/syr/AR5_SYR_FINAL_SPM.pdf [Accessed: 1 October 2018].

Johnson, N. (2013). Native Village of Kivalina v. ExxonMobil Corp: Say Goodbye to Federal Public Nuisance Claims for Greenhouse Gas Emissions. *Ecology Law Quarterly* 40 (2): 557.

Knox, J.H. (2018). Email communication, 8 October 2018, on file with author.

Kravchenko, S. and Bonine, J. (2008). *Human Rights and the Environment: Cases, Law, and Policy*. Durham, NC: Carolina Academic Press.

Parvez, H. and Azfar, A. (2004). Securing Environmental Rights Through Public Interest Litigation in South Asia. *Virginia Environmental Law Journal* 22: 215–47.

Peel, J. and Osofsky, H.M. (2015). *Climate Change Litigation: Regulatory Pathways to Cleaner Energy*. Cambridge: Cambridge University Press.

Puvimanasinghe, S.F. (2009). Sustainable Development in the Courts: Towards a Jurisprudence of Sustainable Development in South Asia: Litigation in the Public Interest. *Sustainable Development Law & Policy* 10: 41.

Rahman, H.A. (2010). Human Rights to Environment in Malaysia. *Health and the Environment Journal* 1(1): 59.

Raman, M. (1997). The 1996 Malaysian High Court Decision Concerning the Bakun Hydro-Electric Dam Project. *Asia Pacific Journal of Environmental Law* 2 (1): 93.

Raymond, N. (2018). Rhode Island sues major oil companies over climate change. *Reuters* 2 July 2018 [Online]. Available at: www.reuters.com/article/us-oil-climatechange-rhode- island/rhode-island-sues-major-oil-companies-over-climate-change- idUSKBN 1JS28M [Accessed: 1 October 2018].

Razzaque, J. (2002). Human Rights and the Environment: The National Experience in South Asia and Africa, Joint UNEP-OHCHR Expert Seminar on Human Rights and the Environment, Background Paper No 4. Geneva, January 2002 [Online]. Available at: http://eprints.uwe.ac.uk/18403/1/Joint%20UNEP%20-%20Razzaque.pdf [Accessed: 5 February 2019].

Tay, M. (2017). Development and Environmental Injustice in Malaysia: A Story of Indigenous Resistance in Sarawak. *EnviroLab Asia* 1(1): Article 7.

UN Environment (2004). Compendium on Human Rights and the Environment: Selected International Legal Materials and Cases [Online]. Available at: http://wedocs.unep.org/bitstream/handle/20.500.11822/9943/UNEP_Compendium HRE. pdf?sequence=1&isAllowed=y [Accessed: 1 October 2018].

UN Environment (2017). *The Status of Climate Change Litigation: A Global Review*. Nairobi: Law Division, UN Environment Programme.

U.S. Global Change Research Program (2017). Climate Science Special Report 10 (Fourth National Climate Assessment) [Online]. Available at: https://science2017.globalchange.gov/downloads/CSSR2017_FullReport.pdf (accessed 10 October 2018).

Wood, M.C. (forthcoming). Atmospheric Trust Litigation: Securing a Constitutional Right to a Stable Climate System. *Colorado Natural Resources, Energy & Environmental Law Review*. Draft available at: https://law.uoregon.edu/images/uploads/entries/Wood-Working_Draft.pdf [Accessed: 1 October 2018].

Cases

Adong bin Kuwau and others v Kerajaan Negeri Johor and Anor (1997). 1 MLJ 418

AP Pollution Control Board II v. MV Nayudu (2000) [Online]. Available at: https://indiankanoon.org/doc/1543623/ [Accessed: 1 October 2018].

Bakun Hydro-Electric Dam project (1996). Referred to in Raman (1997)

Bokhari v. Federation of Pakistan (2003). [Online]. Available at: www.supremecourt.gov.pk/web/user_files/File/C.A._545_2015_dt_8-7-2015.pdf [Accessed: 1 October 2018].

Bulankulama v. Minister of Industrial Development and others (Eppawala case) (2000). S.C. Application No. 884/99 (F/R). [Online]. Available at: www.elaw.org/content/sri-lanka-bulankulama-v-min-industrial-development-eppawala-case-sc-application-no-88499-fr [Accessed: 1 October 2018].

Centre for Environmental Justice v. Ministry of Agriculture, Environment, Irrigation and Mahaweli Development et al. (Bundala National Park Case) (2006). (Sri Lanka Court of Appeal, 23 February 2006). [Online]. Available at: www.elaw.org/node/2070 [Accessed: 1 October 2018].

Case Concerning the Gabcikovo Nagymaros Project (Hungary v. Slovakia) (1996). International Court of Justice

Clean Air Council, SB through his Guardian and BB through his Guardians v. USA, Donald Trump as President and others (2017). Case 2:17-cv-04977-PD [Online]. Available https://cleanair.org/wp-content/uploads/Complaint-17-4977.pdf [Accessed: 1 October 2018].

Communities for a Better Environment v. City of Richmond (2010).

Court of Appeal of California 184 Cal. App. 4th 70 (2010)

Comunidad de Chanaral v. Codelco Division el Salvador (2016)

Demanda Generaciones Futuras v. Minambiente (Future Generations v. Ministry of the Environment and Others) (Colombian youth case) (2018). Original complaint in Spanish. [Online]. Available at: https://cdn.dejusticia.org/wp-content/uploads/2018/01/TutelaCambioClim%C3%A1tico.pdf?x54537&x54537&x54537 [Accessed: 1 October 2018].

Demanda Generaciones Futuras v. Minambiente (Future Generations v. Ministry of the Environment and Others) (Colombian youth case) (2018). Unofficial summary in English by DeJusticia. [Online]. Available at: http://blogs2.law.columbia.edu/climate-change-litigation/wp-content/uploads/sites/16/non-us-case-documents/2018/20180405_11001-22-03-000-2018-00319-00_decision-1.pdf [Accessed: 1 October 2018].

Earthlife Africa Johannesburg v. Minister of Environmental Affairs et al. (2017). Case no. 65662/16

Gbemre v Shell Petroleum Development Company Nigeria Limited and Others (2005). Available at: http://wordpress2.ei.columbia.edu/climate-change-litigation/files/non-us-case-documents/2005/20051130_FHCBCS5305_judgment-1.pdf

Greenpeace Nordic Association v. Norway Ministry of Petroleum and Energy (2017). [Online]. Available at: http://climatecasechart.com/non-us-case/greenpeace-nordic-assn-and-nature-youth-v-norway-ministry-of-petroleum-and-energy/ (including an unofficial translation in English) [Accessed: 1 October 2018].

Gunaratne v. Homagama Pradeshiya Sabha (1998). 2 Sri L.R. 11 [Online]. Available at: www.lawnet.gov.lk/wp-content/uploads/2016/11/002-SLLR-SLLR-1998-V-2-GUNARATNE-v.-THE-HOMAGAMA-PRADESIYA-SABHA-AND-OTHERS.pdf [Accessed: 20 October 2018].

Ioane Teitiota v. The Chief Executive of the Ministry of Business, Innovation and Employment [2015] NZSC 107, New Zealand Supreme Court, 20 July 2015.

In re: AD Immigration and Protection Tribunal, New Zealand (2014). NZIPT 501370-371 (2014) [Online]. Available at: http://blogs2.law.columbia.edu/climate-change-litigation/wp-content/uploads/sites/16/non-us-case-documents/2014/20140604_2014-Cases-501370-371_decision.pdf [Accessed: 1 October 2018]

Irfan v. Lahore Development Authority (2003). No. 25084 (Lahore H.C. 2003). [Online]. Available at: https://elaw.org/content/pakistan-anjun-irfan-v-lahore-development-authority-wp-250841997-dd-20020614-lahore-air-poll [Accessed: 20 October 2018].

Juliana et al. v. USA et al. (April 2016) United States District Court, For the District of Oregon, 6:15-cv-1517-TC, Magistrate Judge Coffin, Order and Findings and Recommendations [Online]. Available at: https://static1.squarespace.com/static/571d109b04426270152febe0/t/576195342fe1316f09d2eb8d/1466012983313/16.04.08.OrderDenyingMTD.pdf [Accessed: 25 September 2018]

Juliana et al. v. USA et al. (2016). United States District Court, For the District of Oregon, 6:15-cv-1517-TC, Judge Aiken, Opinion and Order [Online]. Available at: https://static1.squarespace.com/static/571d109b04426270152febe0/t/5824e85e6a49638292ddd1c9/1478813795912/Order+MTD.Aiken.pdf [Accessed: 25 September 2018]

Karnataka Industrial Areas Development Board v. Sri C. Kenchappa and Others (2006). Case No: Appeal (civil) 7405 of 2000 [Online]. Available at: https://indiankanoon.org/doc/992326/ [Accessed: 1 October 2018]

Lliuya v. RWE AG (Case No. 2 O 285/15 Essen Regional Court)

Massachusetts v. EPA (2007). 549 U.S. 497

Mendoza Beatriz Silva and others v. National Government of Argentina and others in regard to damages suffered (2008). File M. 1569. XL

Metropolitan Manila Development Authority v. Concerned Residents of Manila Bay (2008)

Minors Oposa v. Secretary of the Department of Environment and Natural Resources (1994). 33 ILM 173

Montana Environmental Information Center v. Department of Environmental Quality (USA) (1999). [Online]. Available at: https://caselaw.findlaw.com/mt-supreme-court/1329934.html [Accessed: 20 October 2018].

Mundy v. Central Environmental Authority and others (2003). SC Appeal 58/2003

Narmada Bachao Andolan v. Union of India (2000). Writ Petition (civil) 328 of 2002 [Online]. Available at: https://indiankanoon.org/doc/1642722/ [Accessed: 1 October 2018]

National Audubon Society Inc. v. Davis (2002). 312 F.3d 416 (9th Cir. 2002)

Netherlands v. Urgenda Foundation (Urgenda appeal) (2018) [Online]. Available at: https://uitspraken.rechtspraak.nl/inziendocument?id=ECLI:NL:GHDHA:2018:2610 [Accessed: 1 October 2018]

Petro Flores v. Corporacion Del Cobre, Codelco (1988). Rol 2.052 (Sup. Ct. Chile, 23 June 1988) Translated in *Georgetown International Environmental Law Review* 2: 251–253

Prakash Mani Sharma v. His Majesty's Government Cabinet Secretariat and Others (2003). WP 2237/1990 (2003.3.11) [Online]. Available at: www.elaw.org/content/nepal-prakash-mani-sharma-v-his-majesty%E2%80%99s-government-cabinet-secretariat-and-others-wp-22371 [Accessed: 1 September 2018]

Prakash Mani Sharma and others v. Nepal Drinking Water Corporation and Others (2001) WP 2237/1990 (10 July 2001) [Online]. Available at: www.elaw.org/content/nepal-advocate-prakash-mani-sharma-and-others-vs-nepal-drinking-water-corporation-and-others [Accessed: 1 October 2018]

Salt Miners Labor Union v. Industries and Mineral Development (1993). Supreme Court of Pakistan 1994 SCMR 2061 [Online]. Available at: www.globalhealthrights.org/wp-content/uploads/2013/02/SC-1994-Salt-Miners-v.-Director-Industries-and-Mineral-Development.pdf [Accessed: 1 September 2018]

Senih Ozay v. Ministry of the Environment and Eurogold Madencilik (1997)

Shehla Zia and others v. WAPDA (1994). PLD 1994 SC 693

Subhash Kumar v. State of Bihar (1991). 1991 AIR 420, 1991 SCR (1) 5 [Online]. Available at: https://indiankanoon.org/doc/1646284/ [Accessed 1 October 2018]

Suray Prasad Sharma Dhungel v. Godavari Marble Industries and Others (1992). WP 35/1992

Tan Tek Seng v. Suruhanjaya Perkhidmatan Pendidikan & Another (1996). 1 MLJ 261

Thomson v. Minister for Climate Change Issues (2017). NZHC 733

Urgenda Foundation v. The State of The Netherlands (2015). English Translation, [Online]. Available at: https://elaw.org/system/files/urgenda_0.pdf [Accessed: 20 October 2018].

Vellore Citizens Welfare Forum v. Union of India (1996). [Online]. Available at: https://indiankanoon.org/doc/1934103/ [Accessed: 1 September 2018]

Victor Ramon Castrillon Vega v. Federación Nacional de Algodoneros y Corporation Autonoma Regional del Cesar (1997). Accion de Tutela Case No 4577 (Colom.), discussed in Kravchenko and Bonine, 2008, p.70

Village of Kivalina v. ExxonMobil Corp. (2012). 696 F.3d 849 (9th Cir. 2012)

VZW Klimatzaak v. Kingdom of Belgium (2016). Summary. [Online]. Available at: http://climatecasechart.com/non-us-case/vzw-klimaatzaak-v-kingdom-of-belgium-et-al/ [Accessed: 1 October 2018]

Wolfson v. Brammer (2010). 616 F.3d 1045,1056 (9th Cir. 2010)

Part III

Climate change and human rights

9 From UNFCCC to Paris Agreement

A human rights assessment

Introduction

The climate change legal regime has come a long way since the United Nations Framework Convention on Climate Change (UNFCCC) was first adopted in 1992. Back then, the international community was not sure whether climate change was a real threat, or how much of a threat it was, what to do, what a safe level of temperature increase is, how to address the main greenhouse gases, what to do about developing countries' emissions, etc.

From a legal architecture standpoint, the roller coaster ride that the international community took from the UNFCCC to the Paris Agreement in 2015 is fascinating. The UNFCCC is a framework convention – as indeed, the title itself suggests. It set the broad framework for further action, including establishing an institutional framework and some broad objectives and commitments. Binding obligations came with the Kyoto Protocol, but only for developed countries – this reflects the common but differentiated responsibility (CBDR) principle – and finally, we have the Paris Agreement, which is a hybrid document, i.e. a document that encapsulates voluntary commitments with binding obligations. This approach was necessary to work around the common but differentiated responsibility principle and get buy-in from developing countries. These three very different documents are interspersed with many soft law instruments. So, what is the common but differentiated responsibility principle?

The first time that this principle was applied was in the legal regime governing ozone depletion, although the principle was not specifically identified in it. The principle was formally included in the Rio Declaration on Environment and Development of 1992:

> States shall co-operate in a spirit of global partnership to conserve, protect and restore the health and integrity of the Earth's ecosystem. In view of the different contributions to global environmental degradation, States have common but differentiated responsibilities. The developed countries acknowledge the responsibility that they bear in the international pursuit of sustainable development in view of the pressures their societies place on the global environment and of the technologies and financial resources they command.
>
> (Principle 7)

International law is based on the sovereign equality principle, but international society is anything but equal. The CBDR takes this disparity into account both in relation to the contribution to creating environmental problems and the ability to address them. Climate change is a good example of an environmental issue where the contribution of states varies considerably, and the ability of states to address it also varies. The UNFCCC incorporates the CBDR as a principle that underlies the legal regime. The CBDR principle is based on equity and fairness. While an in-depth analysis of this principle cannot be undertaken here, suffice it to say that it is a revolutionary principle in international environmental law and is an exception to the sovereign equality principle.

This chapter will not discuss each and every document adopted in relation to climate change. Instead, it will highlight key documents that embody human rights commitments (broadly defined). In other words, this chapter is a human rights assessment of the climate change legal documents adopted by the international community.

The establishment of the IPCC and the adoption of the UNFCCC

The Intergovernmental Panel on Climate Change (IPCC) was established by the United Nations Environment Program (UNEP), the World Meteorological Organization (WMO), and the World Health Organization (WHO) in 1988 to synthesize research on climate change. The IPCC is not mandated to carry out its own research. It is an eminent body of over 100 scientists from all over the world. It is regarded as the authority on science relating to climate change. Since 1988, the IPCC has issued five assessment reports, the latest being in 2014 when the IPCC unequivocally asserted that it is extremely likely that the warming of the planet is due to anthropogenic greenhouse gas emissions (IPCC, 2014a). A Special Report titled *Global Warming of 1.5°C* was issued in October 2018. It concluded with high confidence that:

> Human activities are estimated to have caused approximately 1.0°C of global warming above pre-industrial levels, with a likely range of 0.8°C to 1.2°C. Global warming is likely to reach 1.5°C between 2030 and 2052 if it continues to increase at the current rate.
>
> (IPCC, 2018)

The report issued a dire warning about the impending food shortages and worsening wildfires, and that adverse consequences could come as early as 2040:

> If greenhouse gas emissions continue at the current rate, the atmosphere will warm up by as much as 2.7 degrees Fahrenheit (1.5 degrees Celsius) above preindustrial levels by 2040, inundating coastlines and intensifying droughts and poverty. Previous work had focused on estimating the damage if average temperatures were to rise by a larger number, 3.6 degrees Fahrenheit

> (2 degrees Celsius), because that was the threshold scientists previously considered for the most severe effects of climate change.
>
> (Davenport, 2018)

According to the report, to prevent 2.7°F of warming, greenhouse gas emissions must be reduced by 45% from 2010 levels by 2030, and 100% by 2050. Coal as a source of electricity will have to be reduced to 1–6% (almost completely eliminated), while renewable sources will have to increase to 67%. The report acknowledges that while this is technically possible, it may be impossible politically. However, do we have a choice? The future of our children and our grandchildren depends on taking aggressive mitigation measures now.

As noted, the UNFCCC was adopted in 1992 at the Earth Summit at Rio. Its adoption gave rise to many divisions in the international community – North v. South; US v. EU; G 77 v. small islands and other vulnerable states; oil producing states v. others; nations with large forest cover v. Global North, and so on (Hunter et al., 2015). Given the number of divisions, it is a small wonder that a treaty, albeit a framework convention, was adopted at all. Many of these divisions continue to date, with newer alliances such as BASIC, SIDS, and AOSIS being forged.

The Convention embodies several noteworthy features: it specifically includes principles that should guide the parties; obligations are based on the common but differentiated responsibility principle, and states are categorized as all parties, Annex I countries, and non-Annex I countries; and it established an institutional framework to implement the provisions of the convention – a Conference of Parties (COP), a subsidiary body on scientific and technological advice, a subsidiary body for implementation, and a secretariat (Atapattu, 2016a).

The main objective of the Convention is the "stabilization of greenhouse gas concentrations in the atmosphere at a level that would prevent dangerous anthropogenic interference with the climate system" (UNFCCC: Article 2). The Convention does not define "dangerous anthropogenic interference." This limit was not settled in legal texts until the Copenhagen Accord was adopted in 2009, although scientists had been in agreement for some time that global temperature increase should not exceed 2°C. This was affirmed in later documents, and the level of ambition was raised in the Paris Agreement to endeavor to limit the temperature increase to 1.5°C as a result of intense lobbying by small island states.

Despite establishing an institutional framework and identifying initial obligations of states in relation to climate change, the Convention, as a whole, is mostly silent on the impact of climate change on human beings. The only reference it makes to human beings is in the context of adverse effects of climate change, where it refers to human health and the need to protect the climate for the benefit of present and future generations. It also refers to the obligation of Annex I countries to help developing countries and those that are vulnerable to climate change to adapt. Reference is made to states, rather than individuals, and the Convention can be characterized essentially as an environmental treaty (Cullet, 2010), seeking to protect the interests of states, particularly those who are especially vulnerable to climate change. Cullet argues that "the climate change regime needs to move beyond its traditional international environmental law model to encompass consideration of the specific vulnerabilities of individuals and communities" (Cullet, 2010).

The lack of reference to human beings in a treaty that seeks to protect current and future generations from the adverse consequences of climate change is puzzling. Similar to the UNFCCC, the Kyoto Protocol which was adopted in 1997 contains no reference to human rights. As pointed out by Philippe Cullet, "While environmental and economic considerations have been central to the climate change legal regime, the same cannot be said for its human rights aspects" (Cullet, 2010). While vulnerability may provide the conceptual framework to address human rights within Kyoto, vulnerability is mentioned mainly in relation to states, not individuals. Given the close link between climate change and human rights, it is rather surprising that the language of rights is totally absent from the early documents on climate change (Atapattu, 2016a).

The link between climate change and human rights

According to the IPCC, the impacts of climate change are far-reaching and include both sudden events and slow onset events: increased severe weather events both in terms of severity and frequency, water scarcity, food scarcity, desertification, increased wildfires, sea level rise due to melting glaciers and thermal expansion of oceans, ocean acidification, and its consequent impact on small island states, displacement due to climate change both internally and across borders, and potential conflicts over natural resources. Every region of the world will be affected. The worst consequences are yet to materialize and future generations will suffer more than the current generation.

At the beginning, however, the human rights community was slow to recognize this link. Looking at this list of adverse consequences, it is easy to see that most protected rights could be undermined by climate change. Once the link between climate change and human rights became obvious and communities started experiencing the adverse impacts, the human rights community started lobbying to get a human rights provision included in climate documents. States that are especially vulnerable to climate change, such as the small island states that are at great risk of being totally submerged by sea level rise associated with climate change, have been very vocal about recognizing climate change as a human rights issue. The Maldives, Kiribati, Marshall Islands, and Tuvalu have been among the most vocal.

The Malé Declaration on the Human Dimension of Global Climate Change adopted by small island developing states (SIDS) in 2007 was an important milestone that spurred action in this area. It was the first inter-governmental statement to explicitly recognize that climate change has implications for the full enjoyment of human rights, including the rights to life, health, and an adequate standard of living (Malé Declaration, 2007). It requested the UN Human Rights Council to convene a debate on climate change and human rights.

In 2008, the UN Human Rights Council adopted its first resolution on climate change and human rights. In its resolution 7/23, the Council expressed its concern that "climate change poses an immediate and far-reaching threat to people and communities around the world and has implications for the full enjoyment of human rights" (Human Rights Council, 2008a). The resolution also recognized "that the world's poor are especially vulnerable to the effects of climate change, in particular those concentrated in high-risk areas, and also tend to have more limited adaptation capacities," and "further that low-lying and other small island countries, countries with low-lying coastal, arid and semi-arid areas or areas liable to floods, drought and desertification, and developing countries with fragile mountainous ecosystems are particularly vulnerable to the adverse effects of climate change." It requested the Office of the High Commissioner for Human Rights to prepare a detailed analytical study on the link between climate change and human rights (Human Rights Council, 2008b).

While the impacts of climate change are global, every country and every community will not be affected equally. Climate change will have a disproportionate impact on some states and some communities, and the legal regime governing climate change had to grapple with this reality. It is not only the contribution to climate change that varies across states; the impacts vary considerably too.

In its report, the Office of the High Commissioner for Human Rights (OHCHR) reiterated that climate change will have implications for the enjoyment of the full range of human rights. The report identified the following as amongst the main consequences of climate change: contraction of snow-covered areas; sea level rise and higher thermal temperatures; increase in extreme weather events including droughts, heatwaves, and floods; and increase in tropical cyclones (OHCHR, 2009).

In addition to these global impacts, there will be more localized consequences, such as health effects, impacts on food and water availability and supply, changes in cultivation patterns, and weather-resistant crops, pests, and other issues will have an impact on the day-to-day lives of millions of people (OHCHR, 2009). While we tend to focus more on catastrophic events, as they are more visible, we should not lose sight of these smaller, but longer-lasting consequences of climate change which will also impact the enjoyment of human rights. The OHCHR's report recognized the unequal burden of climate change as being reflected in Article 3 of the UNFCCC with some states and communities more impacted than others (Atapattu, 2016).

The Stockholm Declaration of 1972 recognized the interdependence and inter-relatedness of human rights and the environment, as do the treaty bodies, although human rights treaties do not recognize a specific right to a safe and healthy environment. The OHCHR report discussed the impacts on specific rights, both civil and political rights, and economic, social, and cultural rights: the right to life, the right to adequate food, the right to water, the right to health, the right to adequate housing, and the right to self-determination. Furthermore, the report discussed the effects on specific groups such as women, children, and indigenous peoples.

The report noted that according to the first IPCC report, the greatest single impact of climate change could be on human migration, as by 2050, about 150 million people could be displaced by climate change-related events, such as desertification, water scarcity, floods, and storms. The report identified four displacement scenarios: (a) weather-related disasters; (b) gradual environmental deterioration and slow onset disasters; (c) increased disaster risks and relocation of people from high-risk zones; and (d) social upheaval and violence attributable to climate change-related factors (OHCHR, 2009).

The report recognized several situations of displacement to which international law has few or no answer: (a) people moving internally will be entitled to the full range of human rights guaranteed by the state in question; and (b) persons moving across an international border due to an environmental factor have no legal status under international refugee law. While they would be entitled to general human rights guarantees in a receiving State, they will not have a right of entry to that State; (c) with regard to the status of the populations displaced from small island states, due to their eventual submergence, the report recognized that human rights law does not provide a clear answer and will require "adequate long-term political solutions, rather than new legal instruments" (OHCHR, 2009). However, with any legal instruments, it is unclear how the situation of inhabitants of small island states could be addressed, as they will need to be resettled in other countries and their citizenship rights will have to be resolved.

In addition to the impact of climate change consequences on the realization of human rights, the report recognized that mitigation measures that states put in place to address climate change may also have human rights implications. Mitigation is discussed in greater detail in Chapter 12. In brief, mitigation means to take measures to reduce greenhouse gas emissions from various sectors. The impact of agro-fuel production on food security and the impact of increased use of biofuels and the REDD program on the rights of indigenous peoples are examples.

While noting that climate change has "obvious implications" for the enjoyment of human rights, the report stated that it is less obvious whether they can be qualified as human rights violations. It identified several difficulties: (a) the causal link between the human right infringed and climate change is virtually impossible to establish; (b) there are several contributing factors to many of the consequences such as severe weather events and water scarcity; and (c) climate change impacts are often future events, while a human rights framework comes into play after the harm has occurred.

This does not mean that human rights have no role to play vis-à-vis climate change. In this regard, it is important to discuss human rights in the context of: (a) damage caused by historic emissions; (b) mitigation measures; (c) adaptation measures; and (d) damage caused by severe weather events. The assertion made above by the OHCHR that while human rights can be implicated by climate change, it is less clear whether they can be categorized as human rights violations is correct with regard to the damage caused by historic emissions. Consequences of climate change that are manifesting themselves now are due to the historic emissions of greenhouse gases by states, the accumulation of which is giving rise to adverse consequences now. Here it is virtually impossible to trace these impacts to the emissions of a particular state, although we now know who the main emitters are both in terms of states and corporations. Can the "but for" theory apply here – whether in the absence of these emissions, such consequences would have materialized? Are there any collective theories of liability, such as the market share liability principle that we can develop here similar to those at the national level? In the absence of novel theories of liability, it would be impossible to hold major emitters accountable at the international level for damage we are witnessing today (Atapattu, 2009), even though novel legal strategies such as those discussed in Chapter 8 are emerging at the national level.

In the run up to COP15 in Copenhagen in 2009, 20 UN Special Procedures mandate-holders (such as Special Rapporteurs and Independent Experts) issued a joint statement emphasizing that climate change poses a serious threat to the full enjoyment of human rights, and that mitigation and adaptation measures should be developed in accordance with human rights norms, with the participation of affected communities (OHCHR, 2015a). They warned that a weak outcome at COP15 could threaten the enjoyment of protected rights.

It took the international community almost two decades to include a provision on human rights in climate documents. For the first time, a reference was made to human rights at COP16 in Cancun in 2010. Quoting the statements in the Human Rights Council resolution 10/4 that climate change has a range of implications for the effective enjoyment of human rights, the decision stated: "Parties should, in all climate change-related actions, fully respect human rights" (UNFCCC, 2010). Although a non-binding document, it nonetheless signaled a turning point with regard to climate negotiations. The baby-step that was taken at Cancun culminated in the inclusion of a human rights provision in the Paris Agreement in 2015.

At a seminar organized by the UN OHCHR to discuss the Adverse Impacts of Climate Change on the Full Enjoyment of Human Rights in Geneva in 2012, Dr Kumi Naidoo of Greenpeace International pointed out that "climate change is affecting human lives, and is and will cause human rights violations across the world in many forms. Because of climate change, we will see more starvation, more climate forced migration, and more extinction across the world" (Naidoo, 2012). He stated that countries such as Canada and the United States have a responsibility to reduce their emissions and pay compensation for the damage they have caused. He called for the recognition of a right to a healthy environment in the Rio+20 outcome document and urged states to turn the crisis of climate change into an opportunity: "Climate change should bring humanity to its senses. We can use this challenge to break down the divisions between rich and poor nations, developed and developing, North and South" (Naidoo, 2012).

At the seminar, the representative from Bangladesh referred to the urgency of action:

> Climate change is real. It is now and here. We would be contributing to irrevocable damages to the earth and to the humankind if we allow continuation

> of emissions and resultant discriminations and violations. Given its near universal impact, albeit with differing degrees on regions and communities, one can certainly argue that climate change response is a case of predominantly international cooperation, supported by robust national actions. There is a clear requirement of human rights protection and provisions in all aspects of climate change regime (mitigation, adaptation, technology and finance).
> (OHCHR, 2012)

With regard to mitigation measures that states take to address climate change, the situation is different. States are required to ensure that the measures they take to mitigate adverse consequences do not lead to a violation of rights, especially of vulnerable communities. The OHCHR recognized the possibility of mitigation measures infringing upon the rights of people, but did not, unfortunately, elaborate on this. An example of a mitigation measure that could infringe rights of peoples is where a windfarm is located on indigenous lands without consulting or even providing any information to them. Chapter 12 discusses this in more detail.

Similarly, adaptation measures could also lead to a violation of rights if people are not consulted with regard to adaptation plans or the plans are implemented in a discriminatory manner. Participatory rights – access to information and participating in decision-making – are important in this regard. Participatory (procedural) rights were discussed in Chapter 6. Adaptation measures are discussed in Chapter 12.

Finally, with regard to severe weather events, states have particular obligations under human rights law, before, during, and after such events. They have to provide prior warning, temporary shelters, evacuation of people, and emergency relief to those who are affected in a timely manner. States must also provide assistance to the victims to rebuild in the aftermath, taking into account vulnerable people and communities. During the Asian Tsunami in 2004, it took months to reunite children with their families. Special laws had to be passed to ensure that orphaned children were not sold for adoption or trafficked by unscrupulous people.

During severe weather events, low-income and minority communities are often more affected than others, as events such as Hurricane Katrina and Hurricane Irma have demonstrated. Women and children are also susceptible to increased risk, as are the elderly and disabled people. These communities and groups will likely need more assistance than others. The Sendai Framework for Disaster Risk Reduction, 2015–2030 (the follow-up to the Hyogo Framework) notes the link between disasters, sustainable development, poverty alleviation, and climate change. It notes that disasters continue to exact a heavy toll on people and that during the 10-year period between 2005 and 2015:

> Over 700 thousand people have lost their lives, over 1.4 million have been injured and approximately 23 million have been made homeless as a result of disasters. Overall, more than 1.5 billion people have been affected by disasters in various ways, with women, children and people in vulnerable

> situations disproportionately affected. The total economic loss was more than $1.3 trillion. In addition, between 2008 and 2012, 144 million people were displaced by disasters. Disasters, many of which are exacerbated by climate change and which are increasing in frequency and intensity, significantly impede progress towards sustainable development. Evidence indicates that exposure of persons and assets in all countries has increased faster than vulnerability has decreased, thus generating new risks and a steady rise in disaster-related losses, with a significant economic, social, health, cultural and environmental impact in the short, medium and long term, especially at the local and community levels.
>
> (Sendai Framework, 2015)

The Framework embodies several guiding principles, including: the primary responsibility for disaster risk reduction rests on states; there is shared responsibility between the central government and other national entities and stakeholders; relevant policies and plans must be aimed at coherence "across sustainable development and growth, food security, health and safety, climate change and variability, environmental management and disaster risk reduction agendas."

The Framework points out that disaster risk reduction is essential to achieve sustainable development, and that it requires all-society engagement and partnership, paying special attention to those disproportionately affected by disasters:

> It also requires empowerment and inclusive, accessible and nondiscriminatory participation, paying special attention to people disproportionately affected by disasters, especially the poorest. A gender, age, disability and cultural perspective should be integrated in all policies and practices, and women and youth leadership should be promoted. In this context, special attention should be paid to the improvement of organized voluntary work of citizens.
>
> (Sendai Framework, 2015)

Coordinating efforts relating to disaster risk reduction, sustainable development, poverty alleviation, climate adaptation and protecting human rights is crucial to ensure that limited resources are efficiently used. All these issues are inter-related and rather than addressing them in isolation, states need to adopt a coordinated and concerted effort to address them. However, this is not easy, as we are so used to compartmentalizing issues.

Rights at jeopardy

As noted in the OHCHR report (2009), there is a clear link between adverse consequences associated with climate change and the enjoyment of human rights. The enjoyment of many of the rights, if not all, recognized under human rights law can be undermined as a result of the consequences of climate change. Both substantive and procedural rights are at risk. This section will briefly discuss the main rights that could be infringed due to consequences associated with climate change. On the other hand, the human rights framework can be used to avoid some of the infringements with regard to mitigation and adaptation measures that states undertake.

Knox points out that while climate change can drastically affect human rights, the jurisprudence relating to environmental rights was developed in the context of harm that does not cross borders; therefore, this jurisprudence does not easily apply to transboundary harm like climate change. Nonetheless, he points out that: "even without such a broader application, human rights law: (1) requires states to address climate change harm to the extent that it infringes the human rights of people within their own territory; and (2) constrains states' possible responses to climate change" (Knox, 2009).

Substantive rights

The enjoyment of almost all the substantive rights protected under human rights law, from the supreme right to life, protected under Article 6 of the ICCPR, to freedom of movement can be jeopardized as a result of adverse consequences associated with climate change. Whether civil and political rights, economic, social and cultural rights, individual rights, or collective rights, all rights are at risk. Climate change acts as a threat multiplier, exacerbating existing vulnerabilities and further marginalizing those who are in subordinate positions.

While the right to life as formulated in Article 6 was originally adopted to protect citizens from actions of their state, over the years, this provision has been used to protect people's lives from environmental degradation and pollution. We discussed in Chapter 5 how the Human Rights Committee recognized that positive obligations emanated from it by adopting General Comment No. 6 (see Chapter 5). In extreme situations, climate change can infringe the right to life. In the Inuit Petition filed by the Inuit of Canada and the United States against the United States, the complainants invoked, among others, the right to life. We will discuss this case in some detail, because it was the first case to use the human rights framework with regard to a global environmental problem, and because of the range of rights that were alleged to have been infringed. It is also significant for the extensive use of scientific information in the petition.

Similarly, the right to health, protected under Article 12 of the ICESCR can be affected as a result of climate change. According to Article 12(1): "The States parties to the present Covenant recognize the right of everyone to the enjoyment of the highest attainable standard of physical and mental health." In this regard, the reports of the IPCC are insightful. The health impacts of climate change range from increased incidence of vector-borne diseases to health impacts of heatwaves and other severe weather events and states are required to take measures to protect their people from the adverse health impacts associated with climate change.

Other substantive rights that we discussed in Chapter 5 – the rights to food, water, adequate housing, and sanitation – can be jeopardized a result of adverse consequences of climate change. Scientists predict that consequences of climate change, such as sea level rise, increased salinity, severe weather events, wildfires, and increased temperatures will lead to increased water scarcity, land degradation, and desertification which will lead to food scarcity. Unpredictable and erratic changes in weather also lead to crop damage. Many people are already experiencing these adverse consequences.

The Special Rapporteur on the Right to Housing devoted the 2009 report to climate change and its impact on the right to housing. The report discussed the impact of climate change on urban settlements, human mobility, and small islands and low-lying coastal areas and noted that a disproportionate impact will be borne by the poor and slum-dwellers. The report urged that where populations must be relocated, it must be done in consultation with the affected population, with special consideration given to the needs of women. The report stressed that permanent relocation must never result in homelessness (Atapattu, 2016a). It is clear that both slow onset events, such as sea level rise, and sudden onset events, such as hurricanes, can jeopardize the right to housing.

Climate change will affect food security in several ways – it will increase droughts and desertification as well as flooding. It will also result in more severe weather events (Gonzalez, 2011). All these will lead to the destruction of crops, displacement of people, and degradation of land and soil. Despite the availability of food, currently there are about 850 million malnourished people. Climate change will add about 650 million to this number (UNDP, 2007/08). According to the former Special Rapporteur on the Right to Food, Jean Ziegler, "half of the world's hungry people depend for their survival on lands which are inherently poor, and which may be becoming less fertile and less productive as a result of the impact of repeated droughts, climate change and unsustainable land use" (OHCHR, 2008: para. 51). Moreover, special attention should be paid to "vulnerable and disadvantaged groups, including people living in disaster-prone areas and indigenous peoples whose livelihood may be threatened" (OHCHR, 2009: para. 27; Atapattu, 2016a). The current Special Rapporteur on the Right to Food, Hilal Elver, devoted a report to climate change and the right to food (OHCHR, 2015b).

In addition to the rights discussed above, several other rights could be jeopardized as a result of climate change. The right to a livelihood, especially if people have to be relocated, the right to self-determination, especially with regard to the inhabitants of small island states, the right to culture and property, especially with regard to indigenous peoples, and freedom of movement, especially during severe weather events where people are evacuated, are some of the substantive rights that could be affected.

The link between the rights to health, water, and sanitation, and adverse consequences of climate change was recognized by the Special Rapporteur on the Right to Health. He discussed the link between climate change and health and identified three main obstacles to realizing access to safe water and adequate sanitation: poverty, gender inequality, and global warming. As safe drinking water sources diminish, people will resort to unsanitary practices, causing serious health issues. Moreover, the world's poor will be disproportionately affected by climate change. The Special Rapporteur warned of the dire consequences that await us:

> Despite these disturbing trends, the international community has not yet confronted the health threats posed by global warming. The failure of the international community to take the health impact of global warming seriously will endanger the lives of millions of people across the world.
>
> (OHCHR, 2007: para. 102)

Procedural rights

The procedural rights that were discussed in Chapter 6 – right to information, right to participate in the decision-making process, and access to remedies – could also be affected as a result of climate change. People may not be consulted before they are relocated, adaptation plans may be drawn up without the affected people's participation, information may not be provided, and adequate remedies may not be available to victims. These violations can become even more acute in the context of climate change, especially if there are multiple severe weather events, e.g., hurricane after hurricane such as what we saw in the United States during the Fall, 2017. It is sobering to note that even the richest country in the world cannot insulate itself from the adverse consequences of climate change.

Often, in emergency situations, there is no time to consult people, but the provision of reliable information is crucial so that people can exercise caution and move to safer locations. During Hurricane Katrina, many people refused to leave their homes, even though there were strict orders to evacuate. Similarly, during wildfires, some residents do not evacuate despite orders to do so. Should public safety override personal wishes? Can authorities forcibly evacuate people if their safety is at risk? The answer is not very clear-cut.

Mitigation and adaptation measures

Under the climate legal regime, states have pledged to reduce their greenhouse gas emissions by various amounts, using various baselines. Under the Kyoto Protocol, this was a reduction of 5% below the 1990 levels (the EU, however, made a commitment of 7% below 1990 levels). The Doha amendment increased this to 17%, but many Annex I countries did not ratify the amendment. The trend of using voluntary commitments started under the Copenhagen Accord where states pledged many reduction targets using varying baselines. Even though the emission reduction commitments under the Paris Agreement are voluntary (these are called Nationally Determined Contributions (NDCs)), certain binding obligations flow from them. Thus, the Paris Agreement is a hybrid document encompassing both voluntary commitments and binding obligations. In order to give effect to these commitments, states are required to take mitigation measures. These measures could also run the risk of violating rights. For example, in an effort to reduce the reliance on fossil fuels, many states are switching to biofuels. Converting land to grow crops to be used for biofuels could have an impact on food security of people. The REDD program could have an impact on indigenous rights. Similarly, several projects under the Clean Development Mechanism (one of the flexibility mechanisms under the Kyoto Protocol) have led to many human rights violations because guidelines did not include a reference to human rights (Atapattu, 2016a), even though states are required to fulfill their human rights obligations in relation to these projects. It is thus important to apply a human rights framework to mitigation measures to ensure that they do not violate peoples' rights.

Similarly, adaptation plans, and measures must be informed by a human rights framework. States already have human rights obligations that they are required to fulfill to protect their populations from the adverse consequences of climate change. Adaptation measures could run the risk of violating the rights of people if sufficient information is not provided to them or the affected people are not consulted, especially if they are to be relocated from their lands. With regard to indigenous peoples, these procedural requirements are stricter – the standard of free, prior, and informed consent is required if they are to be relocated from their traditional lands.

As the Human Development Report noted as far back as 2007/08: "the world's poor cannot be left to sink or swim with their own resources while rich countries protect their citizens behind climate-defense fortifications. Social justice and respect of human rights demand stronger international commitment on adaptation" (UNDP, 2007/08). The IPCC defines adaptation as "the process of adjustment to actual or expected climate and its effects" (IPCC, 2014), and it has recognized that even most stringent mitigation efforts cannot avoid impacts of climate change which makes adaptation unavoidable (IPCC, 2007). This is because we have locked in a certain level of climate change, not just for the present generation, but for the next generation as well, because of our historic emissions. Thus, learning to cope with the consequences of climate change is crucial. When the climate change legal regime was adopted in 1992, the main focus was on mitigation. Adaptation was not given the attention it deserved until recently. However, at least for the current and the next generation, adaptation is crucial:

> While developing and implementing successful mitigation strategies clearly remains critical in the quest to avoid worst-case climate change scenarios, we have passed the point where mitigation efforts alone can deal with the problems that climate change is creating. Because of "committed" warming – climate change that will occur regardless of the world's success in implementing mitigation measures, a result of the already accumulated greenhouse gases ("GHGs") in the atmosphere – what happens to socio-ecological systems over the next decades, and most likely over the next few centuries, will largely be beyond human control.
>
> (Craig, 2010: 9)

Thus, given the reality of climate change consequences, adaptation assumes more urgency for the present generation than mitigation. Despite this reality, however, adaptation did not receive attention until recently. Wold, Hunter, and Powers point out:

> The increasing focus on adaptation is largely a response to the recognition that many climate impacts are virtually certain at this point and that human suffering and economic losses can be reduced through some adaptive strategies. Mitigation measures are unlikely to prevent climate change before it results in significant environmental and social impacts, and adaptation policies are necessary to alleviate future suffering.
>
> (2015: 95)

Some believe that although equity forms part of the UNFCCC, most debates have referred to fairness in mitigation (allocating the right to emit greenhouse gases), rather than adaptation. Adger, Paavola, and Huq believe that considerations of fairness of adaptation strategies are central to the future of international action:

> Large parts of the world's population already confront significant risks from climate variability in the face of pressing needs for survival, and in the face of insecurity, globalization, and global financial crises. Many of the people who are vulnerable to climate change today live in the developing world. The world's changing climate and our responses to it threaten to exacerbate precisely those trends and pressures that cause present insecurities and that are likely to lead to increased insecurity in the future.
>
> (2006: 2)

Vulnerable groups, particularly those who are dependent on climate-sensitive resources, such as farmers and fishermen, indigenous groups, and women are likely to be disproportionately affected as they are "likely to be at the sharp end of the policy responses to climate change" (Adger, Paavola, and Huq, 2006: 2). These groups are often ignored when policy decisions are made on both mitigation and adaptation.

The Cancun Adaptation framework adopted at COP 16 was the first attempt at establishing a comprehensive framework on adaptation. It stressed that "adaptation must be addressed with the *same priority* as mitigation" (UNFCCC, 2010, emphasis added). An important outcome of this Framework was the creation of a process for the least-developed countries to adopt NAPAs. Another was the creation of an Adaptation Committee. The Framework stressed the need to reduce vulnerability and increase resilience of people in developing countries, especially among those who are particularly vulnerable to climate change (UNFCCC, 2010).

The Cancun Adaptation Framework was also the first time that rights language, albeit limited, was used in the UNFCCC documents. It affirmed that:

> Enhanced action on adaptation should be undertaken in accordance with the Convention, should follow a country-driven, gender-sensitive, participatory

> and fully transparent approach, taking into consideration vulnerable groups, communities and ecosystems, and should be based on and guided by the best available science and, as appropriate, traditional and indigenous knowledge, with a view to integrating adaptation into relevant social, economic and environmental policies and actions, where appropriate.
>
> (UNFCCC, 2010: para. 12)

Furthermore, the Framework requires the preparation of vulnerability and adaptation impact assessments; financial assessments and the economic, social, and environmental evaluation of adaptation plans; disaster risk reduction strategies; and cooperation with regard to displacement. The framework endorsed the common but differentiated responsibility principle in relation to enhanced action on adaptation. Since then adaptation has continued to be one of the four pillars of climate action along with mitigation, finance, and technology. The Paris Agreement added a fifth pillar – the loss and damage mechanism, which will be discussed in greater depth, together with mitigation and adaptation, in Chapter 12.

Vulnerable groups and states

The OHCHR report identifies several groups of people that are more vulnerable to climate change. They "are already in vulnerable situations due to factors such as poverty, gender, age, minority status, and disability" (OHCHR, 2009). The report discusses women, children, and indigenous groups as vulnerable groups, as well as people affected by displacement associated with climate change.

Women

Women in developing countries are especially vulnerable to climate change due to the multiple roles they perform. They are responsible for collecting water and firewood, as well as providing food for the family. With increased food and water scarcity associated with climate change, women have to walk longer distances in search of water. Women are also more vulnerable to severe weather events and face gender-based violence, harassment, and exploitation during periods of displacement associated with severe weather events and conflicts over resources. Women are also engaged in agriculture, but lack tenure or land rights. On the other hand, they are more resilient and have better adaptive capacity. Their knowledge of the land and resources is crucial for adaptation programs. Getting their input is necessary for adaptation programs.

Children

The OHCHR report (2009) highlights that, overall, the health burden of climate change will be borne by children in developing countries. Extreme weather events and increased water stress are already having an impact on malnutrition and infant and child mortality. Girl children are affected by these stresses as they are more responsible for collecting water and firewood. They also have a higher mortality rate due to severe weather events. The Convention on the Rights of the Child requires states parties to take measures to ensure that the rights enshrined there are protected within their jurisdiction. As the future generation, children today are likely to experience worse consequences in the future.

Indigenous peoples

Indigenous people have a special relationship with the land that they live on. Many of them also depend on forests for their daily subsistence, and therefore are especially vulnerable to the adverse consequences of climate change. The petition filed by the Inuit people of the United States and Canada showed how this indigenous community struggled to cope with the adverse consequences of climate change. Similarly, the Athabaskan indigenous group filed an action against Canada for its failure to address pollution by black carbon. This case is currently pending before the Inter-American Commission of Human Rights. The UN Declaration on the Rights of Indigenous Peoples (UNDRIP) is relevant in this regard. It contains several principles that are relevant to the threats posed by climate change.

Climate-related displacement

Another group that is especially vulnerable are those that will be forced to move as a result of climate change-related consequences. As the very first IPCC report noted, the greatest single impact of climate change may be on human migration. It estimated that by 2050, 150 million people could be displaced by climate change-related consequences, such as desertification, water scarcity, and severe weather events. It is likely that such displacement will be primarily internal and will affect poorer regions of the world that are less able to address such displacement. The OHCHR report (2009) discusses four displacement scenarios: (a) weather-related disasters; (b) slow-onset events such as desertification or sea level rise; (c) relocation of people from high-risk zones; and (d) conflicts associated with climate change.

While this category of people is commonly referred to as "climate refugees," it is not a legal term, nor is there a legal framework to govern them. Several proposals have been made to govern different categories of climate refugees – the Peninsula Principles govern climate refugees who are internally displaced, while the Nansen Initiative seeks to cover those who are displaced across borders as a result of severe weather events. Despite the recognition that migration could be the biggest single impact of climate change as far back as 1990, it was only in 2015 at the Paris Conference that the international community established a taskforce to examine the issue. Currently, people who cross an international border due to climate-related consequence are unprotected unless they can satisfy the criteria for a traditional refugee under the Geneva Convention.

Small island states and their inhabitants

While no state in the world will be able to escape the adverse consequences of climate change, some states are more vulnerable to those consequences than others. Small island states are one such category. With sea level rise associated with climate change as well as increased severe weather events, these small islands face a bleak future to the point of total submergence. International law has never dealt with a situation where sovereign nations physically disappear.

Those who are living on small island states face the prospect of losing their state and with it, its protection, if they become submerged as a result of sea level rise associated with climate change. Some believe that these inhabitants will move long before the state disappears because increased severe weather events, frequent flooding, and salt-water intrusion will make these islands uninhabitable. Five islands belonging to the Solomon Islands have already disappeared and people living in the Carteret Islands in Papua New Guinea have already been relocated to the larger island of Bougainville. In Alaska, United States, the communities of Kivalina and Shishmaref – both indigenous – are awaiting relocation, as their current locations are becoming submerged. Efforts are underway to relocate the indigenous community in Isle de Jean Charles in Louisiana, United States, to the mainland. These are only the tip of the iceberg and the worst consequences are yet to materialize. It is not clear how the international community will cope with several million people who will be displaced when they cannot even cope with the current numbers.

Climate litigation

In Chapters 4 and 8, we discussed selected seminal cases involving climate change – these range from compelling governments to take mitigation measures, e.g., the *Urgenda* case in the Netherlands which survived an appeal by the government, to adaptation measures and compensation. Some domestic cases such as the *Juliana* case in the United States and the *Urgenda* case in the Netherlands have withstood challenges from all quarters. The Inuit petition was the first case to use the human rights framework in relation to climate change and is perhaps the most famous case as well. This case was discussed in detail in Chapter 4. As was pointed out there, the Commission dismissed the petition without prejudice stating that the information provided was not sufficient to make a determination (Gordon, 2007), but held a hearing to address matters relating to global warming and human rights (Earthjustice, 2007a). Although the Commission did not give a decision on the merits of the case, the petition broke new ground in international law. It also spurred action at the international level on treating climate change as a human rights issue. Barely a year after the Inter-American Commission dismissed the petition, the Human Rights Council adopted a resolution on the topic, and instructed the OHCHR to compile an analytical report on the relationship between climate change and human rights (OHCHR, 2009).

According to UNEP, "As of March 2017 climate change cases had been filed in 24 countries (25 if one counts the European Union), with 654 cases filed in the U.S. and over 230 cases filed in all other countries combined" (UNEP, 2017: 10), and outside the United States, Australia has seen the highest number of cases at 80 (UNEP, 2017: 11). The report identified several challenges to climate litigation: justiciability, standing, causal link, and separation of powers (UNEP, 2017: chapter 3). The laws that have been invoked range from common law nuisance to human rights law and international law. The UNEP report concludes:

> Litigation has emerged as an important feature of ongoing efforts to promote climate change mitigation and adaptation efforts. This owes in large part to the growing number of national laws that address climate change directly and so provide toeholds for litigants seeking to hold governments and private actors to account for obligations to mitigate or adapt. It also owes to the cohering role played by the Paris Agreement, which puts national laws and policies into a global context and thereby enables litigants to construe governments' commitments and actions as being adequate or inadequate. … In addition to proliferating, climate change litigation also seems to be growing in ambition and effectiveness: cases across the world provide examples of litigants holding governments to account for the actions or inactions that bear upon those litigants' rights amid changes to weather and coastlines.
>
> (2017: 40)

Duty-bearers and rights-holders

One of the challenges of applying a human rights framework to climate change is to identify the "perpetrator." Because climate change is a global issue and every member of the international community is both a contributor and a victim, albeit to different degrees, it is difficult to hold one state responsible for the adverse consequences that people experience. This issue arose in the Inuit petition. Can one state out of 197 states be held accountable for causing climate change? Under the current principles of state responsibility, we cannot. International law does not have collective theories of liability. On the other hand, we know who the main culprits are. For example, the United States is responsible for almost a quarter of all greenhouse gas emissions since the industrial revolution. We also know who the responsible corporations are. So why are we reluctant to hold the main emitters responsible? In a lawsuit filed against 21 major companies, the Governor of Rhode Island, the smallest US state, did just that:

> The state alleges the 21 companies created a public nuisance by inadequately warning customers and regulators of the risks in their products, as well as refuting scientific knowledge and advancing a public message that confused. The state also claims the defendants failed to prevent reasonably foreseeable harm from their products and caused sea level rise to

> encroach on the land. Finally [that] the 21 oil companies violated the state's Environmental Rights Act.
>
> (Maradiaga, 2018)

A similar lawsuit filed by Kivalina, a village in Alaska that is sinking due to climate change, was dismissed in 2012 by the United States Court of Appeals, Ninth Circuit. However, science relating to climate change is more advanced now and people across the globe are experiencing adverse impacts of climate change. Moreover, as we saw above, lawsuits relating to climate change are mushrooming all over the world, so we may need to adopt innovative legal principles to address a global challenge with multiple actors, multiple victims, multiple generations, and multiple sources.

Climate justice versus human rights

Given the disproportionate contribution to climate change by the Global North and the disproportionate impact on the Global South, especially least-developed countries and small island states, many have argued that the human rights framework is limited in scope, as it puts the burden on developing states to protect their populations from the adverse consequences of climate change when their contribution to creating the problem is minimal. It is, therefore, unfair to expect these countries to shoulder a greater burden with limited resources when they have other pressing needs. The Global South believes that the Global North should shoulder a greater burden because they benefited from the development that led to the emission of GHGs since the industrial revolution, whereas developing countries are sharing only the burdens. Bangladesh is among the most vulnerable countries to the consequences of climate change. While noting the relevance of human rights, Bangladesh pointed to the drawbacks of using the human rights framework in relation to climate change, given the disproportionate contribution to its creation by the Global North:

> [i]n dealing with the global problem of climate change, too much emphasis was put on national responsibility. The least developed countries and small island States would be the worst affected by climate change, although they had contributed least to global greenhouse gas emissions. It was not only unfair, but also unjustified, to hold these countries responsible fully for protecting their populations. While human rights based efforts to adaptation and mitigation were useful, such an approach should not only focus on the obligations of national authorities as duty bearers for protecting individuals' human rights.
>
> (Human Rights Council, 2009)

Knox too points to the limitations of the human rights framework in relation to climate change:

> In other words, the environmental human rights jurisprudence is based on the premise that a single polity experiences both the benefits of economic

> development and the environmental harm that it engenders and that it has the responsibility to decide where to strike the balance between the benefits and harms. But climate change is not subject to a single polity; it is inherently a transboundary problem on a global scale. Moreover, human rights law primarily imposes vertical duties, that is, duties owed by a state to those within its own jurisdiction. Extending those duties diagonally, to those outside the state's jurisdiction, may seem necessary in order to apply human rights law to the global aspects of climate change, but such an extension faces serious obstacles, both legal and practical.
>
> (2009)

Given these limitations and the inverse relationship between emissions and consequences, many believe that a justice framework is better suited to climate change. Due to the unequal distribution of benefits and burdens, some argue that a distinction should be made between "luxury emissions" and "subsistence (or survival) emissions" (Shue, 2014: 47). So, what would a justice framework look like in relation to climate change?

The Mary Robinson Foundation – Climate Justice believes that "climate justice links human rights and development to achieve a human-centered approach, safeguarding the rights of the most vulnerable and sharing the burdens and benefits of climate change and its resolution equitably and fairly" (Mary Robinson Foundation Climate Justice, n.d.). It identifies the following core principles that inform climate justice: (a) respect and protect human rights; (b) support the right to development; (c) share benefits and burdens equitably; (d) ensure that decisions on climate change are participatory, transparent, and accountable; (e) ensure gender equality and equity; (f) harness the transformative power of education for climate stewardship; and (g) use effective partnerships to secure climate justice. These principles share elements common with the environmental justice framework discussed in Chapter 1. Thus, participation of relevant stakeholders – including women as a vital component of society and other vulnerable groups – falls within participatory justice while sharing benefits and burdens equitably falls within distributive justice. The right to development also displays characteristics of distributive justice and overlaps with the intra-generational equity component of sustainable development (Atapattu, 2016a). The Mary Robinson Foundation further notes:

> The vast disparity in resources between rich and poor, evident in the gaps between countries in the North and South and also within many countries (both North and South), is the deepest injustice of our age. This failure of resource-fairness makes it impossible for billions of humans to lead decent lives. Climate change both highlights and exacerbates this gulf in equality.
>
> (Mary Robinson Foundation Climate Justice, n.d.)

Distributive, procedural, and remedial justice would be useful in addressing the disparities in the global community as well as disparities within states, particularly in the context of climate change, whereas social justice would be more applicable to communities within states and across generations. The present generation is feeling the impact of decisions taken by our forefathers, while our children and grandchildren will suffer the consequences of our actions today, making climate change a truly inter-generational justice issue (International Bar Association, 2014: 45). Moreover, the need for a justice-centered framework is further highlighted because of the unequal geographic distribution of climate change impacts (Atapattu, 2016a). In this context:

> the most equitable conception of climate change must recognize that while the developed nations have contributed the most to climate change over the past two centuries, it is the developing nations and their peoples who stand to suffer the most extreme consequences of rising sea levels, rising temperatures, and other human-induced environmental shifts.
>
> (International Bar Association, 2014)

Thus, while developed countries have the resources to adapt to climate change consequences, such as building sea walls and storing food against droughts, poor nations and communities do not have such luxuries due to poverty, lack of resources, and historic marginalization (International Bar Association, 2014). Some prefer to apply an environmental justice framework in relation to states, because states are necessarily harmed by damage suffered by individuals (Stillings, 2014: 643).

While the climate change legal regime sought to take the inverse relationship between contribution and damage into account when it adopted the CBDR, the general consensus is that the current legal regime does not go far enough. Thus, a justice-centered approach that introduces ethics into policy-making and which fosters a human rights perspective to climate change is necessary. Furthermore, a climate justice agenda must embrace the implications of climate change on development imbalances, including the fact that the distribution of climate change impacts is inherently unjust (Atapattu, 2016a). The Paris Agreement's reference to climate justice, albeit in a rather half-hearted way, will be the first step towards incorporating climate justice in relation to climate action.

Conclusion

There is a clear link between climate change and human rights, and many of the protected rights would be jeopardized as a result of adverse consequences of climate change:

> It is now "beyond debate" that the adverse effects of climate change will, in their severity, threaten a range of human rights, including the rights to life, health, food, and housing. There is also a dawning recognition that measures taken to mitigate and adapt to climate change have the potential to impinge on human rights.
>
> (Rajamani, 2018: 236)

The obligations that human rights law imposes on states to take measures relating to climate change should be evaluated in relation to four scenarios: (a) mitigation; (b) adaptation; (c) severe weather events; and (d) damage caused by historic emissions. While it is hard to pinpoint liability based on historic emissions, mainly due to archaic legal rules, states do have to take their human rights obligations into account when taking measures in pursuance of their mitigation commitments. This was affirmed by the Paris Agreement. Moreover, irrespective of the contribution to the problem, states have to take reasonable measures to protect their people from these consequences and adopt adaptation measures. Similarly, with regard to severe weather events, states must take measures to protect the people before, during, and after the event – giving them prior warnings and evacuating them if necessary, and providing them with adequate shelter until it is safe for them to return home. It is clear that states have to fulfill their human rights obligations in relation to climate change. International obligations do not operate in a vacuum. We are used to compartmentalizing our obligations into human rights obligations, environmental obligations, trade obligations, etc. (the silo effect), without identifying their interaction or application across disciplines, but climate change is forcing us to get rid of the silo mentality.

DISCUSSION QUESTIONS

1. What are the pros and cons of using a human rights approach to climate change?
2. Should the human rights framework be amended to address a global environmental problem like climate change? If so, how?
3. What remedies should be available to victims of climate change?
4. How should liability principles under international law be revised to address a global phenomenon as complex as climate change which has multiple actors, multiple victims, and multiple sources?
5. How useful is climate litigation?

References

Adger, N., Paavola, J. and Huq, S. (2006). Toward Justice in Adaptation to Climate Change. In: Adger, N., Paavola, J., Huq, A., and Mace, M.J. (eds.), *Fairness in Adaptation to Climate Change*. Cambridge, MA: The MIT Press.

Atapattu, S. (2009). Climate Change, Differentiated Responsibilities and State Responsibility: Devising Novel Legal Strategies for Damage Caused by Climate Change. In: Richardson, B., Le Bouthillier, Y., McLeod-Kilmurray, H., and Wood, S. (eds.), *Climate Law and Developing Countries: Legal and Policy Challenges for the World Economy*. Cheltenham, UK: Edward Elgar, pp.37–62.

Atapattu, S. (2016a). *Human Rights Approaches to Climate Change: Challenges and Opportunities*. New York, NY: Routledge.

Atapattu, S. (2016b). Justice for Small Island Nations: Intersections of Equity, Human Rights, and Environmental Justice. In: Abate, R. (ed.), *Climate Justice: Case Studies in Global and Regional Governance Challenges*. Washington, DC: Environmental Law Institute, p.299.

Atapattu, S. (2018). The Right to a Healthy Environment and Climate Change: Mismatch or Harmony? In: Knox, J. and Pejan, R. (eds.), *The Human Right to a Healthy Environment*. Cambridge: Cambridge University Press, p.252.

Committee on Economic, Social and Cultural Rights (2003). General Comment No 15 on the right to water [Online]. Available at: www2.ohchr.org/english/issues/water/docs/CESCR_GC_15.pdf. [Accessed: 1 October 2018].

Craig, R.K. (2010). Stationarity is Dead – Long Live Transformation: Five Principles for Climate Change Adaptation Law. *Harvard Environmental Law Review* 34: 9.

Cullet, P. (2010). The Kyoto Protocol and Vulnerability: Human Rights and Equity Dimensions. In Humphrey, S., *Human Rights and Climate Change*. Cambridge: Cambridge University Press, p.183.

Davenport, C. (2018). "Major Climate Report Describes a Strong Risk of Crisis as Early as 2040", *New York Times*, 7 October 2018 [Online]. Available at: www.nytimes.com/2018/10/07/climate/ipcc-climate-report-2040.html [Accessed: 1 October 2018].

De Schutter, O. (2009). Joint Statement of the Special Procedure Mandate Holders of the Human Rights Council on the UN Climate Change Conference (Copenhagen, 7–18 December 2009) [Online]. Available at: www.srfood.org/images/stories/pdf/otherdocuments/20091216_statement-climate-change_en.pdf [Accessed 1 September 2018].

Displacement Solutions (2014). The Peninsula Principles on Climate Displacement within States (2013) [Online]. Available at: http://displacementsolutions.org/wp-content/uploads/2014/12/Peninsula-Principles.pdf [Accessed: 20 October, 2018].

Earthjustice (2007a). Inter-American Commission on Human Rights to Hold Hearing on Global Warming [Online]. Available at: http://earthjustice.org/news/press/2007/inter-american- commission-on-human-rights-to-hold-hearing-on-global-warming [Accessed: 1 October 2017].

Earthjustice (2007b). Testimony of Martin Wagner before the Inter-American Commission on Human Rights [Online]. Available at: http://earthjustice.org/news/press/2007/global-warming-human-rights-gets-hearing-on-the-world-stage [Accessed: 1 October 2017].

Earthjustice (2013). Petition to the Inter-American Commission on Human Rights Seeking Relief from the Violations of Rights of Arctic Athabaskan Peoples Resulting from Rapid Arctic Warming and Melting Caused by Emissions of Black Carbon from Canada (2013) [Online]. Available at: http://earthjustice.org/sites/default/files/AAC_PETITION_13-04-23a.pdf [Accessed: 20 October 2018].

Freestone, D. (2012). The International Legal Framework for Adaptation. In M.B. Gerrard and K.F. Kuh, *The Law of Adaptation to Climate Change: US and International Aspects*, Cleveland, OH: American Bar Association, p.610.

Gonzalez, C. (2011). Climate Change, Food Security and Agrobiodversity: Towards a Just, Resilient, and Sustainable Food System. *Fordham Environmental Law Review* 22: 493.

Gordon, J. (2007). Inter-American Commission to Hold Hearing after Rejecting Inuit Climate Change Petition. *Sustainable Development Law & Policy* 7: 2 [Online]. Available at: http://digitalcommons.wcl.american.edu/cgi/viewcontent.cgi?article=1239&context=sdlp [Accessed: 10 January 2015].

Heinrich Böll Stiftung (2009). Climate Change and Right to Food [Online]. Available at: www.srfood.org/images/stories/pdf/otherdocuments/20091216_report-climate- change-summary_en.pdf [Accessed: 1 September 2017].

Human Rights Committee (1982), General Comment 6, The Right to Life, UN Doc HRI/GEN/1/Rev.7, 30 April 1982 at 128 [Online]. Available at: http://hrlibrary.umn.edu/peace/docs/hrcom6.htm [Accessed: 1 October 2017].

Human Rights Council (2008a). Resolution 7/23 on climate change and human rights, A/HRC/RES/7/23, United Nations [Online]. Available at: http://ap.ohchr.org/documents/E/HRC/resolutions/A_HRC_RES_7_23.pdf [Accessed: 1 October 2017].

Human Rights Council (2008b). Report of the Special Rapporteur on the Right to Food, Jean Ziegler, A/HRC/7/5, 10 January 2008 [Online]. Available at: https://documents-dds-ny.un.org/doc/UNDOC/GEN/G08/100/98/PDF/G0810098.pdf?OpenElement [Accessed: 1 October 2018].

Human Rights Council (2009). Report of the Office of the United Nations High Commissioner for Human Rights on the Relationship Between Climate Change and Human Rights. A/HRC/10/61 [Online]. Available at: https://documents-dds-ny.un.org/doc/UNDOC/GEN/G09/103/44/PDF/G0910344.pdf?OpenElement [Accessed: 1 July 2018].

Human Rights Council (2009). Summary of the Human Rights Council's interactive panel on human rights and climate change, June 2009 [Online]. Available at: www.ohchr.org/EN/Issues/HRAndClimateChange/Pages/Panel.aspx [Accessed: 1 October 2017].

Hunter, D., Salzman, J., Salzman, J., and Zaelke, D. (2015). *International Environmental Law and Policy*, 5th edn. St. Paul, MN: Foundation Press.

International Bar Association (2014). Achieving Justice and Human Rights in an Era of Climate Disruption, Climate Change Justice and Human Rights Task Force Report (IBA report), p.35 [Online]. Available at: www.ibanet.org/Article/Detail.aspx?ArticleUid=96b93592-3761-4418-8a52- 54a81b02c5f1 [Accessed: 12 November 2014]

Inuit Circumpolar Council (2005). Petition to the Inter-American Commission on Human Rights Seeking Relief for Violations Resulting from Global Warming Caused by Acts and Omissions of the US [Online]. Available at: www.inuitcircumpolar.com/inuit-petition-inter-american-commission-on-human-rights-to-oppose-climate-change-caused-by-the-united-states-of-america.html [Accessed: 1 September 2018].

IPCC (2007). Working Group II: Fourth Assessment Report. Geneva: IPCC Secretariat, p.747.

IPCC (2014a). Climate Change Synthesis Report, Summary for Policymakers [Online]. Available at: www.ipcc.ch/pdf/assessment-report/ar5/syr/AR5_SYR_FINAL_SPM.pdf [Accessed: 1 October 2017].

IPCC (2014b). Climate Change 2014: Impacts, Adaptation, and Vulnerability, Summary for Policymakers [Online]. Available at: https://ipcc-wg2.gov/AR5/images/uploads/WG2AR5_SPM_FINAL.pdf [Accessed: 1 October 2017].

IPCC (2018). Global Warming of 1.5°C (October 2018). [Online]. Available at: http://report.ipcc.ch/sr15/pdf/sr15_spm_final.pdf [Accessed: 20 October 2018].

Knox, J. (2009). Climate Change and Human Rights Law. *Virginia Journal of International Law* 50(1): 163.

Knox, J. (2013) Mapping Human Rights Obligations Relating to the Enjoyment of a Safe, Clean, Healthy and Sustainable Environment. Individual Report on the International Covenant on Civil and Political Rights, Report No 2, prepared for the Independent

Expert on the Issue of Human Rights Obligations Relating to the Enjoyment of a Safe, Clean, Healthy and Sustainable Environment. Geneva: Office of the High Commissioner for Human Rights.

Mary Robinson Foundation Climate Justice (n.d.). Principles of Climate Justice [Online]. Available at: www.mrfcj.org/pdf/Principles-of-Climate-Justice.pdf [Accessed: 1 October 2017].

Male Declaration on the Human Dimension of Global Climate Change (2007) [Online]. Available at: www.ciel.org/Publications/Male_Declaration_Nov07.pdf [Accessed: 1 July 2018].

Maradiaga, F. (2018). Rhode Island files lawsuit against 21 oil companies for contributing to Climate Change [Online]. Available at: http://turnto10.com/news/local/rhode-island-files-lawsuit-against-21-oil-companies-for-contributing-to-climate-change [Accessed: 1 October 2017].

Naidoo, K. (2012). Seminar to Address the Adverse Impacts of Climate Change on the Full Enjoyment of Human Rights. *Greenpeace International* [Online]. Available at: www.ohchr.org/Documents/Issues/ClimateChange/SpeechKumiNaidoo.pdf [Accessed: 1 October 2017].

OHCHR (2007). Report of the Special Rapporteur on the Right of Everyone to the Highest Attainable Standard of Physical and Mental Health, A/62/214, United Nations [Online]. Available at: http://daccess-dds-ny.un.org/doc/UNDOC/GEN/N07/453/79/PDF/N0745379.pdf?OpenElement [Accessed: 1 October 2017].

OHCHR (2008). Report of the Special Rapporteur on Right to Food A/HRC/7/5, 10 January 2008 [Online]. Available at: http://daccess-dds- ny.un.org/doc/UNDOC/GEN/G08/100/98/PDF/G0810098.pdf?OpenElement [Accessed: 1 October 2017].

OHCHR (2009). Report of the Special Rapporteur on Right to Housing as a Component of the Right to an Adequate Standard of Living and on the Right to Non-Discrimination in this Context, A/64/255, 6 August 2009 [Online]. Available at: http://daccess-dds-ny.un.org/doc/UNDOC/GEN/N09/446/64/PDF/N0944664.pdf?OpenElement [Accessed: 1 October 2017].

OHCHR (2012). Statement of H.E. Dr. Dipu Moni, MP at the seminar on Climate Change and Human Rights [Online]. Available at: www.ohchr.org/Documents/Issues/ClimateChange/DipuMin.pdf [Accessed: 1 October 2017].

OHCHR (2015a). Joint Statement by UN Special Procedures on the Occasion of World Environment Day: Climate Change and Human Rights. [Online]. Available at: www.ohchr.org/EN/NewsEvents/Pages/DisplayNews.aspx?NewsID=16049&LangID=E [Accessed: 1 October 2018]

OHCHR (2015b). Special Rapporteur on the Right to Food, Interim Report. A/70/287. [Online]. Available at: www.ohchr.org/Documents/Issues/Food/A-70-287.pdf [Accessed: 5 February 2019]

Rajamani, L. (2018). Human Rights in the Climate Change Regime: From Rio to Paris and Beyond. In J. Knox and R. Pejan (eds.), *The Human Right to a Healthy Environment*. Cambridge: Cambridge University Press, p.236.

Rio Declaration on Environment and Development (1992). A/CONF.151/26 (Vol. I). [Online]. Available at: www.un.org/documents/ga/conf151/aconf15126-1annex1.htm [Accessed: 10 October 2018].

Shue, H. (2014). *Climate Justice: Vulnerability and Protection*. Oxford: Oxford University Press.

Stillings, Z.L. (2014). Human Rights and the New Reality of Climate Change: Adaptation's Limitations in Achieving Climate Justice. *Michigan Journal of International Law* 35: 637.

The Nansen Initiative (2015). Towards a Protection Agenda for People Displaced Across Borders in the Context of Disasters and the Effects of Climate Change [Online] Available at: www.nanseninitiative.org/global-consultations/ [Accessed: 20 October 2018].

UN (2015). Sendai Framework for Disaster Risk Reduction 2015-2030 [Online]. Available at: www.unisdr.org/files/43291_sendaiframeworkfordrren.pdf [Accessed: 1 July 2018].

United Nations Development Program (2007). Human Development Report 2007/2008, Fighting Climate Change: Human Solidarity in a Divided World [Online]. Available at: http://hdr.undp.org/sites/default/files/reports/268/hdr_20072008_en_complete.pdf [Accessed: 1 September 2018].

United Nations Framework Convention on Climate Change (UNFCCC) (1992). FCCC/INFORMAL/84 [Online]. Available at: https://unfccc.int/resource/docs/convkp/conveng.pdf [Accessed: 1 July 2018].

United Nations Framework Convention on Climate Change (UNFCCC) (2010).Report of the Conference of the Parties on its sixteenth session, Cancun, 29 November–10 December 2010 [Online]. Available at: http://unfccc.int/resource/docs/2010/cop16/eng/07a01.pdf#page=4 [Accessed: 1 October 2017].

UN Environment Programme (UNEP) (2017). The Status of Climate Change Litigation: A Global View (May 2017) [Online]. Available at: http://wedocs.unep.org/bitstream/handle/20.500.11822/20767/climate-change- litigation.pdf?isAllowed=y&sequence=1 [Accessed: 1 October 2018].

Wold C., Hunter, D., and Powers, M. (2013). *Climate Change and the Law*, 2nd ed. Newark, NJ: LexisNexis.

Cases

Juliana et al. v. USA et al. (2016). United States District Court, For the District of Oregon, 6:15-cv-1517-TC, Judge Aiken, Opinion and Order [Online]. Available at: https://static1.squarespace.com/static/571d109b04426270152febe0/t/5824e85e6a49638292ddd1c9/1478813795912/Order+MTD.Aiken.pdf [Accessed: 25 September 2018].

Netherlands v. Urgenda Foundation (Urgenda appeal) (2018). [Online]. Available at: https://uitspraken.rechtspraak.nl/inziendocument?id=ECLI:NL:GHDHA:2018:2610 [Accessed: 25 October 2018].

Urgenda Foundation v. The State of The Netherlands (2015). [Online]. Available at: https://elaw.org/system/files/urgenda_0.pdf [Accessed: 25 September 2018].

Village of Kivalina v. ExxonMobil Corp. (2012). 696 F.3d 849 (9th Cir. 2012).

10 Social movements and civil society

Introduction

In this chapter, we will learn about how local experiences with climate change and climate policies are often taken up by transnationally operating civil society networks to campaign for change at the international level. Many of the institutional achievements that we have today, for example, the inclusion of human rights in the Preamble to the Paris Agreement, have been advocated for by civil society groups for many years. Information about rights infringements at the grassroots level, in relation to the consequences of climate change or climate policy implementation, have been crucial in this advocacy work, because it can be strategically used by these networks. In the following sections, we will learn more about the emergence of the environmental justice movement and the concept of transnational advocacy networks in order to better understand the climate justice movement, including its moderate wing and the more radical wing. We will also learn about environmental human rights defenders (EHRD) and how their rights can be better protected.

The environmental justice movement: key concepts and issues

The environmental justice movement places an emphasis on eliminating discriminatory environmental policies and practices. The movement emerged in the 1960s in the United States, primarily organized by people of color who expressed their concerns about unequal environmental protection in their communities. There were substantial overlaps with the demands of the civil rights movement, which had already raised awareness of public health risks due to inequities in environmental protection. Dr. Martin Luther King, Jr. investigated an incident of environmental and economic injustice in Memphis, Tennessee, where he supported the rights of garbage workers and advocated for better working conditions for them.

In 1982, African-Americans organized non-violent resistance against a polychlorinated biphenyl (PCB) landfill in Warren County, North Carolina. Although the sit-in protests were unsuccessful and more than 500 environmental activists were arrested, the events around Warren County are often understood as the catalyst of the environmental justice movement. It initiated a study of the US General Accounting Office called *Siting of Hazardous Waste Landfills and Their Correlation with Racial and Economic Status of Surrounding Communities*. The study report concluded that three out of four hazardous waste landfills in one particular region (comprising eight states of the South of the United States) were located in predominately African-American communities with a family income often below the poverty level. Thus, the report provided empirical evidence for the movement's claims regarding environmental racism. More empirical support for these claims was delivered by another study of the United Church of Christ Commission on Racial Justice (UCC) investigating the relationship between the location of hazardous waste sites and socio-economic and racial characteristics. The outcome of this study revealed that the most decisive factor in determining the location of these facilities was race – wielding even more influence than other variables, such as poverty, home ownership, or land values (Bullard, 2001: 151).

Calls for environmental justice, both distributive and procedural, are central demands of civil society actors engaged in the environmental justice movement. Civil society is often referred to as the "third sector" of society, distinct from governments and businesses. When we refer to civil society in the context of human rights and the environment, we mean not only non-governmental organizations, but also trade unions, foundations, academics, and, increasingly, religious actors with a keen interest in protecting environmental rights.

In the environmental justice movement, all of the above-mentioned actors come together for a common purpose: to build a social movement with a view to highlighting justice concerns in relation to environmental matters. A social movement is an action system of mobilized networks comprised of groups and organizations that – for a certain period of time and based on a collective identity – aims at initiating, preventing, or reversing social change with various means, sometimes involving protests and violence (Rucht, 1994: 76–7). Very often, actors in a social movement are engaged in conflicts over resources. This means they fight other social actors, often governments, over access to or over exerting control of resources (Diani, 2003). Coming together in a social movement means joining forces and combining strengths to be able to more effectively influence decision-making of states and international organizations to follow collective aims and behavior.

The environmental justice movement became active in the 1980s, criticizing unequal distribution of environmental risks and hazards among poorer and disadvantaged population groups or minorities. Their collective objective is a healthy, clean, and safe environment for everyone. They demand equal distribution of environmental harm and benefits and equal environmental protection from the government regardless of social background, ethnicity, or gender. Other important ideas of the movement are that people affected by environmental issues should be able to meaningfully participate in the decision-making process. Moreover, environmental justice groups respect local communities' way of life, acknowledge cultural differences, and strongly consider local knowledge (Schlosberg, 2007).

Two concepts are at the heart of the activities of the movement: environmental racism and environmental discrimination. Of these, the broader concept is environmental discrimination. It comprises unequal distribution of environmental hazards among certain socio-economic groups or classes, ethnic groups and minorities, and gender. Environmental racism refers to disproportionate exposure to toxic waste, pollution or other environmental hazards or limited access to clean air, water, and other natural resources on a racialized basis. In 1987, for instance, the study *Toxic Waste and Race in the United States* published by the Commission for Racial Justice emphasized that race is the most decisive factor in decision-making on waste facility locations (Godsil, 1991). Warren County, the case described in the introduction to this chapter, where the US State of North Carolina deposited polychlorinated biphenyls (PCB)-contaminated waste in an area where mostly African-Americans live, and environmental groups protested on the basis of justice, became known as the first environmental justice case in the United States (Cutter, 1995). Another case is that of Altgeld Gardens in South Chicago. Altgeld Gardens is a housing project established after the Second World War for African-American veterans returning from the war. With 53 toxic facilities and 90% of the city's landfills surrounding the area, Altgeld Gardens is known as the "toxic doughnut" in the community. In this area, 90% of residents are African-American and many of them live in poverty. Public health studies have revealed that the toxins and pollutants have caused excessive cases of blood, lung, and prostate cancer. Many children were born with brain tumours, foetuses had to be aborted due to brain aberrations, and many other diseases, such as asthma, were very common in the community. Despite this, Altgeld Garden's residents have not been relocated (EJA, 2015).

We find instances of environmental discrimination or racism not only *within societies*, but also *between different countries*. Waste materials or dangerous chemicals are frequently exported from developed countries with higher environmental standards to countries with lower standards that are often in the developing world. This means local population groups in these countries are increasingly exposed to the environmental, economic, social, and health effects of waste, toxins, and other hazards produced in the developed world. Environmental discrimination, however, can also occur between private corporations and local communities. Conflicts involving oil companies, like Shell and Chevron, and the Ogoni people in the Niger Delta in Nigeria are an important example. In the late 1950s, the first Nigerian oil reserves that could be used for commercial purposes were found in Ogoniland, which is home to half a million Ogoni people, an ethnic minority group. Since then, the Nigerian government has forced the Ogoni people to leave their land without consultation, without free, prior, and informed consent, or with inadequate compensation. After a constitutional amendment was introduced in 1979, the Nigerian government took over full ownership and rights pertaining to territory in Nigeria, leaving the Ogoni people with limited entitlements to their traditional lands. Since the 1990s, the conflict between Shell and the native Ogoni has intensified. Oil drilling has led to severe environmental contamination and there is too little economic benefit for the local population. After the

Movement for the Survival of the Ogoni people (MOSOP) initiated a number of protests against Shell and the Nigerian government, one of the leading protestors, Ken Saro-Wiwa, was executed by the government authorities.

Environmental advocacy networks

How affected the local Ogoni people are by the international activities of a multinational corporation is a good example of the world's interconnectedness. As a result, complex problems emerge, and environmental activists often feel that they need to raise public awareness of these issues. Therefore, they link with affected actors at the grassroots level to gather information, share material and resources with other civil society experts working on similar problems, collaborate with international organizations, and advocate for their inclusion in international forums. These network activities have often been the key to the success of the environmental justice movement.

Transnational advocacy networks (TANs) can be defined as forms of network organizations or communicative structures based on common principles, ideas, and norms. In their groundbreaking book *Activists Beyond Borders* (1998), Keck and Sikkink define these networks as:

> nonstate actors that interact with each other, with states, and with international organizations. These interactions are structured in terms of networks, and transnational networks are increasingly visible in international politics. … Others are networks of activists, distinguishable largely by the centrality of principled ideas or values in motivating their formation.
>
> (Keck and Sikkink, 1998: 1)

Network activists use these principled ideas and values to specify criteria for evaluating whether certain actions or policies are just or unjust. In their original work, Keck and Sikkink highlight that environmental advocacy networks differ from human rights advocacy networks because they rather refer to common goods, such as clean air or water, instead of grounding their actions on "rights" (Keck and Sikkink, 1998: 121). However, this has significantly changed since then and the Inuit petition brought before the Inter-American Commission on Human Rights in 2005 is a good example of the interaction between the two groups. In the petition, affected indigenous peoples and environmental advocates framed climate change consequences as a human rights concern. This was critical to introducing rights language in environmental and climate change advocacy.

There are several forms of transnational advocacy networks and they can be composed of different actors engaged at various scales. TANs can comprise international and national NGOs, local social movements, foundations, the media, churches, trade unions, academics, and also regional and international governmental organizations or single governmental representatives (Keck and Sikkink, 1998: 9). If these actors come together and build a network, they create new linkages, multiply access channels to the international system, make material and informational resources available for new actors, and attempt to transform practices of national sovereignty. Important commonalities across transnational advocacy networks are the central role of principles and norms, the idea that individuals can initiate social change, the creative use of information, and the use of sophisticated political strategies within campaigns. The common goal of TANs is to change the policies of states and international organizations.

Keck and Sikkink (1998) have developed a typology of tactics to understand how TANs attempt to achieve this goal. They highlight: (1) *information politics*: understood as strategically using information; (2) *symbolic politics*: to draw on symbols and stories to highlight a situation to a target audience that might be geographically distant; (3) *leverage politics*: network actors being able to gain moral or material leverage over state actors and international organizations (IOs); and (4) *accountability politics*: referring to formerly adopted norms and policies of governmental actors and obligations to comply with them (Keck and Sikkink, 1998: 16). With these tactics, they try to influence the following phases in the decision-making process: (a) problem-definition and agenda-setting; (b) the discursive positions of states and international organizations; (c) institutional procedures; (d) policy recommendations; and (e) the resulting policies and state action (Keck and Sikkink, 1998: 25). Whether TANs can successfully influence the policies of states and international organizations also depends on opportunity structures, i.e., access to (state) institutions and the broader institutional context that can provide opportunities or obstacles (Tarrow, 1994), and mobilization structures, i.e., successful network-building (Joachim, 2003). Mobilization can be short-term and last only a few days, for example, around a political event, or it can have a higher degree of institutionalization and last for years, or even decades. More institutionalized networks may also have common bodies, councils, procedures, and representatives.

In the past, network-building has been key to the success of the environmental justice movement. Networks are not simply non-governmental entities, but are becoming increasingly transnational in nature. This means they bring together non-governmental actors, but can also include governmental actors from various countries and backgrounds. Transnational networks are often marked by complex and decentralized organizational structures; they can comprise local organizations and nation state representatives, but also actors from international organizations or international NGOs. They bring together groups and single actors that are ideologically or normatively motivated and pursue common objectives. Increasingly, networks of the environmental justice movement base their action on human rights norms. Thus, protecting and promoting human rights standards for people affected by environmental challenges can be a strong common objective to pursue. By creating influential dynamics through political and media pressure, environmental networks can contribute to social change and influence the decision-making processes of state actors (della Porta and Diani, 2006). Very often, they refer to creative actions to attract media attention and to raise public awareness for initiating the change they deem necessary.

The climate justice movement

Since the end of the 1990s, the focus of the climate movement has shifted from utilizing a *climate change frame* to employing a *climate justice frame* (della Porta and Parks, 2013: 45). Although climate justice is a fluid framework that is diversely utilized by these networks, it broadly reflects the observation that those people who have contributed the least to climate change and have the fewest resources to adapt are the ones who are affected by it the most (Görg and Bedall, 2013: 88–9). This re-framing from climate change to emphasizing climate injustice comes with at least two advantages for the climate movement. First, it helps to integrate local concerns and more radical groups into the network. Second, it allows for processes of frame-bridging with other movements (della Porta and Parks, 2013: 45), such as those working on human rights, indigenous peoples' rights, and other social justice issues.

The climate justice movement, however, does not function homogeneously, nor does it speak with one voice. Instead, it is characterized by a dominant antagonism: it is divided into a more *moderate wing* accepting capitalism and lobbying of change within the established climate institutions of the UNFCCC, and a more *radical wing*, viewing capitalism and the capitalist way of life as a root cause of climate change that needs to be changed (della Porta and Parks, 2013: 47). Thus, we need to be aware that the climate justice movement comprises more moderate groups, i.e., the reformers, and more radical groups, i.e., the revolutionaries. This leads to more cooperative and more conflictive activities of environmental advocacy networks with state representatives inside and outside of the UNFCCC process (Brunnengräber, 2014).

The moderate wing: claims for justice at the international climate negotiations

At the international level, civil society organizations participate in international climate conferences by advocating, campaigning, and providing expertise (Bernauer and Betzold, 2012: 63). Those organizations with strong ties to and good relations with state delegations have the most advanced access opportunities. They engage with governmental institutions and become part of the official UNFCCC process by acting as accredited organizations (Reitan, 2011: 66). The UNFCCC is particularly open to including non-state observers. These are not only NGOs, but also other third sector groups, such as religious actors, trade unions, and academics. Internationally operating organizations are usually very well organized and act professionally, taking up consultative functions and providing expertise. In some cases, they even obtain the opportunity to speak during the negotiations, especially if they are part of the so-called constituencies. The constituencies at the UNFCCC negotiations are clustered groups of officially registered CSOs sharing certain interests and acting as observers in the process. As of 2014, the UNFCCC Secretariat reported more than 1,600 admitted CSOs that are organized in nine constituencies, among them environmental CSOs (ENGO), business and industry CSOs (BINGO), indigenous people's organizations (IPOs), youth CSOs (YOUNGO), women and gender (WGC), and trade union CSOs (TUNGO), as well as research and independent CSOs (RINGO). Participation in a constituency comes with several advantages: it allows observers to make interventions at certain points in the course of the state negotiation process, it facilitates the use of focal points for better coordination with the UNFCCC Secretariat, and it enhances flexible information-sharing (UNFCCC, 2014).

For achieving concrete social change, it is indispensable for CSOs to interact directly with the established governmental institutions and state representatives. In some cases, civil society actors even become members of national delegations and are "formally granted a 'seat at the table'" (Böhmelt, Koubi, and Bernauer, 2014: 63). This increases their opportunities to influence governmental decisions since it provides them with access to closed sessions, official state documents, and the possibility of presenting their own proposals (Böhmelt, Koubi, and Bernauer, 2014: 19).

Since the Rio Conference on Environment and Development in 1992, civil society participation in decision-making processes within the UN system has been strengthened. Civil society participants are understood to bring expertise and credibility to international negotiations. In open and transparent negotiation processes, states increasingly take up non-state demands (Steffek and Nanz, 2008). CSO participation is considered to improve the democratic development of international organizations (IOs) (Scholte, 2012; Bexell, Tallberg, and Uhlin, 2010). Thus, governmental delegations are also interested in including CSOs, mainly because they might receive additional information and expertise (Betsill and Corell, 2008), and because civil society participation can enhance the legitimacy of states' decisions (Bernauer and Betzold, 2012: 63). CSOs, by contrast, use their close interaction with governments to exert pressure and/or lobby for negotiating, ratifying, enforcing, and complying with international environmental agreements (Bernauer, Böhmelt, and Koubi, 2013). During the past Conferences of the Parties, civil society used the institutional playing field for shaping implementation details (Brunnengräber, 2014). CARE International, the Center for International Environmental Law (CIEL), the Women's Environment and Development Organization (WEDO), and others, for instance, presented recommendations for the negotiations of the Paris Rule Book at a side event of COP 23 in Bonn at the end of 2017 (CIEL et al., 2017). They explicitly proposed suggestions on how to protect human rights while implementing the Paris Agreement. These aspects are mainly relevant for policy-making but when it comes to realizing climate policies, CSOs can also take over consulting, monitoring and control functions, and they can develop certificates for offset-mechanisms (Hein and Garrelts, 2013: 441).

Case study: Human rights and climate change working group

Network-building and first successes

One important example for civil society activism at the climate conferences is the transnational advocacy network *Human Rights and Climate Change Working Group* (HRCCWG). The HRCCWG has been particularly active and successful in their attempt to promote human rights within the climate regime. The idea to launch this group was born in 2008 at COP 14 in Poznan, Poland, and it became operative during the following COP in 2009 in Copenhagen, Denmark. The HRCCWG can be described as a hybrid link of predominantly civil society and some state actors operating at various scales – from the local to the global – with the common objective of promoting and institutionalizing human rights in the climate regime. Among the networks' members are prominent international CSOs, such as the Center for International Environmental Law (CIEL), Earthjustice, Friends of the Earth, Carbon Market Watch, Human Rights Watch, and Amnesty International. They also include local CSOs from various developing countries, gender advocates, indigenous peoples' representatives, academics, representatives from IOs, like the Office of the High Commissioner for Human Rights (OHCHR) and UNICEF, as well as single actors from state delegations. Membership in the network is rather informal; participants can be present at one negotiation meeting joining the group's activities there, and then miss out on the next one. Simultaneously, they can be part of another transnational advocacy networks active at the international climate conferences, including Climate Action Now (CAN), the REDD+ Safeguards Working Group, or the Indigenous Caucus.

At COP 16 in Cancun (Mexico) and at COP 21 in Paris (France), the (so far) most significant successes in promoting and institutionalizing human rights in the climate regime could be achieved. The Human Rights and Climate Change Working Group, among other TANs, was strongly involved in both. We want to look at these successes first before we learn more about the strategies the network is employing to influence state decisions.

First, in the preambulatory clauses of the Cancun Agreement, the UNFCCC member states referred to a Human Rights Council Resolution (HRC 10/4) on human rights and climate change stating that:

> the adverse effects of climate change have a range of direct and indirect implications for the effective enjoyment of human rights and that the effects of climate change will be felt most acutely by those segments of the population that are already vulnerable owing to geography, gender, age, indigenous or minority status, or disability.
>
> (UNFCCC/CP, 2010: 2)

Second, in the operative part of the Cancun Agreement, member states announced that "parties should, in all climate-change related actions, fully respect human rights" (UNFCCC/CP, 2010: 4). These actions, of course, refer to all climate policy initiatives introduced under the legal framework of the UNFCCC and the Kyoto Protocol.

Third, the first procedural rights, also known as safeguards, were institutionalized for the implementation of REDD+ programs in Annex One of the Cancun Agreement. This means that the rights to participation, information, transparency, and free, prior, and informed consent need to be respected for local communities affected by the realization of REDD+ programs. The rights, knowledge and (land) ownership of indigenous peoples are particularly emphasized here (UNFCCC, 2010: Appendix I, paragraph 2).

Fourth, a review of the modalities and procedures of the Clean Development Mechanism (CDM) is underway. During COP 19 in 2013 in Warsaw and at the following Intersessionals in 2014 and 2015 in Bonn, states (and CSOs) discussed stronger stakeholder consultation requirements and several references stating that activities under the CDM have to be carried out in accordance with human rights (Filzmoser, 2013: 2). It can be expected that civil society pressure will be upheld and procedural rights will enter CDM policies in the coming years (Schade and Obergassel, 2014).

Finally, human rights have again been anchored in the preambulatory clauses of the Paris Agreement in 2015. It stipulates that state parties should:

> when taking action to address climate change, respect, promote and consider their respective obligations on human rights, the right to health, the rights of indigenous peoples, local communities, migrants, children, persons with disabilities and people in vulnerable situations and the right to development, as well as gender equality, empowerment of women and intergenerational equity.
> (UNFCCC, 2015: Preamble)

Members of the HRCCWG consider these developments as progressively evolving successes in their effort to institutionalize human rights in the climate regime.

In the following section, we want to take a closer look at what strategies were used by the Human Rights and Climate Change Working Group as a transnational advocacy network at the international climate negotiations to promote human rights institutionalization in climate agreements.

Strategies employed by networks at climate conferences

The most successful strategy the HRCCWG has employed for achieving its objectives was to build friendly relations with the decision-makers at the climate negotiations, i.e., with the state representatives. Making use of these receptive relations, the network receives access to negotiating texts and attempts to include human rights language in respective drafts, asking state parties to introduce these drafts in closed negotiation sessions.

Relations with state parties are particularly well-established if the negotiators are open to human rights arguments due to their own liberal democratic state identity, increased pressure by domestic CSOs, or if they rely on the network's expertise and capacities as a number of developing countries do. If the latter is the case, network members provide their expert knowledge and attend sessions on behalf of small delegations that cannot afford to travel to the negotiations with a large number of staff. This means civil society actors become officially registered as members of state delegations and draft texts for them that can be introduced in meetings officially closed to non-state observers.

To justify the use of human rights language in climate agreements, the HRCCWG frequently works with case studies emphasizing the adverse effects of climate change and climate policies on local people from Asia, Latin America, and Africa. At the climate negotiations, case studies are usually presented during so-called "side events" that run parallel to the meetings of state parties and are accessible to all governmental delegations as well as to non-state observer groups, including civil society, the business community, representatives from religious organizations, and academics. Cases of rights violations are sometimes presented by locally affected people such as indigenous or agro-pastoralist groups themselves. In some cases, the international partners of the HRCCWG sponsor representatives of local groups so that they are able to travel to where the COP takes place, join these international meetings, and bring forward their experiences and corresponding demands. Civil society networks help to process these cases, demonstrating that there are severe drawbacks in the body of rules and regulations of climate policies, such as REDD+ and the CDM, leading to rights infringements in the implementation process. They use these groups to voice their demands for change in the modalities and procedures of climate instruments, and to support the pleas of affected people for more information, transparency, and participation opportunities. In this way, local claims are brought into the international negotiation process.

Whereas the initial focus of the HRCCWG was on framing emission reductions as a human rights obligation, there is a clear emphasis on response measures and their impact on local populations in developing countries now. Thus, procedural human rights, and especially the right to participation, lie at the heart of the network's activities. Altogether, we can find a two-way process here: local advocates inform the policy-making process of states within the forum of an international organization, and their decisions can, in return, change climate policy implementation at the local level.

The Inter-Constituency Alliance to Promote Human Rights in the Paris Agreement

Prior to COP 21 in Paris, the two coordinators of the Human Rights and Climate Change Working Group initiated an inter-constituency alliance in order to combine the strengths of several civil society networks with constituency status, including women's groups, environmental CSOs, indigenous peoples' organizations, trade unions, and others, for the negotiations of a new climate treaty. This inter-constituency alliance can be understood as a "super-network," i.e., a network above several networks, sharing the same objectives in shaping the Paris Agreement.

Prior to the Paris negotiations, the inter-constituency alliance was established due to the fact that most observer organizations shared some common concerns. Among them were the protection and fulfillment of human rights, in particular, the right to health, indigenous peoples' rights, and sustainable development, a just transition of the workforce and the creation of decent jobs, gender equality and equal participation of women, and inter-generational equity. During the Intersessionals in Bonn in June 2015, the alliance managed to include all of these issues in the operative part of the negotiation text in Article Two. This was quite meaningful because Article Two was to determine the objective of the Paris Agreement. Thus, including human rights there would have meant acknowledging that the purpose of the Agreement was the protection and fulfillment of basic rights in the face of a changing climate and through the implementation of climate policies. However, the text of Article Two was still in brackets after Bonn and during almost the entire Paris conference. This means that this part of the text was not agreed on and still subject to further negotiations. Two days before the agreement was adopted at COP 21, the reference to human rights was removed from the operative part of the text, and only remained in the Preamble. This happened despite strong advocacy around the common objectives of the "super-network." Nevertheless, human rights and all the other related aspects the inter-constituency alliance advocated for remained part of the Preamble. This can be seen as a success of the network that can be traced back to the strong collaborative effort of the "super-network," which coordinated its activities and harmonized its strategies to then go back into the sub-networks to lobby with their specific state partners.

Several legal experts of the inter-constituency alliance, as well as the former UN Special Rapporteur on Human Rights and the Environment, John Knox, counted the institutionalization of human rights in the Preamble to the Paris Agreement as a success because it reiterated that all climate-relevant action has to be in accordance with existing human rights obligations, and this can be understood as an advancement compared to previous climate treaties.

Enhanced outreach on the road to Paris

Besides initiating and coordinating the inter-constituency alliance, there were also other changes to the HRCCWG leading to a strengthening of its advocacy efforts. After they had commenced working mainly with environmental law organizations, representatives from indigenous peoples' and women's rights' CSOs, and prominent non-governmental human rights players, Human Rights Watch (HRW) and Amnesty International (AI), came on board at the COP in Lima in 2014. A common Greenpeace–Amnesty statement, as well as a widely-spread common press release by HRW and AI, marked the arrival of the major human rights players in the network. The joint statement by Greenpeace and Amnesty International noted:

> Amnesty International and Greenpeace International call on governments at COP 21 to protect human rights by including an explicit reference to human rights in Article 2 as well as agreeing to phase out fossil fuels and deliver 100% renewables for all by 2050.
>
> (Amnesty International, 2015)

In addition to these new non-governmental partners from the human rights world, state actors joined the working group as well. Among them were mainly international organizations, like the OHCHR and UNICEF. Both took an active part in the network, further developed its strategy of coordinating meetings, and engaged in awareness-raising activities of state delegates.

Besides IOs, there were also some states taking an active part in the promotion of human rights in the climate agreement. Prior to the negotiations, 18 governments took action and initiated the *Geneva Pledge for Climate Action*, calling for enhanced institutional interaction between the UNFCCC and the OHCHR, and emphasizing that human rights obligations need to be observed in all climate-relevant actions. Among the committed states were mostly Latin American countries (e.g., Mexico, Peru, and Costa Rica), many small island states (e.g., the Maldives, Kiribati, and Samoa), as well as few European nations (e.g., France, Sweden, and Ireland).

Another great push for human rights in the climate regime came from increased media attention and a successful Twitter campaign. Under #Stand4Rights, the HRCCWG and several of their partners, disseminated information regarding new versions of the negotiating text including human rights, spread the word on further awareness-raising actions, and put pressure on governments who argued against rights in the climate agreement during the negotiations. An interesting example of the latter is the use of the #Stand4Rights to tweet a joint press release by Amnesty International and Human Rights Watch that blamed and shamed those countries that were severely blocking human rights language in the climate treaty at that point, namely the United States, Saudi Arabia, and Norway. The press release quoted Ashfaq Khalfan, Law and Policy Programme Director at Amnesty International, with the following sentences:

> Norway, Saudi Arabia, United States are at risk of being labelled "human rights deniers" in addressing climate change. … Norway is claiming to be a bridge builder on the issue of human rights, but rather seems intent on blowing up an essential bridge between environmental protection and human rights.
>
> (Amnesty International and Human Rights Watch, 2015)

Everyone on the mailing list of the HRCCWG was asked to re-tweet this press release, and the wide public attention around this led to immediate reactions of the opponents. Norway, for instance, released an official statement emphasizing that it supports human rights in the operative part of the treaty, but not in Article Two (which sets out the purpose of the agreement).

Thus, both prior to and in Paris, the network was profiting from an enhanced outreach including combined strengths in the inter-constituency alliance and active participation of international organizations and state actors, as well as increased press and social media coverage. All of these aspects led to a partial success – an institutionalization of human rights in the Preamble of the new climate agreement.

The Human Rights and Climate Change Working Group is just one example of a moderate climate justice network. Others include the Climate Action Network (CAN) and Stop Climate Chaos (della Porta and Parks, 2013).

The radical wing: claims for justice outside established institutions

There is no single understanding of climate justice or of the strategies to be employed in order to achieve justice in the context of climate change. During the COP14 climate negotiations in Bali in 2007, the network Climate Justice Now was formed. This network is convinced that many solutions to the climate crisis promoted by governments, with the help of international organizations, financial institutions, and multinational corporations are wrong. They oppose many of the solutions discussed at the international climate negotiations, including forest carbon markets, agrofuels, and carbon offsetting. Moreover, they claim that industrialized nations are very often unjustifiably pressuring developing countries into commitments for greenhouse gas emission reductions without offering adequate financial support. They also criticize developed nations for not living up to the legal and moral obligations they have made in the international climate regime by continuing to maintain their unsustainable energy, consumption, and production patterns (Carbon Trade Watch, 2008). Thus, in contrast to more moderate groups like the Human Rights and Climate Change Working Group, radical networks often oppose the underlying ideological, normative, and economic foundations that build the basis of the climate regime. They politicize climate change and criticize historic social and political relations between states as the root causes of climate injustice. The main argument they bring forward is that, both historically and currently, industrialized nations have been mainly responsible for greenhouse gas emissions, and have exploited the resources of developing nations, especially during times of colonization but also after that, to accumulate wealth.

Now industrialized countries promote market-based solutions that prevent a real change in the unjust world system and do not lead to a substantial decrease in greenhouse gas emissions, but reproduce asymmetric economic and power relations. These market-based solutions include mechanisms introduced under the Kyoto Protocol, which operationalized the UNFCCC, such as International Emissions Trading (IET), the Clean Development Mechanism (CDM), and Joint Implementation (JI). The problem with the CDM, for instance, was that industrialized countries and companies from the Global North could continue to pollute, if they bought credits from sustainable development projects in the Global South that are meant to decrease emissions. Through purchasing these offsets, developed countries could achieve their emissions reduction targets determined under the Kyoto Protocol. To qualify for accreditation by the CDM Board, projects need to demonstrate "additionality," i.e., developers have to prove additional greenhouse gas reductions that would be achieved with the respective project. Many of these projects, however, did not meaningfully prove this "additionality," nor did they contribute to development, but were large-scale carbon-offset projects that resulted in land grabbing, environmental degradation, and social rights violations. Offsetting leads to a delay in meaningful climate action in developed countries. Faced with the increasing energy demands in developing countries, it is questionable whether offsets reduce or actually increase greenhouse gas emissions overall (Carbon Trade Watch, 2018). Most CDM projects, among them many hydroelectric dams, were implemented in China and India and thus, did not have an impact in the least-developed countries, e.g., in Sub-Saharan Africa.

Many groups of the more radical wing of the climate justice movement do not just suggest the reform of market-based mechanisms, they view them as a completely false solution to the problems posed by climate change. They highlight that market schemes lead not only to exacerbated climate injustice between developed and developing countries, but also to increased inequalities within societies – often at the expense of indigenous peoples, pastoralist groups, and other minorities. They claim that we need to profoundly change our economic and political system, first and foremost capitalism, to initiate a real change towards more climate justice. Thus, if you see a demonstration where people hold up posters and placards with the slogan "system change not climate change," you might be seeing members of a more radical climate justice network suggesting that moderate policy changes are not enough to address climatic challenges, but that a system change hand-in-hand with altering production and consumption patterns is required. Important examples of this more radical wing are Rising Tide, Klimacamp, and Climate Justice Action Network (della Porta and Parks, 2013).

Environmental human rights defenders

In recent years, the situation of environmental human rights defenders (EHRD) has received increasing attention. We can understand environmental human rights defenders as either individual persons or civil society groups who are working on protecting the environment, but who are facing increasing threats and intimidation. The Special Rapporteur on the Situation of Human Rights Defenders, Michel Forst, differentiates between "land and environmental rights defenders," "environmental rights defenders," and "just environmental activists" (OHCHR, 2016). All of these categories of activists refer to the close link between environmental and human rights issues and the strong relationship between land rights and environmental rights (OHCHR, 2016). EHRD can be lawyers and journalists, but very often they are local people who might not even be aware of acting in this role. Their main concern is that their traditional lands, communities, customary practices, and their way of life are threatened by large-scale projects, including dams, logging, mining, or oil extraction (Knox, 2017). They aim to protect the environment in order to ensure that the basic human rights that depend on it, such as the right to water, food, adequate housing, or the right to life, are still guaranteed at the grassroots level today and for future generations.

State governments have to ensure that environmental human rights defenders (EHRDs) can exercise their human rights to freedom of association and expression, to information, and to participation in decision-making for them to be able to pursue activities protecting the environment. However, this is not always possible. Many of them are being threatened or harassed, or are even facing violence and death. A recent study confirms that across the world three EHRDs are killed every week (Global Witness, 2018). In a comprehensive mapping, Global Witness confirmed that between 2002 and 2013, at least 908 people were killed who were seeking to protect their land and the environment from unsustainable exploitation. Global Witness believes that the actual number is probably much higher, as they can only count cases if they have sufficient information and evidence based on certain criteria. More than 90% of these killings took place in nine countries, among them Brazil, Honduras, and the Philippines (Global Witness, 2018). Because of fierce competition over access to natural resources, the lack of legal recognition of these groups, and dysfunctional institutions in some countries, these numbers are expected to increase in the future. Thus, it is imperative to develop protection mechanisms and raise awareness of this situation. EHRDs themselves have reported that support from well-known international NGOs has helped them to successfully campaign against environmental abuse, transnational pressure by NGOs and international organizations has helped activists to be freed from detention, and environmental assessments conducted by UN agencies have confirmed their concerns regarding unsustainable resource exploitation (Knox, 2017: 21).

States need to further develop their national legislation with a view to protecting EHRDs. This should be on the basis of the main principles of the UN Declaration on the Right and Responsibility of Individuals, Groups and Organs of Society to Promote and Protect Universally Recognized Human Rights and Fundamental Freedoms (UN Declaration on Human Rights Defenders) adopted by the United Nations General Assembly in 1998. In addition, there is a Model National Law on Human Rights Defenders, developed by the International Service for Human Rights (ISHR) and 28 leading experts in the field, including current and former UN Special Rapporteurs on Human Rights Defenders. States can refer to these existing norms and models when developing domestic protection mechanisms (Knox, 2017: 21). Adopting a constitutional human right to a healthy environment and complying with human rights obligations relating to the environment and the rights of indigenous peoples are also crucial steps in the protection of environmental human rights defenders.

Conclusion

In this chapter, we observed how civil society has taken up the link between human rights and the environment for advocacy and litigation purposes. Environmental human rights defenders as individual advocates, as well as transnational advocacy groups, as strongly interconnected networks do not treat environmental and human rights concerns separately anymore. On the contrary, transnationally operating networks – that often rely on information from the grassroots level – have raised awareness among state actors, such as climate negotiators, and called upon them not to ignore the link between environmental degradation and rights infringements.

We also have to acknowledge how advocacy at the intersection between the environment and human rights leads to the emergence of new actor partnerships, such as the inter-constituency alliance. In these new constellations, actors combine their strengths to change the practices of states and international organizations with a view to protecting the environment and related human rights.

DISCUSSION QUESTIONS

1. What are the strategies transnational advocacy networks use in advocating for policy change within international organizations? Which of these strategies, from your perspective, are most successful/useful?
2. How has the Human Rights and Climate Change Working Group in the case study presented above used information politics, symbolic politics, leverage politics, and accountability politics?
3. Do you think lobbying for climate justice outside the established institutions, such as the UNFCCC, can be successful?
4. Does civil society engagement make climate conferences too complex to come to clear-cut solutions?
5. How important are environmental human rights defenders in initiating change at the local level? Could they link up with Transnational Advocacy Networks to increase pressure on state governments?

References

Amnesty International (2015). Protection of Human Rights from Climate Change Requires Urgent Shift to 100% Renewable Energy for All: Joint Statement: Greenpeace – Amnesty International [Online]. Available at: www.amnesty.org/en/documents/pol30/3033/2015/en/ [Accessed: 17 March 2018].

Bernauer, T. and Betzold, C. (2012). Civil Society in Global Environmental Governance. *The Journal of Environment and Development* 21 (1): 62–6.

Bernauer, T., Böhmelt, T., and Koubi, V. (2013). Is There a Democracy-Civil Society Paradox in Global Environmental Governance? *Global Environmental Politics* 13 (1): 88–107.

Betsill, M.M. and Corell, E. (eds.) (2008). *NGO Diplomacy: The Influence of Nongovernmental Organizations in International Environmental Negotiations*. Cambridge, MA: MIT Press.

Bexell, M., Tallberg, J., and Uhlin, A. (2010). Democracy in Global Governance: The Promises and Pitfalls of Transnational Actors. *Global Governance* 16(1): 81–101.

Böhmelt, T., Koubi, V., and Bernauer, T. (2014). Civil Society Participation in Global Governance: Insights from Climate Politics. *European Journal of Political Research* 53 (1): 18–36.

Brunnengräber, A. (2014). Between Pragmatism and Radicalization: NGOs and Social Movements in International Climate Politics. In: Dietz, M. and Garrelts., H. (eds.) *Routledge Handbook of the Climate Change Movement*. Abingdon, UK: Routledge, pp.269–79.

Bullard, R.P. (2001). Environmental Justice in the 21st Century: Race Still Matters. *Phylon* 49 (3/4): 151–71.

Carbon Trade Watch (2008). Climate Justice Now! Statement [Online]. Available at: www.carbontradewatch.org/index.php?option=com_content&task=view&id=227&Itemid=95 [Accessed: 17 March 2018].

Carbon Trade Watch (2018). Carbon Offsets [Online]. Available at: www.carbontradewatch.org/issues/carbon-offsets.html [Accessed: 3 April 2018].

CIEL, CARE, WEDO et al. (2017). Delivering on the Paris Promises: Combating Climate Change while Protecting Rights Recommendations for the Negotiations of the Paris Rule Book [Online]. Available at: www.ciel.org/wp-content/uploads/2017/05/Delivering-On-Paris-Web.pdf [Accessed: 17 October 2018].

Cutter, S. (1995). Race, Class, and Environmental Justice. *Progress in Human Geography* 19 (1): 111–22.

Della Porta, D. and Parks, L. (2013). Framing Processes in the Climate Movement: From Climate Change to Climate Justice. In: Dietz, M. and Garrelts., H. (eds.) *Routledge Handbook of the Climate Change Movement*. Abingdon: Routledge, pp.19–30.

Della Porta, D. and Diani, M. (2006). *Social Movements: An Introduction*. Malden: Blackwell-Wiley Publishing.

Diani, M. (2003). Networks and Social Movements: A Research Programme. In: Diani, M. and McAdam, D. (eds.) *Social Movements and Networks*. Oxford: Oxford University Press, pp.299–318.

EJA – Environmental Justice Atlas (2015). Chicago's Toxic Doughnut, USA. [Online]. Available at: https://ejatlas.org/conflict/chicagos-toxic-doughnut-usa [Accessed: 16 October 2018].

Filzmoser, E. (2013). Clean Development Mechanism. In Human Rights and Climate Change Working Group (ed.), *Summary of Rights-Related Developments at COP 19*. Geneva: CIEL.

Global Witness (2018). At What Cost? Irresponsible Business and the Murder of Land and Environmental Defenders in 2017 [Online]. Available at: www.globalwitness.org/en/campaigns/environmental-activists/at-what-cost/ [Accessed: 17 October 2018].

Godsil, R.D. (1991). Remedying Environmental Racism. *Michigan Law Review* 90(2): 394–427.

Görg, C. and Bedall, P. (2013). Antagonistische Positionen: Die Climate-Justice-Koalition vor dem Hintergrund der Theorie gesellschaftlicher Naturverhältnisse. In: Dietz, M. and Garrelts., H. (eds.), *Die internationale Klimabewegung: Ein Handbuch*. Wiesbaden: VS Springer, pp.75–105.

Hein, J. and Garrelts, H. (2013). Ambiguous Involvement: Civil Society Actors in Forest Carbon-Offsets – the Case of the Climate Community and Biodiversity Standards (CCB). In: Dietz, M. and Garrelts., H. (eds.), *Routledge Handbook of the Climate Change Movement*. Abingdon, UK: Routledge, pp.319–33.

HRCCWG (2018). Human Rights and Climate Change Working Groups. [Online]. Available at: http://climaterights.org/ [Accessed: 19 December 2018].

Human Rights Watch (2015). Human Rights in Climate Pact Under Fire. [Online]. Available at: www.hrw.org/news/2015/12/07/human-rights-climate-pact-under-fire [Accessed: 17 June 2016].

Joachim, J. (2003). Framing Issues and Seizing Opportunities: The UN, NGOs, and Women's Rights. *International Studies Quarterly* 47 (2): 247–74.

Keck, M.E. and Sikkink, K. (1998). *Activists Beyond Borders: Advocacy Networks in International Politics*. Ithaca, NY: Cornell University Press.

Knox, J. (2017). *Environmental Human Rights Defenders: A Global Crisis. Policy Brief.* Versoix, Switzerland: Universal Rights Group

OHCHR (2016). Report of the Special Rapporteur on the Situation of Human Rights Defenders, UN Doc. A/71/281, para. 7, 3 August 2016. Geneva: Office of the High Commissioner for Human Rights.

Petition to the Inter-American Commission on Human Rights (2005). Petition to the Inter-American Commission on Human Rights Seeking Relief from Violations Resulting From Global Warming Caused by Acts and Omissions of the United States. [Online]. Available at: http://blogs2.law.columbia.edu/climate-change-litigation/wp-content/uploads/sites/16/non-us-case-documents/2005/20051208_na_petition.pdf [Accessed: 19 December 2018].

Reitan, R. (2011). Coordinated Power in Contemporary Leftist Activism. In Olesen, T. (ed.) *Power and Transnational Activism.* London: Routledge, pp.51–71.

Rucht, D. (1994). *Modernisierung und Neue Soziale Bewegungen: Deutschland, Frankreich und USA im Vergleich.* Frankfurt: Campus.

Schade, J. and Obergassel, W. (2014). Human Rights and the Clean Development Mechanism. *Cambridge Review of International Affairs* 27 (4): 717–35.

Schlosberg, D. (2007). *Defining Environmental Justice: Theories, Movements, and Nature.* Oxford: Oxford University Press.

Scholte, J.A. (2012). A More Inclusive Global Governance? The IMF and Civil Society in Africa. *Global Governance*. 18(2): 185–206.

Spitulnik, D. (2011). Small Media Against Big Oil (Nigeria). In: Downing, J. (ed.), *Encyclopedia of Social Movement Media*. Thousand Oaks, CA: SAGE Publications.

Steffek, J. and Nanz, P. (2008). Emergent Patterns of Civil Society Participation in Global and European Governance. In Steffek, J., Kissling, C. and Nanz, P. (eds.), *Civil Society Participation in European and Global Governance*. Basingstoke, UK: Palgrave Macmillan, pp.1–29.

Tarrow, S. (1994). *Power in Movement: Social Movements, Collective Action and Politics*. Cambridge: Cambridge University Press.

UNFCCC (2014). Nongovernmental Organization Constituencies. [Online]. Available at: https://unfccc.int/files/parties_and_observers/ngo/application/pdf/constituency_2011_english.pdf [Accessed: 24 June 2016].

UNFCCC (2015). Paris Agreement. Adopted at the 21st Conference of the Parties (COP), 12 December 2015, Paris.

UNFCCC/AWG-LCA (2010). United Nations Framework Convention on Climate Change/Ad Hoc Working Group on Long-term Cooperative Action under the Convention: Negotiating Text, FCCC/AWGLCA/2010/14.

UNFCCC (2010). United Nations Framework Convention on Climate Change/Decision 1/CP.16. The Cancun Agreements: Outcome of the Work of the Ad Hoc Working Group on Long-term Cooperative Action under the Convention, Appendix I.

United Christ of Church Commission of Social Justice (1987). Toxic Waste and Race in the United States. [Online]. Available at: www.nrc.gov/docs/ML1310/ML13109A339.pdf [Accessed: 18 December 2018].

United Nations (1998). UN Declaration on the Right and Responsibility of Individuals, Groups and Organs of Society to Promote and Protect Universally Recognized Human Rights and Fundamental Freedoms. New York: UN General Assembly.

United States General Accounting Office (1983). Siting of Hazardous Waste Landfills and Their Correlation with Racial and Economic Status of Surrounding Communities. [Online]. Available at: http://archive.gao.gov/d48t13/121648.pdf [Accessed: 18 December 2018].

11 Vulnerability and climate change

Introduction

Adverse consequences of climate change, and sometimes even climate policies, can exacerbate situations of injustice between countries, within societies and among generations. Certain population groups – often those who are already disadvantaged or treated unequally – are more vulnerable to the adverse consequences of climate change than others. Vulnerability can be understood as diminished capacities of individuals, communities, or groups in coping with, adapting to, and recovering from hazards, such as environmental degradation or climate change. Often, but not exclusively, economically less well-off people or socially marginalized communities from developing countries are more vulnerable than others. Their exposure to certain risks differs according to a variety of factors, including their social group, gender, ethnic identity, or other forms of identity, and age. If people manage to build capacities and resources to cope with these hazards, they can reduce their vulnerability (IFRC, 2018).

In this chapter, we will take a closer look at societal groups that require special protection in the face of climatic challenges, in particular, women, children, indigenous peoples, and migrants. We will analyze and discuss the situation of these groups in-depth and reflect on the effectiveness of protective mechanisms that have been introduced to ensure that these groups can enjoy their human rights in the face of climatic challenges. We will also explore the role of vulnerable groups in developing appropriate mitigation and adaptation strategies. Mitigation policies can be understood as programs to reduce or prevent greenhouse gas emissions. Adaptation is a response to climate change impacts with the aim of reducing the vulnerabilities of the above-mentioned social groups.

Vulnerable groups

Women

Climate change impacts as well as policy responses to these impacts can exacerbate gender disparities. Due to already existing inequalities, women in certain cultures are often more vulnerable than men when experiencing extreme weather events, like storms or floods, or other disasters. This is especially true for elderly women and young girls. In situations of natural catastrophes or extreme weather events, for instance, many women do not leave their homes, as they are trained to comply with local gender-specific norms and are the main caregivers for children and elderly people. This means they more frequently become victims of floods, storms, or other extreme events, especially if they are staying in their home (BRIDGE and DFID, 2008: 6). Sometimes when women are trying to escape from natural calamities, they are, exposed to violence and aggression. There is empirical evidence that female climate migrants trying to find a safe place or staying in emergency shelters or in temporary accommodation are increasingly threatened by domestic or sexual violence (BRIDGE and DFID, 2008: 7–8). Often, women's vulnerability to the consequences of climate change increases because of structural inequalities, such as inability to own property, discrimination in the distribution of food and medicine, exclusion from decision-making processes, and restricted access to information and financial resources (United Nations, 2009: 16; BRIDGE and DFID, 2008: 3). This can be understood as a situation of intra-generational injustice.

Apart from risks associated with the impacts of climate change, it is also important to take gender-specific division of labor into account when analyzing the situation of women. In times of a changing climate, women are confronted with increased pressure with respect to the work tasks they have to complete. Worldwide, girls and women are still mainly responsible for the major part of reproductive activities, including taking care of old and sick family members, as well as children. In the context of climatic challenges, support and caregiving work can even increase, for example, due to growing health issues. This means women might even need to take over more reproductive work and have less time to generate an income. This can lead to higher poverty risks. Moreover, women can find themselves exposed to situations of severe physical and psychological stress, if they, in addition to caring for the elderly and sick, need to walk longer distances to fetch water or firewood. Additional care responsibilities will lead to time constraints, which can prevent women in participating in decision-making within their communities pertinent to managing the impacts of climate change (BRIDGE and DFID, 2008: 3–4).

In many parts of the world that are already severely affected by a changing climate, women play an important role in agriculture and food security. Throughout Sub-Saharan Africa and in parts of Asia, women cultivate maize, rice, vegetables, and other food crops, and sell them at the local market. Female family members are the ones supplying water for the household and looking after livestock (BRIDGE and DFID 2008: 4–5). Droughts and floods especially can lead to crop shortfalls and endanger the provision of food for women and their families. Droughts can make it much more difficult for them to fetch water, because watering holes may be even more remote, with much longer distances to walk there. These long walks cost time and expose women to the risk of physical and sexual violence, especially in conflict-prone areas. Lack of water, on the other hand, constitutes a serious risk to the right to health. Similarly, heavy rainfalls and floods increase women's workload and make it more difficult for them to maintain their household. Extreme weather events and natural catastrophes can seriously endanger, *inter alia*, the right to adequate housing (BRIDGE and DFID, 2008: 5). Because of their substantial role in agriculture and food provision, it is of utmost importance that women have a say in respective adaptation strategies and can play a meaningful role in decision-making processes in this regard.

In summary, women and girls are more vulnerable with respect to the consequences of climate change. However, this does not imply that we understand them as passive victims. Because women perform so many important and different roles within the family, community, and society, they should be able to participate in climate adaptation and mitigation decisions in a meaningful way. This is also relevant when considering the relationship between climate change and human rights. The realization of many social rights, including the rights to food, water, adequate housing, and health, depend on the contribution and labor of women and girls. Climatic impacts enhance stress and the burden on females. Due to increased household work and further need for reproductive care, girls often refrain from attending school, and women have less time for education and income-generating activities in the labor market. As a consequence of their responsibilities, women and girls have fewer opportunities to participate in climate-relevant decisions, for instance, regarding adaptation and mitigation strategies, within their communities and beyond. Against this background, climate change can be understood as a "threat multiplier" (IFAD, 2010), deepening already existing gender disparities and inequalities.

The Committee on the Elimination of Discrimination against Women that monitors states' implementation of the Convention on the Elimination of all Forms of Discrimination against Women (CEDAW, 1979) has made several relevant recommendations in relation to climate change. The Convention obliges states to take all necessary measures for the full development and advancement of women. This is particularly relevant in relation to climate change because women are often more severely affected by climate change consequences or even by climate policies that lack a gender dimension. Thus, in 2010, the Committee made recommendations with respect to the rights of older women, emphasizing that climate change as well as disaster response measures need to be gender-sensitive, and need to take the vulnerabilities of older women into account (CEDAW, 2010). Another recommendation adopted in 2016 addresses the situation of rural women. The protection of rural women is anchored as a specific obligation in Article 14 of CEDAW. Women living in rural areas are often particularly exposed to climate induced-risks and they also heavily depend on the use of natural resources, especially those who are poor. The Committee has also stressed that a part of states' obligation not to discriminate against women is to mitigate climate change, as it disproportionally affects women (CEDAW, 2016a; CIEL and GI-ESCR, 2018). In 2018, the Committee adopted a General Recommendation on the Gender-related Dimensions of Disaster Risk Reduction in a Changing Climate (CEDAW, 2016b; Bonner, 2018). It emphasizes that three principles of CEDAW are particularly relevant in the context of natural disasters and pertinent climate action. These are: substantive equality and non-discrimination, participation and empowerment, and accountability and access to justice (CEDAW, 2016b). It also draws attention to certain areas of rights protection that are of special concern in the context of disaster risk reduction, including gender-based violence, migration and forced displacement, the right to health, the right to an adequate standard of living, the right to education and information, the right to work, and the right to social protection (CEDAW, 2016b; CIEL and GI-ESCR, 2018). In all of these recommendations, the Committee highlighted the importance of gender mainstreaming in climate policy responses, as well as meaningful participation of women in decision-making and climate policy implementation. Gender mainstreaming means to continuously assess and pay attention to the different implications policies have on men and women. Furthermore, the committee emphasized climate education, called for the reduction of greenhouse gas emissions, as climate change negatively impacts on women, and suggested that investment in renewable energy can be explored with respect to employment opportunities for women (CIEL and GI-ESCR, 2018: 4). In addition, as a result of lobbying by women's groups, a Women and Gender Constituency was established under the UNFCCC.

Children

According to the Convention on the Rights of the Child (CRC), all legal, political, and welfare decisions made with regard to children should be primarily guided by the best interests of the child. All children should be protected and cared for, and they should be able to develop and grow while they are provided with food, water, adequate housing, and health care (CRC, 1989). The consequences of climate change, however, seriously put the future of children at risk. Nevertheless, children are only very rarely included in debates about the changing climate and appropriate policy responses.

Because children are still in the process of physiological and psychological development, they are more vulnerable to harm caused by environmental and climate hazards. They are more susceptible than adults to air pollution, to infection and diseases, to malnutrition, and to lack of clean drinking water and sanitation. Climatic challenges today and in the future will have severe effects in the least developed countries where the largest number of under-18 year olds live. Many of these children are already facing severe poverty, which will be exacerbated by climate impacts. The Stern Review, commissioned by the UK government in 2006, for instance, estimated that by 2100, an additional 165,000–250,000 children (as compared to a world without climate change) per year could die in South Asia and Sub-Saharan Africa alone, mainly due to malnutrition but also as a consequence of the spread of diseases (Stern, 2006: 55).

Natural disasters can destroy important infrastructure, preventing children from accessing schools and health care services. Climate change also leads to lower income levels within the family, which again can prevent children, and particularly girls, from attending school. As certain household tasks become more difficult and time-consuming, such as collecting water or caring for older or sick family members, girls may need to help in the domestic sphere for longer hours, leaving them with less time for school or doing their homework. Girls who often have to combine school and work, have less time to engage in both paid employment in the marketplace and education. Faced with the harsh reality of climate change, they will have considerably less time for play, leisure, and recreational activities as outlined in Article 31 of the Convention on the Rights of the Child (CRC, 1989).

Malnutrition will severely increase in the context of droughts, floods, and changes in precipitation. This will have a serious impact on children's health, will weaken their immune systems, and decrease their resistance to infectious diseases. Both morbidity and mortality rates will increase due to malnutrition. The occurrence of respiratory diseases, diarrhea, malaria, and other vector-borne illnesses caused by parasites, viruses, and bacteria is sensitive to weather conditions and can increase with a changing climate. Poor health and hunger affect children's ability to learn and can impact school achievements. Decreased education levels have long-term effects, including regarding information relating to climate change. Young adults will find it more difficult to understand new information on climate adaptation and might become even more prone to the negative effects of climate change. In some cases, families will be forced to sell their scarce assets in order to survive after extreme weather events, droughts, or floods. This can mean that they will lose their land, cattle, or livestock. Thus, they cannot raise their level of income over time, accumulate wealth, or increase consumption levels. They may also not be able to send children to primary or secondary schools. Particularly in developing countries, families have fewer adaptive capacities. They lack access to important infrastructure, including clean water and sanitation, financial resources, and social services. This makes families, and especially children, even more vulnerable to the consequences of climate change (Stern, 2006: 97–114).

People in the least-developed countries (LDCs) already are, and will be, the most severely affected by the negative impacts of climate change – and these countries usually have very large child populations. Under-18 year olds account for about half of the population in LDCs, and the population in these countries is growing considerably faster than those in industrialized countries. This puts a strain on these countries, especially in education and health sectors, and with regard to providing adequate nutrition. Thus, it will become increasingly difficult for the LDCs to achieve the Sustainable Development Goals in the face of adverse consequences of climate change. Scarce resources and adaptive capacities also lead to higher risks of violent conflicts due to increased economic, social, and political stress. In this context, children can experience psychosocial trauma, they can become victims of displacement, forced migration, or family separation, or they can become exposed to trafficking and exploitation. Sometimes they are forcibly recruited as child soldiers.

By causing anthropogenic climate change, human beings have created multiple layers of climate injustice. If we recall the three dimensions of climate injustice elaborated on in the introduction, we can see that children, especially those in developing countries, are affected by several of them at the same time. Children in developing countries experience *international injustice*. This means that after industrialized countries have extensively utilized carbon-intensive industries to foster growth and wellbeing in the developed world, developing countries shall not be able to do the same, as the "cumulative carbon budget" has been exhausted (Shue, 2014: 324–7). At the same time, families in developing countries lack the capacities to adapt to severe climate impacts. In addition to this, children also experience *intra-societal injustice*, as they are more vulnerable to emissions, pollution, extreme weather events, and droughts than other population groups. They are also more susceptible to the risks of malnutrition and vector-borne diseases and other human impacts resulting from climate change. Furthermore, children in developing countries experience *inter-generational justice*. This means that they suffer from decisions on unsustainable energy and consumption patterns made by past and current generations of political leaders. In order to ensure that future generations of children will be able to enjoy their basic rights, we need to take inter-generational responsibility seriously, substantially reduce greenhouse gas emissions, and initiate a transition process to a low-carbon economy (UNICEF, 2010). These multiple injustices in relation to children in developing countries have not yet been sufficiently understood and considered – in research, as well as in policies.

Inter-generational equity and justice have, so far, been discussed from a normative perspective. There are still too few attempts at practicing it. One idea for providing future generations with a voice is to establish the position of a Commissioner for Future Generations. Countries like Israel and Hungary have realized this by asking their Commissioner to review decisions with reference to environmental effects and the situation of future citizens (UNICEF, 2010). One important advantage of this approach is that the Commissioner has raised awareness among the public regarding the concerns of future generations. Another idea UNICEF has proposed is to mainstream inter-generational equity across all governmental sectors and departments, similar to what has been practiced with respect to gender mainstreaming (UNICEF, 2010). Other important ideas are to raise children's own awareness and let them have a voice in their community's approach to climate adaptation and mitigation. This would enhance capacities of children and the entire community to deal with climate change. Views of children should be heard and they should be seen as active protagonists in promoting environmentally sustainable lifestyles and the interests of future generations – instead of considering them merely as a passive vulnerable group (UNICEF, 2015).

In a landmark constitutional climate lawsuit called *Juliana v. United States*, 21 youths and children filed a lawsuit which asserted that action of the US government has caused climate change and led to the violation of the constitutional rights to life, liberty, and property, and failure to protect public trust resources. The lawsuit was filed against the US government in the Federal District Court of Oregon in 2015. Earth Guardians is a civil society plaintiff in this case and another NGO, Our Children's Trust, acts as a supporter (Our Children's Trust, 2018). The plaintiffs emphasize that there is only a very short window of opportunity to phase out reliance on fossil fuels in order to reduce greenhouse gas emissions. Therefore, they are seeking a declaration confirming that their constitutional rights and public trust rights have been violated, and a court order that halts these violations and directs the government to develop a plan for substantially reducing greenhouse gas emissions (Aiken, 2016: 2). Initially, the US government – in partnership with representatives from the fossil fuel industry – tried to have the case dismissed. This led to several interesting developments in this case, such as a recommendation to deny both motions to dismiss issued by US Magistrate Judge Thomas Coffin and upheld by US District Court Judge Ann Aiken who released an historic opinion and order in November 2016. Therein, she held: "Exercising my 'reasoned judgement', I have no doubt that the right to a climate system capable of sustaining human life is fundamental to a free and ordered society" (Aiken, 2016: 32). Many activists of the climate movement have interpreted this as the first time ever a fundamental right to a safe climate has been recognized. The case was discussed in Chapter 8.

Indigenous peoples

Indigenous peoples live in harmony with nature, in a close and mutually beneficial relationship with land, water, animals, and other components of an ecosystem. Not only do their material livelihoods depend on intact ecosystems but also their cultural, spiritual, and medical practices. Because of this close relationship with the environment and natural resources, indigenous peoples are often among the first people within societies that are affected by the consequences of climate change. Some of the areas where indigenous communities live, such as small island states, tropical forests, or the polar regions are particularly vulnerable to environmental degradation and climatic changes (Chazalnoel and Carini, 2018). There are many examples from different world regions illustrating this. In the Himalayas, for instance, glacial melts put hundreds of millions of indigenous peoples at risk, because changes in seasonal water flows lead to increased water in the short-term, but reduced water in the long-term. In the Amazon region, deforestation and forest fragmentation as a result of climate change accelerate carbon release into the atmosphere, exacerbating further changes, such as droughts. Thus, rainforests are replaced by savannahs, resulting in substantial impacts on the livelihoods of indigenous peoples. In the Arctic, global warming and weather changes threaten the survival of certain species. This leads to risks in food security for indigenous peoples who have traditionally been fishing, gathering, and hunting polar bears, seals, or walruses. Changes in these ecosystems not only challenge the local economy, but also hamper cultural and spiritual practices and traditions (United Nations, 2008). In addition to this, indigenous communities very often belong to the poorest segments of society, which puts them even more at risk when they face environmental challenges, as poverty further reduces their adaptive capacities (ILO, 2017: ix).

This interference with indigenous peoples' traditional livelihoods constitutes a situation of intra-societal injustice. Historically, indigenous communities have contributed considerably little to greenhouse gas emissions. Quite on the contrary, they are an integral part of ecosystems and take an active role in preserving them. In order to do so, they draw on their valuable ancestral knowledge, which has helped them to adapt to climatic changes over many generations. In Vietnam, for instance, communities planted mangroves along the coast in order to diffuse strong waves from tropical storms. And in Guyana, people are moving from the savannah to the forest during periods of drought, and have started to plant crops, such as cassava, on moist flood plains. This knowledge is recognized in the 2015 Paris Agreement as highly relevant for appropriate climate adaptation (UNFCCC, 2015).

However, indigenous communities are, due to their close relationship with land and natural resources, among the first societal groups that are confronted with the impacts of climate change, like droughts, reduced water levels in rivers or lakes, floods, and changes in precipitation. In addition to this, they are very often adversely affected by our political responses to climate change, i.e., by adaptation and mitigation programs. In the case study presented in Chapter 12, we will learn how indigenous peoples have been violently relocated in the context of governmental plans to build the hydroelectric Gibe III dam in Ethiopia. The government argues that it is taking an important step towards energy transition and green growth through hydropower generation to protect the rights of future generations. At the same time, however, they do not adequately compensate local communities for taking away their traditional land, disrupting their livelihoods, and interfering with their cultural practices. In this way, climate change and climate policies are exacerbating already existing social inequalities (Vanderheiden, 2004; Humphreys, 2014) in Ethiopia and elsewhere. Some governments even use climate policies, such as hydroelectric dams, to realize their vision of "nation-building" (Gadgil and Guha, 1994: 110) and development as modernization by purposefully changing traditional lifestyles of indigenous peoples. One crucial example is the creation of jobs for relocated indigenous peoples, such as in the case of the hydroelectric dam Gibe III discussed in Chapter 12, in agricultural initiatives that may reduce greenhouse gases in the atmosphere, but lead to an increase of monoculture crops and plantations, possibly fostering a decline in biodiversity (United Nations, 2008). This can put food security and health of indigenous peoples at risk. They might also decide to leave their traditional lands, either as a consequence of environmental challenges, natural disasters, or because they are forced away by climate or development policies. Deforestation, for instance, has prompted native communities to move into the cities for economic reasons, but very often they end up in poor urban areas and slums.

The United Nations' program to tackle deforestation, called Reducing Emissions from Deforestation and Forest Degradation (REDD+), however, has in many cases also led to infringements of indigenous peoples' rights – despite containing safeguards that have to be observed (Faustino and Furtado, 2014). If indigenous peoples become migrants, they are particularly vulnerable and often face double discrimination, as migrants on the one hand, and as indigenous peoples, on the other hand (United Nations, 2008).

Two international legal instruments are relevant for protecting the rights of indigenous peoples in the context of climatic challenges and respective policy responses: the 2007 United Nations Declaration on the Rights of Indigenous Peoples (UNDRIP, 2007) and the 1989 ILO Indigenous and Tribal Peoples Convention (C169) (ILO, 1989). UNDRIP, in its Preamble, directly refers to the historical injustice indigenous peoples have experienced as a result of colonization and dispossession of their traditional lands and resources. The Declaration also makes reference to the need to respect indigenous knowledge as it contributes to sustainable and equitable development and conservation of natural resources. In its operative clauses, UNDRIP determines that indigenous peoples shall not be forcibly removed from their lands and that relocation cannot take place without free, prior, and informed consent (Article 10), indigenous peoples have the right to participate in decision-making in all matters affecting their life (Article 18), they have the right to determine their own priorities and strategies with respect to exercising their right to development (Article 23), indigenous peoples have the right to their traditional lands, territories, and resources, and they have the right to redress, including restitution or compensation, if their traditional territories have been taken away or used without their free, prior, and informed consent (Articles 26 and 28), and states have to consult indigenous peoples and need to obtain their free, prior, and informed consent for any project affecting their lands, territories, and resources (Article 32) (UNDRIP, 2007).

The ILO Indigenous and Tribal Peoples Conventions contains very similar provisions with a view to protecting indigenous peoples from being relocated or forced to leave their traditional lands, policies that interfere with their self-determination, cultural and spiritual practices, or the use of their natural resources. In Article 6, for instance, it is emphasized that governments shall ensure that indigenous peoples can participate in decisions regarding policies and programs that directly affect them (Article 6). Governments also need to conduct social, environmental, cultural, and spiritual impact assessments of all planned development activities (Article 7). Rights of land ownership and use shall be protected for indigenous peoples, including nomadic peoples (Article 14). Natural resources have to be safeguarded, and cannot be exploited without consulting indigenous peoples, who have the right to participate in respective benefits (Article 15). People shall not be removed from their lands without their free, prior consent and if they, in exceptional circumstances, need to be relocated, they have to be adequately compensated (Article 16).

The right to be consulted, the need to obtain free, prior, and informed consent, and to conduct environmental and social impact assessments has, at least in some cases, stopped the construction of mega-dams that would negatively affect indigenous peoples, their livelihoods, and cultural practices. After a massive mobilization campaign, Federal Judge Ilan Presser ordered the Brazilian administration to refrain from issuing an environmental license for the São Luiz do Tapajós dam in the absence of prior consultation with indigenous people and other threatened communities. In Malaysia, continued protests by indigenous peoples first led to a moratorium of the Baram dam, and then to the return of land to indigenous peoples who own the land according to native customary rights. In the case of *Sarayaku v. Ecuador* (2012), the Inter-American Court of Human Rights awarded the Sarayaku people 1,340,000 US Dollars in damages after Ecuador's State Petroleum Company entered into a contract with a private company seeking to exploit crude oil in the Amazon region on land that, for a large part, was traditionally occupied by the Sarayaku People. The Argentinean Company commenced seismic survey activities, which led to the destruction of caves, waterfalls, and underground rivers used as drinking water sources and destroyed at least one site of particular spiritual importance to the Sarayaku. It also planted explosives on the people's traditional lands for further exploring oil resources. The Court found Ecuador liable not only for failing their obligation to consult with the Sarayaku people, but also for breaching their right to communal property, cultural identity, and, by agreeing to explosives being planted, their right to life.

At the United Nations, indigenous peoples are organized in the Permanent Forum on Indigenous Issues. At the international climate conferences, the caucus that brings indigenous peoples participating in the UNFCCC process together is the International Indigenous Peoples Forum on Climate Change (IIPFCC). Representatives of the IIPFCC have been actively engaged in the inter-constituency alliance addressed in Chapter 10, successfully advocating for principles of climate justice to be included in the 2015 Paris Agreement. Thus, today the preamble of the Paris Agreement contains a paragraph emphasizing that the rights of indigenous peoples must be respected, promoted, and considered in all climate-relevant action. We discussed the petition brought by the Inuit peoples in Chapters 4 and 9 (Inuit Petition, 2005).

Environmental migrants

Today, human mobility in many parts of the world is caused by environmental degradation and weather calamities. With increasing and intensified weather-related disasters and respective impacts on peoples' livelihoods, we can expect even more people to be on the move (IPCC, 2014). The International Organization for Migration (IOM) proceeds on the assumption that between 25 million and 1 billion people will be environmental migrants by 2050 (IOM, 2014: 38). Migration in the context of environmental and climatic challenges is a complex issue, and it is difficult to find coherent policy responses. There is no officially recognized definition of "environmental refugees" or "climate refugees" under international refugee law or other international legal frameworks yet. One reason for the lack of a clear-cut definition is that usually different factors combined with climate impacts or the consequences of environmental degradation prompt people to move away from their home. An example can illustrate this: Climatic challenges can lead to changes in precipitation, droughts, and reduced access to water. This can severely damage the local economy and disrupt people's livelihoods. If they lack the capacity to adapt to this situation, people could decide to move from rural areas into the city in order to search for a different livelihood. In this case, however, a combination of factors strongly related to climate change has prompted their movement (Randall, 2018). If slow and complex processes play a role, it becomes even more difficult to single out the environment or climate change as the most important factor causing people to move.

The International Organization for Migration (IOM) has developed a working definition capturing the complexities at stake. It defines environmental migrants as

> persons or groups of persons who, predominantly for reasons of sudden or progressive change in the environment that adversely affects their lives or living conditions, are obliged to leave their habitual homes, or choose to do so, either temporarily or permanently, and who move either within their country or abroad.
>
> (IOM, 2011: 33)

Environmental migration, according to the IOM, can take different forms. It can be temporary or permanent, voluntary or forced, and it can be internal or international. As climate impacts exacerbate already existing vulnerabilities and can lead to decreases in household income, certain segments of the population might also want to move, but are unable to do so due to lack of resources. These people are called "trapped populations." Some areas, such as coastal areas or low-lying island states, might become uninhabitable as a consequence of rising sea levels. Here, migration can also be understood to be an inevitable adaptation strategy enhancing people's ability to cope with environmental challenges and climate change. Decreasing natural resources can enhance tensions within and between societies, possibly leading to different forms of conflict and as a consequence, to displacement (IOM, 2014: 38). Some contend that severe weather events associated with climate change may compel inhabitants of small island states to move long before the islands themselves become submerged (McAdam, 2012).

Currently, there is no reliable data on migrants who are on the move due to gradual environmental degradation processes. The Internal Displacement Monitoring Centre (IDMC), however, has published annual estimates suggesting that 165.9 million people were displaced between 2008–2013, with over 90% of displacement relating to natural disasters in the last four years of this time period. In 2013, for instance, approximately 7.3 million people were displaced by climate and weather-related calamities, such as Typhoon Haiyan in the Philippines. In comparison, there have been 327,000 people displaced due to violence and conflict during the same time period in Asia. In 2017, more than 2.5 million people were displaced by disasters in the Philippines alone, as compared to 645,000 people who migrated in the context of conflict (IDMC, 2018). Even though reliable estimates on slow processes of environmental degradation and related migration movements are still lacking, we can assume that in the long-term, many more people will migrate as a consequence of these slow climatic challenges than in relation to natural calamities (IOM, 2014: 39). Moreover, migration as a reaction to slow environmental degradation tends to be permanent, whereas temporary migration as a response to environmental stressors can also be circular, i.e., on a recurrent basis. Enhanced frequency and intensity of extreme weather events means that there is less time for recovery and rehabilitation, and this means climate change will, in the long run, lead to increased permanent migration (IOM, 2014: 40).

People in least-developed countries, coastal areas, and small island states will be more exposed to climate impacts due to their vulnerable geographic location. Thus, we can expect migratory movements and displacements mainly from these countries. Within these societies, there are people who have already been socially and economically marginalized; they are even more susceptible to these challenges as they have less information, fewer resources and capacities to adapt, and only very limited alternative livelihood strategies. This can also mean that they lack the material means needed to migrate, which leaves them trapped in their vulnerable situations. Those who have the capacity to migrate usually remain within their country and move from rural to urban areas. Post-disaster displacement often means that affected population groups move to the closest safe destination. Only very few migrants will dispose of sufficient resources to travel long distances or even move from one continent to another. Cross-border migration in the context of environmental challenges can already be observed in East Africa, related to droughts, and in East Asia, related to floods. Environmental migrants use existing migratory routes and prefer to move to places where they have pre-existing family ties and historic or cultural roots.

Environmental migrants are not yet protected under international legal frameworks. According to the United Nations Convention Relating to the Status of Refugees 1951, refugees are defined as those fleeing persecution due to the grounds identified in the Convention. This means there is no legislative basis guiding international cooperation programs and there is no legal precedent informing the process of relocation from one territory to another when it comes to environmental migration. Recently, there have been attempts to address these shortcomings by drafting two international instruments. The Global Compact for Safe, Orderly and Regular Migration was adopted in July 2018 by the United Nations and is a non-legally binding international framework proposing cooperative measures to regulate different forms of migration. It contains a section on "Natural disasters, the adverse effects of climate change, and environmental degradation" (Global Compact for Migration, 2018). Despite its non-binding character, the Compact is meant to encourage improved cooperation in managing the root causes of migration, including environmental and climatic challenges, as well as the people who migrate. Ideally, it should prompt states to initiate national legal processes and policies to collaborate with other states and find appropriate solutions for migrants. One example could be New Zealand whose government in 2017 presented a plan to issue 100 special refugee visas to Pacific Islanders. Quite contrary to this, the United States has already left the drafting process in December 2017, due to fears that the main ideas of the Compact do not correspond with the immigration agenda brought forward by the Trump administration (McDonnell, 2018).

The other international instrument, the Global Compact on Refugees, mentions climate change and natural disasters solely as factors interacting with refugee movements, but does not consider them as sole causes (Global Compact on Refugees, 2018). Therefore, it will not significantly change the international community's policy responses in that respect. In addition to the two Compacts, however, there are also other initiatives relevant to this area, such as a United Nations "Task Force on Displacement" set up under the 2015 Paris Agreement presenting recommendations for integrative approaches to minimize and address displacement related to climate change in late 2018 (UNFCCC, 2018).

Elderly and disabled people

Although advocacy and awareness-raising in relation to climate justice often focuses on children, women, and indigenous peoples, there are more societal groups that are disproportionally affected by climate impacts. These include persons with disabilities and older people (OHCHR, 2009). Their concerns and needs have to be mainstreamed into adequate climate adaptation and mitigation policies in order to develop just climate practices.

As weather and precipitation change and the availability of natural resources diminishes, the rights to food, water, and health will be at risk. Climate adaptation programs need to make sure that distribution sources for food, water, and healthcare are accessible to persons with disabilities, even after severe weather events, and that there are mechanisms in place to offset climate impacts on poverty reduction for disabled people in developing countries. Occupations, especially in the agricultural sector, will change due to climate change and there need to be appropriate policy responses to ensure that barriers to agricultural occupations are removed for disabled people, even if these multiply in the context of climate change. Furthermore, alternative livelihood trainings should be available, as changes in occupation, or even relocation, may be a necessary adaptation strategy (GPDD and World Bank, 2009).

During extreme weather events, people with disabilities are more likely to be left behind, their social networks interrupted, and they experience discrimination. In addition, they may not be able to access social and healthcare services. Therefore, disabled people need personal emergency plans and access to information related to climate change, including through alternative formats (Walker, 2013). Emergency shelters as well as early warning systems need to be accessible. In order to realize this, humanitarian and disability organizations need to build strong partnerships and promote inclusive practices. The Convention on the Rights of Persons with Disabilities explicitly addresses situations of humanitarian emergencies and natural disasters. In Article 11, it states that

> States Parties shall take, in accordance with their obligations under international law, including international humanitarian law and international human rights law, all necessary measures to ensure the protection and safety

> of persons with disabilities in situations of risk, including situations of armed conflict, humanitarian emergencies and the occurrence of natural disasters.
> (CRDP, 2006: Article 11)

Moreover, the Convention stresses the obligation of states to cooperate, as well as the need for inclusive and accessible international development programs in collaboration with international organizations, regional organizations, and civil society (Ansell, 2009). However, there is still a need for further research and case studies on how this can work in practice – especially in the context of climate change and increased risk of natural disasters.

Older people constitute another societal group that is physically, emotionally, and financially less resilient than others when faced with adverse consequences of climate change. Their vulnerability can be explained by heightened physiological susceptibility combined with socio-economic factors, such as smaller social networks, less economic means, or poverty. Older people are often less mobile, have pre-existing health conditions, and restricted access to resources (Filiberto et al., 2011). The elderly might lack awareness regarding environmental risks, and thus lack the capacity to adapt to a changing climate. They are also at particular risk when it comes to health-related effects, especially in the context of heatwaves, changes in precipitation, or weather-related disasters. In order to reach out to older people and increase their awareness on climatic impacts, innovative dissemination methods might be necessary. As many older people understand climate change as a concern of younger generations, it is important to engage the elderly in relevant discussions and include them in processes of policy-making to strengthen adaptation and mitigation approaches (Haq, Brown, and Hards, 2010).

This chapter discussed different vulnerabilities in the context of climate change by considering the situation of women, children, indigenous peoples, environmental migrants, people with disabilities, and elderly people. One lesson to be learned from analyzing these different situations is that it is not enough to reflect on injustices between developed and developing countries. In order to design appropriate policy responses, we need to look inside countries and discuss the needs of particular groups, communities, and individuals within societies. Another important aspect to consider is that one community or an individual group can be affected by multiple vulnerabilities. For instance, an indigenous community can be forced to leave their traditional lands and become environmental migrants. Or a group of disabled women might have to deal with an extreme weather event. Thus, it is important to provide access to information, include the voices of these societal groups and their multiple concerns, and to ensure participation in adaptation and mitigation policy-making. Traditional knowledge of indigenous peoples, for example, can help to improve adaptation approaches, as they have successfully adapted to environmental changes over years and developed sustainable livelihood strategies in intact ecosystems. In this way, we do not only view vulnerable groups as passive victims but as active decision-makers in the context of climate change.

DISCUSSION QUESTIONS

1. How can climate injustice within societies be diminished?
2. How can capacities of women and children be enhanced in order to reduce their vulnerability to climate change?
3. Is it possible to include all vulnerable groups in decision-making processes? What are the advantages and disadvantages of doing this?
4. What can we learn from indigenous peoples' traditional knowledge? How can we use this knowledge to design better adaptation and mitigation policies?
5. Do we need new international legal instruments to cover environmental migration?
6. Do we need social workers and care workers who are climate experts in order to develop personal emergency and adaptation plans for the elderly and persons with disabilities?

References

Aiken, A. (2016). *Opinion Order. Kelsey Cascada Rose Juliana et al. vs. the United States of America et al. Case No. 6:15-cv-01517-TC.* United States District Court for the District of Oregon, Eugene Division.

Ansell, K. (2009). *Where Disability Meets Climate Change.* [Online]. Available at: www.bbc.co.uk/ouch/features/where_disability_meets_climate_change.shtml [Accessed: 8 October 2018].

Bonner, O. (2018). *Tackling Gender Disparity at the Intersection of Human Rights and Climate Change.* [Online]. Available at: www.ciel.org/gender-disparity-intersection-human-rights-climate-change/ [Accessed: 21 September 2018].

BRIDGE and DFID (2008). *Gender and Climate Change: Mapping the Linkages: A Scoping Study on Knowledge and Gaps.* University of Sussex: Institute of Development Studies.

CEDAW (1979). Convention on the Elimination of All Forms of Discrimination Against Women. 18 December 1979: New York, NY.

CEDAW (2010). General Recommendation No. 27 on Older Women and Protection of their Human Rights. Committee on the Elimination of All Forms of Discrimination Against Women.

CEDAW (2016a). General Recommendation No. 34 on the Rights of Rural Women. Committee on the Elimination of All Forms of Discrimination Against Women.

CEDAW (2016b). Draft General Recommendation No. 37 on the Gender-related Dimensions of Disaster Risk Reduction in a Changing Climate. Committee on the Elimination of All Forms of Discrimination Against Women.

Chazalnoel, M.T. and Carini, F. (2018). *Environmental Migration and Indigenous Peoples: What is at Stake?* [Online]. Available at: https://medium.com/@IOMatUN/environmental-migration-and-indigenous-peoples-what-is-at-stake-1c127228fb67 [Accessed: 5 October 2018].

CRPD (2006). Convention on the Rights of Persons with Disabilities (A/RES/61/106). Adopted at the General Assembly on 13 December 2006 at United Nations Headquarters in New York.

CRC (1989). Convention on the Rights of the Child. Adopted and Opened for Signature, Ratification and Accession by General Assembly Resolution 44/25 of 20 November 1989, entry into force 2 September 1990.

CIEL and GI-ESCR (2018). States' Obligations under the Convention on the Elimination of All Forms of Discrimination Against Women, in the Context of Climate Change. Center for International Environmental Law and The Global Initiative for Economic, Social and Cultural Rights.

Faustino, C. and Furtado, F. (2014). The Green Economy, Forest Peoples and Territories: Rights Violations in the State of Acre, Fact-finding and Advocacy Mission Preliminary Report. Rio Branco, Uruguay: World Rainforest Movement.

Filiberto, D., Wethington, E., Pillemer, K., Wells N.M., Wysocki, M. and Parise, J.T. (2011). Older People and Climate Change: Vulnerability and Health Effects. [Online]. Available at: www.asaging.org/blog/older-people-and-climate-change-vulnerability-and-health-effects [Accessed: 8 October 2018].

Gadgil, M. and Guha, R. (1994). Ecological Conflicts and the Environmental Movement in India. *Development and Change* 25 (1): 101–36.

Global Compact for Migration (2018). Global Compact for Safe, Orderly and Regular Migration. [Online]. Available at: www.un.org/pga/72/wp-content/uploads/sites/51/2018/07/180713_Agreed-Outcome_Global-Compact-for-Migration.pdf [Accessed: 7 October 2018].

Global Compact on Refugees (2018). Global Compact on Refugees. [Online]. Available at: www.unhcr.org/gcr/GCR_English.pdf [Accessed: 7 October 2018].

GPDD and World Bank (2009). The Impact of Climate Change on People with Disabilities. Report of the e-Discussion hosted by Global Partnership for Disability and Development and World Bank, 8 July 2009, CBM International.

Haq, G., Brown, D., and Hards, S. (2010). *Older People and Climate Change: The Case for Better Engagement.* Stockholm: Stockholm Environment Institute.

Humphreys, S. (2014). Climate Justice: The Claim of the Past. *Journal of Human Rights and the Environment* 5 (0): 134–48.

IDMC (2018). 2017 Internal Displacement Figures by Country. [Online]. Available at: www.internal-displacement.org/database/displacement-data [Accessed: 7 October 2018].

IFAD (2010). *Climate change/strategy: Enabling Poor Rural People to Overcome Poverty.* Rome: International Fund for Agricultural Development.

IFRC (2018). World Disasters Report 2018. International Federation of Red Cross and Red Crescent. [Online]. Available at: https://media.ifrc.org/ifrc/world-disaster-report-2018/ [Accessed: 19 December 2018].

ILO (1989). Convention Concerning Indigenous and Tribal Peoples in Independent Countries. Adopted at the 76th Session of the International Labour Conference, 27 June 1989.

ILO (2017). *Indigenous Peoples and Climate Change: From Victims to Change Agents Through Decent Work.* Geneva: International Labour Office.

IOM (2011). Glossary on Migration, International Migration Law No. 25, 2nd edn. Geneva, IOM [Online]. Available at: http://publications.iom.int/bookstore/index.php?main_page=product_info&cPath=56&products_id=1380 [Accessed: 7 October 2018].

IOM (2014). *IOM Outlook on Migration, Environment and Climate Change*. Geneva: International Organization for Migration.

IPCC (2014). *Climate Change 2014: Synthesis Report. Contribution of Working Groups I, II and III to the Fifth Assessment Report of the Intergovernmental Panel on Climate Change*. Pachauri, R.K. and Meyer, L.A. (eds.). Geneva: IPCC.

Kichwa v. Ecuador (2012). Merits, Reparations and Costs, Judgment, Inter-Am. Ct. H.R. (ser. C) No. 245, 58 (June 27, 2012).

McAdam, J. (2012). *Climate Change, Forced Migration and International Law*. Oxford: Oxford University Press.

McDonnell, T. (2018). The Refugees the World Barely Pays Attention To [Online]. Available at: www.npr.org/sections/goatsandsoda/2018/06/20/621782275/the-refugees-that-the-world-barely-pays-attention-to?t=1538866632395&t=1538940699329 [Accessed: 7 October 2018].

OHCHR (2009). Report of the Office of the United Nations High Commissioner for Human Rights on the Relationship between Climate Change and Human Rights, A/HRC/10/61.

Our Children's Trust (2018). *Juliana v. US – Climate Lawsuit*. [Online]. Available at: www.ourchildrenstrust.org/us/federal-lawsuit/ [Accessed: 3 October 2018].

Petition to the Inter-American Commission on Human Rights (2005). Petition to the Inter-American Commission on Human Rights Seeking Relief from Violations Resulting From Global Warming Caused by Acts and Omissions of the United States. [Online]. Available at: http://blogs2.law.columbia.edu/climate-change-litigation/wp-content/uploads/sites/16/non-us-case-documents/2005/20051208_na_petition.pdf [Accessed: 19 December 2018].

Randall, A. (2018). Climate Refugees Definition: Can We Define a Climate Refugee? [Online]. Available at: http://climatemigration.org.uk/climate-refugees-definition/ [Accessed: 7 October 2018].

Shue, H. (2014). *Climate Justice: Vulnerability and Protection*. Oxford: Oxford University Press.

Stern, N. (2006). *The Economics of Climate Change: The Stern Review*. Cambridge: Cambridge University Press.

UNDRIP (2007). United Nations Declaration on the Rights of Indigenous Peoples. Adopted at 107th Plenary Meeting of the United Nations General Assembly, 13 September 2007.

UNFCCC (2015). Paris Agreement. Adopted at the 21st Conference of the Parties (COP), 12 December 2015, United Nations Framework Convention on Climate Change, Paris.

UNFCCC (2018). Task Force on Displacement [Online]. Available at: https://unfccc.int/wim-excom/sub-groups/TFD#eq-3 [Accessed: 7 October 2018].

UNICEF (2010). *A Brighter Tomorrow: Climate Change, Child Rights and Intergenerational Justice*. London: UNICEF UK.

UNICEF (2015). *Unless We Act Now: The Impact of Climate Change on Children*. New York: United Nations' Children's Fund.

United Nations (1951). *Convention Relating to the Status of Refugees*. Geneva: United Nations General Assembly.

United Nations (2008). Climate Change and Indigenous Peoples [Online]. Available at: www.un.org/en/events/indigenousday/pdf/Backgrounder_ClimateChange_FINAL.pdf [Accessed: 3 October 2018].

Vanderheiden, S. (2004). Justice in the Greenhouse: Climate Change and the Idea of Fairness. *Social Philosophy Today* 19: 89–101.

Walker, R. (2013). Understanding the Impacts of Climate Change and Adverse Weather Events on People with a Disability and Their Carers [Online]. Available at: www.cakex.org/sites/default/files/documents/Enliven%20Disability%20&%20Climate%20Change%20Report.pdf [Accessed: 8 October 2018].

12 Mitigation, adaptation, and loss and damage

Introduction

In addition to climate change consequences, such as sea level rise, extreme weather events, floods, and droughts, the political responses to address these impacts, i.e., climate policies, can also have adverse effects on human rights. Practices that neglect or even violate rights can be identified in several climate mitigation and adaptation policies, especially in relation to lack of access to information and justice procedures, and lack of opportunities provided to participate in decision-making. This is still a deficit in policy-making, because the voices of those who are affected by climate impacts need to be considered more strongly in policy responses and we know, for instance, that local knowledge and experiences of indigenous peoples can be fruitfully used to design and improve adaptation policies. Thus, many scholars and practitioners emphasize the need to include human rights considerations in a coherent and consistent way in climate policy-making.

It is important to realize that policy planning for mitigation and adaptation as well as loss and damage is closely connected: the degree of mitigation determines the necessity to adapt, and if mitigation and adaptation can be successfully conducted, loss and damage will be limited (Harmeling, 2018). In this chapter, we will first take a closer look at rights concerns in the context of mitigation projects before moving on to learning more about climate adaptation as well as loss and damage.

Climate change mitigation from a human rights perspective

Climate change mitigation comprises actions to reduce or prevent greenhouse gas emissions. Mitigation strategies can be complex and range from switching to renewable energy or employing innovative technologies to managing forests and changing consumer behavior. Examples of mitigation policies are activities under the Clean Development Mechanism (CDM), Reducing Emissions from Deforestation and Forest Degradation (REDD+) programs, and domestic endeavors under the umbrella of Green Economy (GE) strategies.

Clean Development Mechanism

The Clean Development Mechanism was established under the Kyoto Protocol. It combines two main objectives: emission reductions and sustainable development. At the same time, it provides industrialized countries with some flexibility on how to meet their binding emission reduction targets. Annex I countries under the Kyoto Protocol, i.e., industrialized countries, can meet their commitments by investing in emission reduction projects and by buying Certified Emission Reduction units (CERs). The projects are implemented in non-Annex I countries, i.e., developing countries, and are supposed to contribute to their sustainable development, e.g., by enhancing access to energy for the domestic population.

The negative human rights consequences of CDM project activities have been subject to discussion among civil society organizations and academics, and at the UN. A compilation of case studies by the NGO Carbon Market Watch (cooperating with other organizations in the Carbon Market Watch Network) has revealed that the local realities of CDM projects often go hand-in-hand with constraints in the realization of substantive and procedural rights. Empirical evidence for these rights constraints can be found mainly in Asia and in Latin America (Carbon Market Watch, 2013). One relevant example is the Adani's Mundra coal power project in India. It was registered in 2009 as the first coal power project to receive carbon credits. The stated purpose behind this was to reduce fossil fuel (coal) consumption through higher power generation efficiency (Joshi, 2013: 4). It has been reported, however, that the project harmfully affects the local environment and conflicts with India's national environmental laws. A human rights analysis revealed that the project's emissions, especially the flying ash, have detrimental consequences on the right to health and the right to means of subsistence of the local community. In addition, regular lethal incidents were reported to occur at the industrial site, and force had been used to close down a creek for project implementation which had been used by several fishermen. Moreover, the livelihoods of fishermen and farmers, i.e., their right to food and to water, were endangered by contaminated fish and the withdrawal of the grazing land of 14 villages. In summary, Adani's Mundra coal power project neither operated in accordance with the mandate of the Clean Development Mechanism, nor did it contribute to sustainable development (Joshi, 2013: 5).

Another example is Panama's Barro Blanco hydroelectric dam project. It is currently under construction damming the Tabasará River in the Chiquiri Province of Panama. Construction of the dam commenced in 2011 and during the same year, Barro Blanco was registered as a project under the Clean Development Mechanism. The dam project is developed by a company named Generadora del Istmo S. A. (GENISA), which was created especially to build Barro Blanco. Financial support is granted by two European development banks – the German Investment Corporation and the Netherlands Development Finance Company – and the Central American Bank for Economic Integration (CABEI). Prior to project implementation, GENISA commissioned an environmental impact assessment study and consulted the affected indigenous Ngäbe and Buglé communities. In 2007, GENISA and representatives of the Ngäbe and Buglé signed a cooperation agreement including the protection of safeguards and indigenous peoples' rights. Based on this study, the Panamanian Environmental Authority approved the dam project and a validation team by the consulting team AENOR confirmed for the UN CDM Executive Board that free, prior, and informed consultation had taken place.

In 2009, suggestions to increase Barro Blanco's capacity from 19 megawatts to 28.8 megawatts were discussed. This raised critical questions about the impact assessment studies that had been conducted for a smaller dam. At the same time, there were disagreements about mining projects under a new legislation proposed by the Panamanian government that would severely affect indigenous territories. These new proposals triggered conflict between the affected indigenous communities and the government of Panama. Social mobilization against the dam was organized within the Movimiento 10 de Abril (M-10) and reached out to transnational advocacy networks. With the help of legal experts from environmental NGOs, a series of court proceedings was launched to prevent the project (Hofbauer and Mayrhofer, 2016). Protests culminated in the blocking of the Pan–American highway for a week during January 2012 (Spain, 2012). Demonstrations increased pressure and prompted the Panamanian government to enter into negotiations with the indigenous leader Silvia Carrera (Evans, 2014).

In 2012 and 2013, the international protest campaign continued and led to an inspection conducted by the United Nations Development Programme (UNDP), and subsequently to a follow-up complaint to an Independent Experts Panel of the involved international development banks. UNDP's assessment report found major flaws in the initial consultation process and confirmed that the continuation of this dam project will flood indigenous peoples' homes as well as cultural sites, and also turn the Tabasará River into a stagnant lake ecosystem, adversely affecting the means of subsistence of the Ngäbe communities. Thus, the economic, social, and cultural rights of the communities would be violated as a result of the dam construction and dam operation. Finally, a temporary suspension order was issued for the project in 2015, which was later overruled by Panama's Supreme Court.

Despite continuing protests often culminating in violent confrontations with the police and the killing of one protestor, Onesimo Rodriguez, allegedly by a masked policeman after a rally, dam construction continued. In 2016, Panama formally withdrew support for Barro Blanco and cancelled its registration as a CDM project (Carbon Market Watch, 2016). In December 2016, Panama's Supreme Court ruled in favor of the project declaring that it was in the "public's interest" (Giraldo, 2017), despite opposition by the Ngäbe communities. As the Supreme Court's decisions cannot be appealed, the dam was expected to become operative in 2017. At the time of completing this book in October 2018, there was no information available confirming that the dam has started operating yet.

The human rights violations in relation to Barro Blanco, including deficient consultations, lack of free, prior, and informed consent by affected indigenous communities, infringements on social and cultural rights, as well as violent police reaction to protests, make Barro Blanco an important case in discussions on how to improve CDM projects. Many advocacy organizations argue that the example of Barro Blanco demonstrates that the modalities and procedures of the CDM have to be reformed, and need to include strong environmental and social safeguards on the basis of human rights. In their rhetoric, even environmental NGOs that seek to improve CDM projects have started to use human rights language and argue from the perspective of climate justice (Kuchler, 2017).

Reducing Emissions from Deforestation and Forest Degradation (REDD+)

Reducing Emissions from Deforestation and Forest Degradation or REDD+ can be understood as a voluntary climate mitigation approach as part of the UN Framework Convention on Climate Change (UNFCCC). It has been developed against the background that 17% of global net emissions result from deforestation and forest degradation. Under the REDD+ framework, countries that take action to reduce deforestation and forest degradation will be financially rewarded according to their achieved emission reductions. Since 2010, not only emission reductions from deforestation and forest degradation, but also the conversion and enhancement of forest carbon stocks, and the sustainable management of forests have become components of REDD+.

At COP16 in 2010, procedural criteria for the realization of REDD+ programs were introduced into the Cancun Agreements (UNFCCC, 2010). The institutionalization of these social and economic "safeguards" comprised recommendations including respect for the knowledge and rights of indigenous peoples and local communities in consideration of the 2007 United Nations Declaration on the Rights of Indigenous People (UNDRIP, 2007) and other obligations anchored in international law. Moreover, the complete and effective participation of all affected people, particularly indigenous peoples and local communities, with reference to their free, prior informed consent has been emphasized (UNFCCC, 2010: Appendix I, paragraph 2).

In an empirical study on the implementation of REDD+ in Peru, Johanna Steudtner (2012) has demonstrated how program realization – despite these safeguards – can go hand-in-hand with severe infringements of rights of local communities. These often occur in the context of conflicts around property, land, and resources (Steudtner, 2012: 124). Forest protection and management measures can affect the people who live on the territories at stake and who use the forest as a source of subsistence. Often, these are indigenous population groups whose right to self-determination (ICCPR, 1966; ICESCR, 1966: Article 1) conflicts with the forest management measures under REDD+ and cannot be guaranteed anymore. Similarly endangered (and closely related) is their right to their own means of subsistence, also incorporated in the International Covenant on Economic, Social and Cultural Rights (ICESCR, 1966: Article 1.2). Additionally, other targeted rights – adopted with the aim of explicitly protecting indigenous peoples – can be threatened by the activities initiated through REDD+. These rights are anchored in the 1989 ILO Convention concerning Indigenous and Tribal Peoples in Independent Countries (ILO, 1989) and the 2007 United Nations Declaration on the Rights of Indigenous People (UNDRIP, 2007).

The CDM and REDD+ are examples of mitigation approaches under the UNFCCC that can have adverse human rights effects. At the national level, rigorous climate policy implementation can lead to very serious human rights violations that severely affect those who are the most vulnerable. This situation of climate injustice is exacerbated if climate policies are implemented in the context of an authoritarian state, as the following case study on Ethiopia demonstrates.

Case study: Green economy and the Gibe III hydroelectric dam in Ethiopia

The Federal Democratic Republic of Ethiopia today is known as an authoritarian government which is "not free" according to FHI 360, and "least free" regarding political rights (FHI, 2018). In 2009, the government adopted a Charities and Societies Proclamation which entered into force in 2010, heavily restricting and controlling activities and publications of civil society in the country, in particular with respect to human rights and minority issues. Moreover, Ethiopia ranks fourth among the most censured countries according to the Committee to Protect Journalists (CPJ, 2015), imposing severe restrictions on the press and the media, and arresting journalists and researchers who raise critical questions and concerns.

Also in the course of 2010, the Ethiopian government adopted a five-year Growth and Transformation Plan (GoE, 2010). Within the framework of this development plan, the government formulated the ambitious objective of becoming a middle-income country in the next decade between 2020–2023. In order to achieve this, the Ethiopian government launched a green economy strategy in 2011 entitled the Climate Resilient Green Economy (CRGE) initiative (GoE, 2011a).

Ethiopia's CRGE

The CRGE strategy was formulated by the former Environmental Protection Authority (EPA) under the leadership of the Prime Minister's Office. Researchers of the governmental Ethiopian Development Research Institute and Addis Ababa University were involved in the strategic planning process from an early stage. The CRGE vision envisages that Ethiopia not only adapts to climate change, but also takes it as an opportunity to transform its economy to attract investments and to become an African leader in low-carbon growth (GoE, 2011b).

Economic development and commercial benefit played a key role in this planning process, in which the Prime Minister's Office was advised by the private consultancy firm, McKinsey & Co. The CRGE policy document stipulates (a) fostering economic development and growth; (b) abatement and avoidance of future emissions; and (c) improving resilience to climate change as the three main objectives (GoE, 2011a: 19). To achieve these objectives, different sectors are envisaged to be transformed, including energy, agriculture, industry, forestry, and transport. Whereas the EPA was initially responsible for policy implementation, a new Ministry of Environmental Protection and Forestry, established by the end of 2013, took over this task.

With its CRGE strategy, the government focused on becoming a "green economy-frontrunner" (GoE, 2011a: 1). Other social aspects of development, such as public health, access to water and food security are clearly formulated to be "co-benefits" (GoE, 2011a: 2). Hydropower generation is a key priority in the CRGE. An illuminating example for the neglect of social issues in transforming the energy sector within the framework of the green economy strategy is the Gilgel Gibe III hydroelectric dam. Gibe III is currently the biggest and most representative Ethiopian dam project and is considered to be a "milestone for building Ethiopia's green economy" (Wubete, 2017).

Gibe III dam

Gibe III is currently the biggest African dam project in operation.[1] It is meant to deliver electricity to more than 80 million people, even to regions that have not yet had access to electricity at all (Human Rights Watch, 2012). The dam is located about 300 kilometers southwest of the capital, Addis Ababa, on the Omo river. Its construction began in 2006, and it commenced power generation in 2015. All generators became operational at the end of 2016, producing 1,870 megawatts of electricity. The dam will more than double Ethiopia's current capacity, so that it will be able to sell energy to neighboring countries, including Sudan, Kenya, and Djibouti. Project development and oversight are in the hands of the Ethiopian Electric Power Corporation (EEPCo). Salini Impregilo, an Italian construction company, is constructing the dam. The largest funder is the Export-Import Bank of China, whereas the Ethiopian government spent US $572 million of its national budget to finance the dam project. After carrying out preliminary evaluation studies, the World Bank (WB), the European Investment Bank (EIB), and the African Development Bank (AfDB) announced in 2010 that they were no longer considering funding Gibe III, due to serious doubts relating to adverse social and environmental impacts (International Rivers, 2011).

Governmental and non-governmental actors refer to the potential green economy policies, including the Gibe III hydroelectric dam project, have for future social development. A representative from the Ethiopian Ministry of Water and Energy stresses that the electricity generated will help to improve the country's infrastructure, including schools and hospitals. Furthermore, it will enhance people's access to clean and safe drinking water. According to another representative of the same Ministry, increased access to electricity can be viewed as a precondition for realizing several social rights.

Some NGOs share this view. They see the potential of green economy policies for promoting human rights, particularly in the long run. Some, however, explain that Ethiopia's government does not necessarily think in terms of individual rights, but rather focuses on development aspects from which society as a whole can profit.

Trade-offs in transition

Despite the potential attributed to realizing human rights and social development in the future, the rigorous transition path followed by the government has led to serious rights infringements against indigenous peoples who depend on the river Omo for their living. Substantial empirical evidence for this can be found in reports published by Human Rights Watch (2012, 2015) and International Rivers (2011, 2015), press releases and articles issued by Survival International (2015, 2016), and interviews with governmental and nongovernmental experts in Ethiopia.

According to these reports and documents, several human rights violations have been committed in the course of establishing Gibe III and associated plans for irrigated agricultural development. Human Rights Watch estimates that about 200,000 residents of the Lower Omo and another 300,000 people living around Lake Turkana in Kenya are affected (Human Rights Watch, 2012). Indigenous communities have been forced by the government and the military to leave their ancestral territories. Harassment, violence, and arbitrary arrests were used to coerce these people. Agro-pastoral and pastoral communities who used the river Omo for their subsistence were resettled in different places. They had to observe their grazing lands being flooded, were forced to reduce their cattle, and were informed that they will most probably lose access to the river (Human Rights Watch, 2012).

There have been a number of incidences in which the government provided investors with land that was vacant because the people who traditionally occupied it were pastoralists moving with their herds according to seasons, depending on rain and grazing land. On their return, their ancestral lands had – in many cases – been taken away and issued to private investors. Land rights are strongly interlinked with other social rights. If people practice subsistence agriculture and farming, their right to means of subsistence, including food, water, adequate housing, or even health, can be adversely affected if access to their territories and water sources is denied. In addition, their right to self-determination and cultural rights are infringed upon.

Thus, in the case of Gibe III, green economy transition endangers indigenous peoples' rights to food and to means of subsistence as embodied in the International Covenant on Economic, Social and Cultural Rights (ICESCR) ratified by the Government of Ethiopia in 1993 (UNTC, 2016). Moreover, consultations with the local communities have not been held, free, prior, and informed consent has not been obtained, and land losses – due to relocations – have not been appropriately compensated. This means that land rights stipulated in the ILO Indigenous and Tribal Peoples' Convention and rights to natural resources pertaining to these ancestral lands (ILO, 1989: Articles 13–15) are being violated with the building of this dam. Although Ethiopia has not ratified the ILO Convention, it is acting contrary to customary international law as reflected in the UN Declaration on the Rights of Indigenous Peoples and the Declaration on the Right to Development. It is also in breach of binding regional instruments, such as the 1981 African Charter on Human and Peoples' Rights (ACHPR, 1981). The ACHPR, ratified by Ethiopia in 1998, includes important provisions in this respect. These comprise the right to information (Article 9), the right to property (Article 14), participation in cultural life (Article 17), peoples' right to existence and self-determination (Article 20), free disposal of peoples' wealth and natural resources (Article 21), and peoples' right to development (Article 22), as well as the right to a general satisfactory healthy environment (Article 24). None of these rights were guaranteed to affected citizens in the course of green economy transition in Ethiopia.

Ethiopian NGOs are concerned about this situation. They highlight the ambitious character and rigorous implementation strategies in the context of the CRGE. A number of critical observers claim that the government is not particularly interested in climate resilience or social development. The fact that social matters are neglected in Ethiopia's green economy strategy was voiced by a representative of UNEP in an informal interview. Instead, state actors have made an extremely smart move politically to secure financial aid available in the climate policy area. The government understands green economy as a window of opportunity to secure funds for dealing with the consequences of climate change caused by developed countries and to foster economic development. Relying on financial support and investments, the government is interested in a swift economic transformation. Focusing on overall development and progress, state actors have accepted severe constraints for indigenous peoples.

The transnational dimension

Field research carried out by Human Rights Watch (2015) and International Rivers (2015) illuminates the transnational dimension to this scenario. Reports from Turkana County in North-western Kenya reveal that Ethiopia's massive hydroelectric and irrigation projects, including Gibe III, also have a devastating impact on human rights across borders. The local population, indigenous peoples relying on livestock, heavily depend on the water supply of Lake Turkana which receives 90% of its water from the Omo River. Ethiopia's dam projects, irrigation canals, and cotton and sugar plantations have drastically reduced the water supply for Turkana people, constraining access to water for the local population in a serious way. Further consequences are food scarcity and competition over grazing lands. This directly affects the economic, social, and cultural rights of the Turkana people, including the rights to water, food, and health, and increases the potential for conflict. Thus, there are also indirect effects on civil and political rights, like the right to personal integrity and security. In a report published by International Rivers, a pastoralist from Turkana County points to this conflict situation: "We will fight with whoever is coming close to the lake. We have our children here. They will destroy that electricity project. Have you heard me clearly? A person from another land can't come and close our lake" (International Rivers, 2015: 13).

A representative from the Ministry of Pastoral Economy and Fisheries of the Turkana County Government emphasized: "We cannot eat electricity. What we require is food and income for the Turkana community" (International Rivers, 2015: 9).

Therefore, Gibe III as a green economy endeavor not only leads to direct and indirect adverse consequences for social groups within Ethiopia, but also in Kenya, revealing a transboundary effect. As a consequence of these severe rights violations, a situation of conflict is emerging (Carr, 2012).

Governmental ambitions

Confronting governmental representatives with these rights concerns during field research, most of them responded with a very clear prioritization. None of them denied that rights infringements in the context of green economy transition are taking place. Instead, they claimed that these are "costs" one has to take into account when implementing these projects.

Interviewees from the Ministry of Water and Energy and from governmental research institutes seem to take these detriments into account because they affect only certain societal groups. They believe that so-called "primitive" indigenous peoples do not fit into the CRGE vision of modernization and rapid economic growth. Traditional pastoralist and agro-pastoralist lifestyles are hampering the government's vision of development. These research results are in line with Human Rights Watch's Report on the Lower Omo Valley and concerns raised by a representative of an international NGO, who explained the government's climate policy comprises further restrictions on pastoralists and agro-pastoralists.

Climate change adaptation from a human rights perspective

Adaptation policies are of utmost importance because the consequences of climate change, including extreme weather events, floods, changes in precipitation and droughts have severe impacts on the lives of many people, especially in developing countries. Sea level rise, for instance, can lead to the loss of land, lack of clean drinking water, damage to coastal infrastructure, homes and properties, loss of agricultural lands, damage to beaches, and threats to tourism. From a human rights perspective, this means that in severe cases, the right to life and the right to self-determination can be affected, as well as the rights to water, health, adequate housing, means of subsistence, culture, and property (ICCPR, 1966; ICESCR, 1966). Coastal areas, low-lying island states, and the Arctic region are the most affected by these climate change impacts resulting from sea level rise. Temperature increases affect many regions in the world, but are most severely felt in Sub-Saharan Africa, Northern Africa, South Asia, Latin America, and the Middle East (OHCHR, 2009). In those regions, the rise in temperature leads to the spread of disease, changes in fisheries and agriculture, loss of biodiversity, and threats to tourism. Mostly affected are the rights to life, health, and an adequate standard of living, including means of subsistence (ICCPR, 1966; ICESRC,

1966; Orellana and Johl, 2013). A report published by the Office of the High Commissioner for Human Rights (OHCHR) identified women, children, and indigenous peoples, disabled people, and the elderly – in all of these regions – as the most vulnerable population groups with respect to climate change impacts (OHCHR, 2009). It is important to acknowledge that these impacts severely differ between a scenario of 1.5°C and 2°C global warming – a temperature increase of 1.5°C will already severely threaten human rights (OHCHR, 2015: 12). The difference between a 1.5°C and a 2°C temperature increase is also addressed in the 2018 report by the Intergovernmental Panel on Climate Change (IPCC, 2018). The IPCC report, which also directly refers to human rights, states that the difference between both scenarios will be dramatic and that "rapid, far-reaching and unprecedented changes in all aspects of society" are necessary to protect human and ecosystem health. To keep the temperature increase below 1.5°C, further commitment that even goes beyond the level of ambition agreed upon in the 2015 Paris Accord will be necessary (IPCC, 2018).

Adaptation is a response to these climate change impacts with the aim of reducing the vulnerabilities of the above-mentioned social groups (and entire ecosystems), minimizing the effects of climate change. The problem for many local population groups from developing economies is that those who are the least responsible for climate change impacts are often the ones most affected, but they have the fewest resources and least capacity to adapt. From a perspective of climate justice, this makes it even more important to include these people in policy-making, to ensure meaningful participation in decision-making processes, and develop their capacities to adapt. The OHCHR has, in a submission to the Conference of the Parties prior to the Paris negotiations, highlighted capacity-building, especially for vulnerable communities:

> States must build adaptive capacities in vulnerable communities, including by recognizing the manner in which factors such as discrimination, and disparities in education and health affect climate vulnerability, and by devoting adequate resources to the realization of the economic, social and cultural rights of all persons, particularly those facing the greatest risks.
>
> (OHCHR, 2015: 2)

According to the Intergovernmental Panel on Climate Change (IPCC), adaptive capacity in relation to climate impacts is very closely linked to social and economic development (IPCC, 2007). The IPCC itself does not use a human rights-based approach in relation to adaptation (even though the newest IPCC report launched in October 2018 makes several references to human rights), but it has linked its scientific findings of Working Group II (Impacts, Adaptation, and Vulnerability) to the concept of human security. It reveals that there is ample evidence to suggest that human security will be severely threatened by the impacts of climate change. Moreover, it asserts that cultural values, which are necessary for individual and community wellbeing, are at risk. This is in line with empirical studies emphasizing damaging effects of a changing climate on cultural heritage, and thus, cultural rights (Maus, 2014). The IPCC also points to migration movements caused by climate impacts and compromising human security. Finally, one of the strongest arguments for adopting a human rights-based approach to climate adaptation is robust scientific evidence for the fact that indigenous, local, and traditional knowledge and experiences can serve as a major resource for adaptation (IPCC, 2007; Harmeling, 2018). Thus, by meaningfully developing and employing a rights-based approach, and enabling local population groups to participate in policy-making, adaptation can be considerably strengthened in many parts of the world.

First attempts at a human rights-based approach in adaptation

There is still an obvious policy gap in linking human rights with international or national adaptation planning and there have, so far, only been few attempts to include human rights in adaptation. The UNFCCC Adaptation Committee, set up with the Cancun Agreements, is the main institution responsible for developing a coherent adaptation framework. Harmeling (2018) points out that the Adaptation Committee could, in the future, provide guidance to the state parties in relation to adaptation planning, paying stronger attention to human rights protection. This is relevant with respect to the exchange of information regarding vulnerable communities, sectors or regions. But the focus on the SDGs and their relationship to human rights is also a strong entry point – as sustainable development goals and respective indicators could be linked to national adaptation goals and indicators (Harmeling, 2018: 94).

At COP16 in Cancun, UNFCCC parties agreed on the establishment of a process for designing and supporting National Adaptation Plans (NAPs). The guidelines for this process, adopted at COP17 in Durban, do not reflect a comprehensive human rights-based approach for adaptation, even though the need for participatory consultations was emphasized (Harmeling, 2018: 97). An analysis by Lottje, Hirsch, and Minninger (2017) suggests several criteria for designing NAPs following a human rights-based approach to development: (a) including a human rights audit in the NAP process to make sure that basic rights, like the rights to water, health and food, are provided for, and the main principles such as non-discrimination, participation, and transparency, are respected; (b) identifying vulnerabilities according to sector and region and particularly vulnerable population groups; (c) including civil society and vulnerable groups in all stages of the planning process, and enabling meaningful participation in decision-making.

Prior to the Paris negotiations, governments were asked to submit their Intended Nationally Determined contributions (INDCs) to reducing greenhouse gas emissions and, in most cases, also for adaptation actions. In these INDCs, 24 states mentioned human rights and 17 states including Bolivia, Brazil, Ecuador, Malawi, Mexico, the Philippines, and South Sudan, emphasized the need to include human rights in climate action. Seven more states, including Cuba, Indonesia, Venezuela, and Zimbabwe, referred to human rights in their domestic context (HRCCWG, 2015). This is an example that demonstrates how certain states take up the language of human rights and include it into their mitigation and adaptation commitments.

The Adaptation Fund, the Least Developed Countries Fund, and the Green Climate Fund have increasingly begun to take human rights into consideration. The Environmental and Social Policy of the Adaptation Fund, for instance, states that: "Projects/programmes supported by the Fund shall respect and where applicable promote international human rights" (Adaptation Fund, 2016). An emphasis is placed on women's empowerment, gender equality, and the protection of vulnerable communities. Moreover, stakeholder consultations on the basis of human rights are highlighted, as well as the compliance of all projects implemented under the Adaptation Fund with the core international human rights treaties. Where these instruments have not been ratified by a state, program realization has to be in accordance with the Universal Declaration of Human Rights. The Least Developed Countries Fund (LDCF) is less advanced in its human rights approach than the Adaptation Fund. Since it is operated by the Global Environment Facility (GEF), certain environmental and social safeguards have to be respected when implementing projects funded by the LDCF. Consultations with affected population groups, especially indigenous peoples, are particularly highlighted in this regard. The Green Climate Fund (GCF) has now become fully operational, and allocates about 50% of its resources to adaptation action in vulnerable states, among them Least Developed Countries (LDC), Small Island Developing States (SIDS), and African countries. The GCF applies the Performance Standards of the International Financial Corporation – that are highly relevant from a human rights perspective – on an interim basis. These include environmental and social risk assessments, consultations with affected local community groups, community health and safety, participation and consent of indigenous peoples, and protection of cultural heritage (IFC, 2012). The new consultation draft of the GCF contains human rights as the guiding principle in all activities financed by the GFC. This also means that new monitoring entities will need to be established (Harmeling, 2018: 103).

The relevance of a human rights-based approach to adaptation becomes even clearer when looking at an example. In his submission to the UNFCCC, the Special Rapporteur for Human Rights and the Environment emphasized the need to provide access to information and ensure public participation in decision-making in relation to climate change and climate change policies. In this regard, the Special Rapporteur highlighted several country cases as good practices. One example is an adaptation project in Honduras, funded by UNDP, with the objective of reducing climate change risks to water resources in the country. Honduras is one of the Latin American countries that are most vulnerable to climate change and its continuous population growth has led poorer population groups to settle in areas that are prone to landslides and flooding from weather events related to climate change. The project was carried out in Tegucigalpa City and the surrounding areas, and consisted of three major components. The first step was to acknowledge climate change risks in water resource laws and corresponding planning processes. In a second step, adaptation measures to safeguard water supply were taken. Capacities within a certain number of households were built to collect rainwater, which was then diverted to a water tank of a public school. This reduced the water infiltration into the soil after precipitation and, therefore, minimized the risk of landslides. At the same time, more drinking water was available to the schoolchildren. The third step was an awareness-raising component, and training decision-makers and the public in understanding climate change impacts and taking initiatives to reduce vulnerabilities. This is an excellent example that illustrates the need for a human rights-based approach to climate adaptation. It demonstrates how state governments and communities can respond to climate change impacts and reduce risks by protecting basic human rights, like the rights to water, food, and health. At the same time, this example emphasizes that relevant information on impacts and protection measures need to be made transparent, and the public needs to be able to meaningfully participate in decision-making and in all actions taken.

Inspirations from the Paris Agreement

In the Paris Agreement, states acknowledged that they need to "respect, promote and consider their respective obligations to human rights" (UNFCCC, 2015) in all climate-relevant action they take. This includes mitigation as well as adaptation policies. Some countries argued against including human rights in the Paris Agreement, because the main responsibility for implementing human rights remains with state governments. African countries, for example, opposed this link between climate change and human rights in the agreement, because they fear that, in face of the devastating consequences of climate change, they might not be able to respect, protect, and fulfil human rights on their state territory anymore. They are also concerned that developed states will impose conditionalities – on the basis of human rights standards – in the climate arena when financing certain policies (Wallbott and Schapper, 2017).

Article 7 of the Paris Agreement can be understood as a door-opener to a human rights-based approach to adaptation action for several reasons. First, parties acknowledge that adaptation should follow a "country-driven, participatory and fully transparent approach" that should be "based on and guided by, as appropriate, traditional knowledge, knowledge of indigenous peoples and local knowledge systems" (UNFCCC, 2015: Article 7). Second, there is a strong link to the Sustainable Development Goals (SDGs). This is viewed by some scholars as an indirect human rights dimension in the operative part of the agreement, because the SDGs are very strongly linked to core human rights (Ferreira, 2016; Harmeling, 2018). Third, international cooperation in adaptation is emphasized. Article 7(6) states that the needs of developing countries and particularly vulnerable population groups have to be taken into account. The obligation of international cooperation is a key human rights principle that is anchored in the International Covenant on Economic, Social and Cultural Rights (ICESRC, 1966). In one of his reports, the UN Special Rapporteur on Human Rights and the Environment states that international cooperation is particularly relevant with regard to "global environmental threats to human rights, such as climate change" (Knox, 2014: 18).

Loss and damage from a human rights perspective

Considering loss and damage means acknowledging that not all harm resulting from climate change can be avoided through mitigation and adaptation measures. Loss and damage has been defined as "negative effects of climate variability and climate change that people have not been able to cope with or adapt to" (Warner et al., 2012: 20). The term "damage" refers to monetary harm, whereas the term "loss" is used to take non-monetary harm into account (Adelman, 2016: 33). The latter includes not only physical, but also social, cultural, and psychological harm. These negative effects had been emphasized by Small Island Developing States (SIDS) since the beginnings of the UNFCCC negotiations. Despite this, it took more than 20 years for the Warsaw International Mechanism on Loss and Damage to be established. The Paris Agreement included the loss and damage mechanism as the fifth pillar of climate action. Under Article 8, state parties acknowledged:

> the importance of averting, minimizing and addressing loss and damage associated with the adverse effects of climate change, including extreme weather events and slow onset events, and the role of sustainable development in reducing the risk of loss and damage.
>
> (UNFCCC, 2015)

It is significant that both extreme weather events and slow onset events are covered by the loss and damage mechanism. Many of the aspects we have discussed in the context of adaptation are also relevant with respect to loss and damage, first and foremost the obligation to cooperate, and the protection of vulnerable regions, sectors, and population groups. Areas that will be prioritized in terms of cooperative action relating to loss and damage include, among others, early warning systems, emergency systems, comprehensive risk assessment and management, risk insurance, and non-economic losses, as well as resilience of communities, livelihoods, and ecosystems (UNFCCC, 2015: Article 4).

By devoting an entire article to it in the 2015 Paris Agreement, loss and damage has become an important pillar of the international climate regime, next to mitigation, adaptation, technology, and finance. It has also received considerable attention in the (I)NDCs, with 44% of the SIDS and 34% of the Least Developed Countries mentioning loss and damage (Kreienkamp and Vanhala, 2017: 2). The fact that not a single industrialized country has referred to it demonstrates how differently developed and developing states are affected by climate-related impacts and harm, and the difficulty of cooperating in this regard. The Paris decision accompanying the 2015 Agreement, for example, explicitly excludes the use of Article 8 as a "basis for any liability or compensation" (Harmeling, 2018: 99). Compensation and effective remedies for the harm caused, however, is exactly what is relevant when we consider loss and damage from a human rights perspective.

A submission by the OHCHR to the UNFCCC in 2016 highlights that those who suffer from harm caused by climate change must have access to meaningful remedies, such as judicial and other redress mechanisms. In the context of climate change, states have an obligation to protect rights-holders, and they are responsible for harm that occurs inside and outside their territory. Moreover, they are obliged to regulate businesses under their jurisdiction in order to prevent further harm being caused (OHCHR, 2016). Scholars like Adelman (2016) argue that developed countries have an ethical obligation to compensate SIDS for loss and damage caused by climate change. He suggests that compensation can be granted without admitting liability, from which many developed states have shied away in the past. However, the principles of climate justice provide an ethical justification for compensation as a measure of corrective justice (Adelman, 2016). Today, there are still strong disagreements between developed and developing countries regarding loss and damage. But the issue of compensation and effective remedies is at the heart of a human rights-based approach to climate change, and will be further advocated for by vulnerable countries and the climate justice movement.

DISCUSSION QUESTIONS

1. Why do some states hesitate to adopt procedural safeguards for the Clean Development Mechanism?
2. Do authoritarian countries like Ethiopia fear further democratization processes when implementing procedural rights, like access to information, participation, and remedies, in the context of green economy transition?
3. What can be done if green economy projects, like the Gibe III dam in Ethiopia, have transboundary human rights effects?
4. How can adaptation projects promote international human rights?
5. What role does human rights play in relation to the loss and damage mechanism?

Note

1 This will change once the Grand Renaissance Dam in Ethiopia becomes operational.

References

Adaptation Fund (2016). Environmental and Social Policy. [Online]. Available at: www.adaptation-fund.org/document/environmental-and-social-policy-approved-in-november-2013/ [Accessed: 19 October 2018].

Adelman, S. (2016). Climate Justice, Loss and Damage and Compensation for Small Island Developing States. *Journal of Human Rights and the Environment* 7(1): 32–53.

Carr, C.J. (2012). Humanitarian Catastrophe and Regional Armed Conflict Brewing in the Transborder Region of Ethiopia, Kenya and South Sudan: The Proposed Gibe III Dam in Ethiopia. Africa Resources Working Group.

Carbon Market Watch (2016). In Landmark Decision, Panama Withdraws UN Registration for Barro Blanco Hydrodam Project. [Online]. Available at: https://carbonmarketwatch.org/2016/11/10/press-statement-in-landmark-decision-panama-withdraws-un-registration-for-barro-blanco-hydrodam-project/ [Accessed: 19 October 2018].

Carbon Market Watch (ed.) (2013). Local Realities of CDM Projects: A Compilation of Case Studies. Brussels: Carbon Market Watch.

CPJ (Committee to Protect Journalists) (2015). Ten Most Censured Countries. [Online]. Available at: https://cpj.org/2015/04/10-most-censored-countries.php [Accessed: 25 August 2016].

Evans, K. (2014). Tabasará Libre: A Case Study of Development and Indigenous Rights. [Case study of Barro Blanco]. Independent Study Project Collection 1875. (unpublished). [Online]. Available at: https://digitalcollections.sit.edu/isp_collection/1875. [Accessed: 19 December 2018].

Ferreira, P.G. (2016). Did the Paris Agreement Fail to Incorporate Human Rights in Operative Provisions? Not If You Consider the 2016 SDGs, CIGI Papers No. 113. [Online]. Available at: www.cigionline.org/sites/default/files/documents/Paper%20no.113.pdf [Accessed: 19 October 2018].

FHI (2018). Freedom House: Ethiopia. [Online]. Available at: https://freedomhouse.org/report/freedom-world/2018/ethiopia [Accessed: 24 April 2018].

Giraldo, C.M. (2017). Panama's Barro Blanco Dam to Begin Operation, Indigenous Pleas Refused. [Online]. Available at: https://news.mongabay.com/2017/03/panamas-barro-blanco-dam-to-begin-operation-indigenous-pleas-refused/ [Accessed: 17 July 2018]

GoE (2010). *Growth and Transformation Plan*. Addis Ababa: Government of the Federal Democratic Republic of Ethiopia.

GoE (2011a). *Ethiopia's Climate-Resilient Green Economy: Green Economy Strategy*. Addis Ababa: Government of the Federal Democratic Republic of Ethiopia.

GoE (2011b). *Ethiopia's Vision for a Climate Resilient Green Economy*. Addis Ababa: Government of the Federal Democratic Republic of Ethiopia.

Harmeling, S. (2018). Climate Change Impacts: Human Rights in Climate Adaptation and Loss and Damage. In: Duyck, S., Jodoin, S., and Johl, A. (eds.), *The Routledge Handbook on Human Rights and Climate Governance*. London and New York: Routledge, pp.90–109.

Hofbauer, J. and Mayrhofer, M. (2016). Panama "Barro Blancho" Case Report. COMCAD Working Papers, 144 [Online]. Available at: www.ssoar.info/ssoar/bitstream/handle/document/50667/ssoar-2016-hofbauer_et_al-Panama_Barro_Blancho_Case_report.pdf?sequence=1 [Accessed: 19 October 2018].

HRCCWG (2015). Integrating Human Rights into the Paris Commitments – (I)NDCs. [Online]. Available at: http://climaterights.org/our-work/unfccc/human-rights-in-indcs/ [Accessed: 19 December 2018].

Human Rights Watch (2012). *What Will Happen if Hunger Comes? Abuses Against the Indigenous Peoples of Ethiopia's Lower Omo Valley*. New York, NY: Human Rights Watch.

Human Rights Watch (2015). *There is No Time Left: Climate Change, Environmental Threats, and Human Rights in Turkana County, Kenya*. New York, NY: Human Rights Watch.

IFC (2012). Performance Standards on Environmental and Social Sustainability. International Financial Corporation, World Bank Group [Online]. Available at: www.ifc.org/wps/wcm/connect/115482804a0255db96fbffd1a5d13d27/PS_English_2012_Full-Document.pdf?MOD=AJPERES [Accessed: 19 October 2018].

International Covenant on Economic, Social and Cultural Rights (ICESCR) (1966). Adopted and Opened for Signature, Ratification and Accession by General Assembly Resolution 2200 A (XXI) of 16 December 1966.

International Covenant on Civil and Political Rights (ICCPR) (1966). Adopted and Opened for Signature, Ratification and Accession by General Assembly Resolution 2200 A (XXI) of 16 December 1966.

ILO (1989). *Convention Concerning Indigenous and Tribal Peoples in Independent Countries*. Adopted at the 76th Session of the International Labour Conference on 27 June 1989. Geneva: International Labor Organization.

International Rivers (2011). *Ethiopia's Gibe III Dam: Sowing Hunger and Conflict*. Berkeley, CA: International Rivers.

International Rivers (2013). Chronology of Events for Barro Blanco Dam (Panama) [Online]. Available at: www.internationalrivers.org/chronology-of-events-for-barro-blanco-dam-panama [Accessed: 19 October 2018].

International Rivers (2015). *Come and Count our Bones: Community Voices from Lake Turkana on the Impacts of Gibe III Dam*. Berkeley, CA: International Rivers.

IPCC (2007). Adaptation and Mitigation Options. In: Pachauri, R.K. and Reisinger, A. (eds.), *Summary for Policymakers. Climate Change 2007. Synthesis Report*. Contribution of Working Groups I, II and III to the Fourth Assessment Report of the Intergovernmental Panel on Climate Change. Geneva, Switzerland: Intergovernmental Panel on Climate Change.

IPCC (2018). Global Warming of 1.5°C. [Online]. Available at: https://report.ipcc.ch/sr15/pdf/sr15_spm_final.pdf [Accessed: 19 December 2018].

Joshi, F. (2013). Adani's Mundra Coal Power Project India. In Carbon Market Watch (ed.), *Local Realities of CDM Projects: A Compilation of Case Studies*. Brussels: Carbon Market Watch, pp.4–5.

Knox, J. (2014). *Report of the Independent Expert on the Issue of Human Rights Obligations Relating to the Enjoyment of a Safe, Clean, Healthy and Sustainable Environment*. Geneva: Human Rights Council.

Kreienkamp, J. and Vanhala, L. (2017). *Climate Change Loss and Damage*. Policy Brief, March 2017. London: Global Governance Institute.

Kuchler, M. (2017). The Human Rights Turn: Engos' Changing Tactics in the Quest for a More Transparent, Participatory and Accountable CDM. *Environmental Politics* 26(4): 648–668.

Lottje, C., Hirsch, T. and Minninger, S. (2017). Poverty Orientation and Civil Society Participation in National Adaptation Plans (NAP) [Online]. Available at: www.germanclimatefinance.de/2017/02/08/poverty-orientation-civil-society-participation-national-adaptation-plans-nap/ [Accessed: 19 October 2018].

Maus, S. (2014). Hand in Hand against Climate Change: Cultural Human Rights and the Protection of Cultural Heritage. *Cambridge Review of International Affairs* 27(4): 699–16.

OHCHR (2009). Report of the Office of the United Nations High Commissioner for Human Rights on the Relationship between Climate Change and Human Rights A/HRC/10/61, 15 January 2009. Geneva: Human Rights Council.

OHCHR (2015). Key Messages on Human Rights and Climate Change [Online]. Available at: www.ohchr.org/Documents/Issues/ClimateChange/KeyMessages_on_HR_CC.pdf [Accessed: 19 October 2018].

OHCHR (2016). Response to the Request of Ad Hoc Working Group on the Paris Agreement (APA) [Online]. Available at: https://unfccc.int/sites/default/files/722.pdf [Accessed: 19 October 2018].

Orellana, M.A. and Johl, A. (2013). *Climate Change and Human Rights: A Primer.* Washington, DC: The Center for International Environmental Law.

Organization of African Unity (1981). African Charter on Human and Peoples' Rights. [Online]. Available at: www.achpr.org/instruments/achpr/ [Accessed: 19 December 2018].

Spain, T. (2012). Panama: Violent Repression of Indigenous Protest. Report Submitted to the Latin American Bureau. [Online]. Available at: https://lab.org.uk/panama-violent-repression-of-indigenous-protests/ [Accessed: 19 September 2017].

Steudtner, J. (2012). Die Einflussnahme der Menschenrechte auf das Klimaregime in den Post-Kyoto Verhandlungen unter besonderer Berücksichtigung der Mitigationsmaßnahme REDD+ – Eine kritische Illustration am Beispiel Peru. Master's Thesis, University of Vienna.

Survival International (2015). Ethiopia: Tribe Starves as Dam and Land Grabs Dry up River. [Online]. Available at: www.survivalinternational.org/news/10691 [Accessed: 26 August 2016].

Survival International (2016). Survival Reports Italian Corporation to OECD over Dam Disaster. [Online]. Available at: www.survivalinternational.de/news/11172 [Accessed: 26 August 2016].

UNDRIP (2007). United Nations Declaration on the Rights of Indigenous People. [Online]. Available at: www.un.org/esa/socdev/unpfii/documents/DRIPS_en.pdf [Accessed: 19 December 2018].

UNFCCC (2010). *The Cancun Agreements: Outcome of the Work of the Ad Hoc Working Group on Long-term Cooperative Action under the Convention*, Appendix I, paragraph 2. Cancun: UNFCCC.

UNFCCC (2015). Paris Agreement. Adopted at the 21st Conference of the Parties. [Online]. Available at: http://unfccc.int/files/essential_background/convention/application/pdf/english_paris_agreement.pdf [Accessed: 10 January 2017].

UNTC (2016). United Nations Treaty Collection. International Covenant on Economic, Social and Cultural Rights. [Online]. Available at: https://treaties.un.org/Pages/ViewDetails.aspx?src=TREATY&mtdsg_no=IV-3&chapter=4&lang=en [Accessed: 20 March 2016].

Wallbott, L. and Schapper, A. (2017) Negotiating By Own Standards? The Use and Validity of Human Rights Norms in UN Climate Negotiations. *International Environmental Agreements: Politics, Law and Economics*. 17 (2): 209–28.

Warner, K., van der Geest, K., Kreft, S., Huq, S., Harmeling, S., Kusters, K., and De Sherbinin, A. (2012). Evidence from the Frontlines of Climate Change: Loss and Damage to Communities Despite Coping and Adaptation Loss and Damage in Vulnerable Countries Initiative. Policy Report. Report No. 9. United Nations University Institute for Environment and Human Security (UNU-EHS), Bonn [Online]. Available at: www.ehs.unu.edu/ [Accessed: 12 April 2014].

Wubete, F. (2017). Ethiopia: Gibe III – Another Milestone for Building Green Economy. [Online]. Available at: http://allafrica.com/stories/201701040514.html [Accessed: 29 June 2017].

Part IV

Emerging issues related to environmental rights

13 Extraterritorial application of environmental rights

Introduction

Whether human rights law operates extraterritorially has become a hotly debated issue in recent times. With more and more adverse consequences of climate change emerging, victims of these adverse consequences are increasingly contemplating the remedies they have under human rights law to seek redress. Since these threats originate from outside the territory, holding the victim's state responsible for damage is not viable. Thus, the questions that arise in this context are:

- Do states have obligations under international human rights law to protect the human rights of nationals in neighboring states who are affected by transboundary environmental damage due to action taken within the states' territory?
- If human rights are violated by transboundary environmental harm, what kind of recourse do victims have?

An example will make these questions clearer. States A and B are neighboring states with a shared border. There is a steel manufacturing plant in State A near the border with State B. Because of the operations of the steel plant in State A, there is air pollution in State B and citizens of State B are experiencing health problems, including asthma. Does State A have obligations under international human rights law to protect the rights of citizens in State B? In this scenario, what is the recourse that citizens of State B have under international human rights law to obtain redress? In other words, can human rights law operate diagonally, although the general principle is one of vertical application?

A lesson from history

A dispute that took place in the late 1930s between the United States and Canada that led to arbitration in 1941 provides an example of a real-life situation of transboundary damage (*Trail Smelter case*, *US v. Canada*, 1938, 1941). It is one of the first air pollution cases. A smelter in British Columbia near the Canadian/

US border was causing damage by fumes to the people and property across the border in the State of Washington (the case does not refer to the "environment" for obvious reasons). The parties referred the dispute to arbitration. Not surprisingly, the arbitral tribunal could not find any international laws applicable to the situation and turned to US domestic law for guidance. The tribunal stated:

> Under principles of international law as well as under the law of the United States, no State has the right to use or permit the use of its territory in such a manner as to cause injury by fumes in or to the territory of another or the properties *or persons therein*, when the case is of serious consequence and the injury is established by clear and convincing evidence.
>
> (1965, emphasis added)

On the question whether an injunction should be issued, the arbitral tribunal stated, quoting from the Supreme Court decision in *Georgia v. Tennessee Copper Co.* (1907) 206 U.S. 230:

> It (the State) has the last word as to whether its mountains shall be stripped of their forests and its inhabitants shall breathe pure air … It is not lightly to be presumed to give up quasi-sovereign rights for pay and … if that be its choice, it may insist that an infraction of them shall be stopped. This court has not quite the same freedom to balance the harm that will be done by an injunction against that of which the plaintiff complains, that it would have in deciding between two subjects of a single political power. Without excluding the considerations that equity always takes into account … it is a fair and reasonable demand on the part of a sovereign that the air over its territory should not be polluted on a great scale by sulphurous acid gas, that the forests on its mountains, be they better or worse, and whatever domestic destruction they may have suffered, should not be further destroyed or threatened by the act of persons beyond its control, that the crops and orchards on its hills should not be endangered from the same source … Whether Georgia, by insisting upon this claim, is doing more harm than good to her own citizens, is for her to determine. The possible disaster to those outside the State must be accepted as a consequence of her standing upon her extreme rights.

This is quite remarkable language considering that this was the pre-human rights and pre-environmental protection era. If this dispute were to arise today, it is quite clear that different terminology and different legal principles would be used in relation to this dispute. In particular, human rights language is likely to be invoked in relation to the damage caused to the people due to transboundary air pollution. Would Canada incur responsibility under human rights law for the damage suffered by US citizens? What is the responsibility of the United States to protect its citizens from transboundary air pollution? These are the questions that we will examine in this chapter.

The general principle applicable to human rights is that they operate vertically between States as duty-bearers and citizens as rights-holders. States must fulfill their human rights obligations within their territory. As a general principle, the human rights framework "assumes a single polity that experiences both the benefits and the costs of environmental policies" (Knox, 2009). In other words, human rights law applies territorially. Since "territory" includes not only the land area but also the territorial sea and the air space above that, human rights law operates in all areas over which states exercise sovereignty or control.

International environmental law, on the other hand, specifically recognizes the extraterritorial dimension of state action: states are under an obligation to refrain from engaging in activities that could cause transboundary harm, as pollution does not respect political boundaries. According to Principle 21 of the Stockholm Declaration of the United Nations Conference on the Human Environment (1972), the wording of which is reminiscent of the decision in the *Trail Smelter* arbitration case:

> States have, in accordance with the Charter of the United Nations and the principles of international law, the sovereign right to exploit their own resources pursuant to their own environmental policies, and the responsibility to ensure that activities within their jurisdiction or control do not cause damage to the environment of other States or of areas beyond the limits of national jurisdiction.
>
> (Stockholm Declaration, 1972: Principle 21)

This principle is considered to have entered the realm of customary international law governing the environment (Sands and Peel, 2012: 191). Under principles of state responsibility, international responsibility would attach if two criteria are satisfied: if the state

(a) knew, or should have known, that there was a real and immediate risk to protected rights (e.g., the right to life), and failed to take the necessary measures that would have been reasonably expected to prevent such risk, and
(b) if there is a casual link between the significant harm to the environment and the human rights impacts.

(Draft Articles on State Responsibility, 2011)

Some human rights treaties establish the territoriality principle: for example, the ICCPR provides that each state party "undertakes to respect and to ensure to all individuals *within its territory* and subject to its jurisdiction, the rights recognized in the present Covenant" (Article 2, emphasis added). However, the ICESCR has no such restriction and encourages international co-operation: "each party undertakes to take steps, individually and through international assistance and cooperation … to achieve progressively the full realization of rights recognized in the Covenant" (Article 2), The Committee on Economic, Social and Cultural

Rights has taken the view that states have extraterritorial obligations in four situations:

(a) refraining from interfering with the enjoyment of human rights in other countries;
(b) taking measures to prevent third parties over which they exercise influence from interfering with the enjoyment of human rights in other countries;
(c) taking steps to facilitate fulfilment of human rights in other countries, including disaster relief, emergency assistance, and assistance to refugees and displaced persons; and
(d) ensuring that human rights are given due attention in international agreements

(CESCR, 1999a, 1999b, 2000, and 2002)

The UN Human Rights Committee, established under the ICCPR to oversee the implementation of the treaty, elaborated on the nature of obligations under Article 2: states must ensure that rights are applied to *all persons within its territory* and *to all persons subject to its jurisdiction*. In addition, rights are extended to those within the *effective control of the state*, even if they are outside the territory, thus recognizing a limited application of the extraterritoriality principle.

Maastricht Principles on Extraterritorial Obligations of States in relation to Economic, Social and Cultural Rights

A group of 40 international law experts was convened by Maastricht University and the International Commission of Jurists to draft Principles on Extraterritorial Obligations of States in relation to Economic, Social and Cultural Rights (ETO Consortium, 2011). While these are non-binding, several emminent scholars participated in this exercise. The Principles noted in the Introduction:

> Despite the universality of human rights, many States still interpret their human rights obligations as being applicable only within their own borders. This attempt to limit obligations territorially has led to gaps in human rights protection in various international political processes and a lack of regulation for the protection of human rights.
>
> (ETO Consortium, 2011)

The report notes that these gaps have become severe due to globalization. They include: (a) lack of human rights regulation of transnational corporations; (b) absence of human rights accountability of intergovernmental organizations, particularly financial institutions; (c) ineffective application of human rights law to investment and trade law policies and disputes; and (d) lack of implementation of duties to protect economic, social, and cultural rights abroad.

According to the Maastricht principles, "extraterritorial obligations (ETOs) are a missing link in the universal human rights protection system. Without ETOs, human rights cannot assume their proper role as the legal basis for regulating globalization and ensuring universal protection of all people and groups" (ETO Consortium, 2011). The principles stress that they do not establish new elements of human rights law and merely clarify extraterritorial obligations of states under current international law.

The Preamble recognizes that human rights of individuals and groups are affected by the extraterritorial acts and omissions of States. Globalization in particular has given rise to states and other actors exerting considerable influence on the realization of ESC rights across the world. Referring to the Vienna Declaration and Program of Action, the Preamble notes that despite the repeated commitment to the realization of ESC rights for everyone, countless people around the world are unable to enjoy their ESC rights. These principles draw from international law and seek to clarify the content of extraterritorial obligations to realize ESC rights. They complement and build on the Limburg Principles on the Implementation of the International Covenant on Economic, Social and Cultural Rights (ESCR, 1986) and the Maastricht Guidelines on Violations of Economic, Social and Cultural Rights (Dankwa, Flinterman, and Leckie, 1997).

According to the principles, ETOs encompass:

(a) obligations relating to the acts and omissions of a state, within or beyond its territory, that have effects on the enjoyment of human rights outside of that state's territory; and
(b) obligations of a global character that are set out in the UN Charter and human rights instruments to take action through international cooperation to realize human rights universally.

(ETO Consortium, 2011: Principle 8)

The document embodies a set of seven general principles. These are:

1 All human beings everywhere are born free and equal in dignity and are entitled without discrimination to human rights and freedoms;
2 States must observe the principles of non-discrimination, equality, including gender equality, transparency and accountability;
3 All States have obligations to respect, protect and fulfill human rights, including civil, cultural, economic, political and social rights, both within their territories and extraterritorially;
4 Each state has the obligation to realize ESC rights for all persons within its territory to the maximum of its ability. All States have extraterritorial obligations to respect, protect and fulfill ESC rights as set forth in these principles.
5 All human rights are universal, indivisible, interdependent, and interrelated. The present principles elaborate ETOs in relation to ESC rights without excluding their applicability to other rights.
6 ESC rights and corresponding territorial and extraterritorial obligations are contained in the sources of international human rights law, including the UN Charter, Universal Declaration of Human Rights, International Covenant on Economic, Social and Cultural Rights and other universal and regional instruments.

7 Everyone has the right to informed participation in decisions that affect their rights. States should consult with relevant national mechanisms, including civil society groups in designing and implementing policies relevant to their obligations in relation to ESC rights.

(ETO Consortium, 2011)

With regard to the scope of the obligations, the Principles provide that states have obligations to respect, protect and fulfill ESC rights in:

(a) situations over which it exercises authority or effective control, whether or not it is in accordance with international law;
(b) situations over which States' acts or omissions bring about foreseeable effects on the enjoyment of ESC rights, within or outside its territory;
(c) situations in which the state is in a position to exercise decisive influence or to take measures to realize ESC rights extraterritorially, in accordance with international law.

(ETO Consortium, 2011)

The Principles further provide that a State's obligation to respect, protect and fulfill economic, social and cultural rights extraterritorially does not authorize a State to act in violation of the UN Charter and general international law (Principle 10).

States must desist from acts and omissions that create a real risk of nullifying or impairing the enjoyment of economic, social and cultural rights extraterritorially. The responsibility of States is engaged where such nullification or impairment is a foreseeable result of their conduct. Uncertainty about potential impacts is not a defense (Principle 13). Borrowing from international environmental law, the Principles provide that states must conduct prior assessment, with public participation, of risks and potential extraterritorial impacts of their laws, policies, and practices on the enjoyment of ESC rights. In addition, states must use the assessment to inform the measures it plans to take to prevent violations or ensure their cessation and provide effective remedies (Principle 14).

The principles provide that states must refrain from both direct and indirect interference with the enjoyment of ESC Rights: "All States have the obligation to refrain from conduct which nullifies or impairs the enjoyment and exercise of economic, social and cultural rights of persons outside their territories" (Principle 20). Indirect interference includes conduct that impairs the ability of another state or an international organization to comply with their obligations with regard to ESC rights or coerces another state or international organization to breach their obligations with regard to ESC rights (Principle 21).

The principles further elaborate on obligations to protect and fulfill, and provide that states must ensure the availability of effective mechanisms to provide for accountability in the discharge of their ETOs. These include procedures for monitoring compliance with human rights obligations. With regard to the remedies, the Principles provide that:

> States must ensure the enjoyment of the right to a prompt, accessible and effective remedy before an independent authority, including, where necessary, recourse to a judicial authority, for violations of economic, social and cultural rights. Where the harm resulting from an alleged violation has occurred on the territory of a State other than a State in which the harmful conduct took place, any State concerned must provide remedies to the victim.
>
> (ETO Consortium, 2011: draft Articles 37 and 38)

These principles were further elaborated on by the commentary provided by De Schutter et al. (2012). They observe that many of the principles embodied in there reflect the provisions in human rights treaties. With regard to the provisions on state responsibility, the commentary points out that they are based on the International Law Commission's (ILC) draft articles on state responsibility, as well as the pronouncement by the International Court of Justice (ICJ) in its Advisory Opinion on Nuclear Weapons that states have a general obligation to ensure that activities within their jurisdiction or control respect the environment of other states. Of course, the non-harmful use of territory is a well-established principle of international environmental law and derives directly from the principle of sovereignty itself. The first articulation of this in the environment field was Principle 21 of the Stockholm Declaration. Similarly, the concept of foreseeability is also addressed by the ILC and provides a good yardstick and "by ensuring that a state shall not be surprised with claims of responsibility for unforeseeable risks that are only remotely connected to its conduct" (De Schutter et al., 2012: 1113). Foreseeability and causality are well-established principles of law.

It is interesting that the commentary also relied on international environmental law principles such as the precautionary principle, the legal status of which is not quite certain, although it is steadily gaining ground. In the *Pulp Mills* case (2010), the ICJ noted that a precautionary approach may be relevant in the interpretation and application of the Statute. Similarly, the Seabed Disputes Chamber of the ITLOS noted in its Advisory Opinion on the responsibilities and obligations of states with respect to the activities in the Area that the precautionary approach has been incorporated into a growing number of international treaties: "In the view of the Chamber, this has initiated a trend towards making this approach part of customary international law" (ITLOS, 2011).

Prior assessment of impacts and public participation are, of course, established principles of international environmental law. As the ICJ recognized in the *Pulp Mills* case, the practice of EIA

> has gained so much acceptance among states that it may now be considered a requirement under general international law to undertake an environmental impact assessment where there is a risk that the proposed industrial activity may have a significant adverse impact in a transboundary context, in particular, on a shared resource.
>
> (*Pulp Mills* case, 2010)

Knox discusses the possibility of extending human rights obligations extraterritorially under two prongs: with regard to the first prong that requires imposing strict procedural requirements, a state will have to conduct an EIA of transboundary harm originating in its territory and give all those affected, including non-residents, an opportunity to participate in the decision-making process (2009). As discussed in the *Pulp Mills* case (2010), the ICJ said that this requirement is now part of customary international law. What the ICJ did not say was whether this requirement extends to giving all those affected the ability to participate in the decision-making process. While it is required under most national laws to provide information and give an opportunity to participate in the decision-making process to those affected within its territory, the logical conclusion with a transboundary process should be to give all those who are affected, regardless of nationality, a similar opportunity. In other words, states cannot just enjoy benefits and pass the costs to the citizens of another state. The ICJ unfortunately did not go as far as to say that the affected public in the neighboring state should be given the opportunity to participate in the decision-making process.

With regard to the second prong which requires giving deference to substantive decisions, a state could argue that its decision with regard to transboundary harm should be entitled to the same level of deference as its decisions regarding internal harm, "because extending the procedural safeguards ensures that the state takes into account the full range of costs and benefits of its actions" (Knox, 2009). With regard to the question whether human rights law provides a legal basis for such an extraterritorial extension of environmental rights jurisprudence, the answer is not very clear-cut.

The test that most human rights bodies have adopted is whether the state in question has *effective control* (or authority, as the Inter-American Commission has articulated, De Schutter et al., 2012: 1103) in deciding whether it has jurisdiction, but the test these bodies have applied with regard to transboundary harm has not been identical. As will be discussed in the next section, this has been expanded by the Inter-American Court. According to Knox who discusses whether extraterritorial application of human rights could be made in relation to climate change:

> Clearly, not all transboundary harm could qualify as "effective control." Or the jurisdictional limits on states' obligations under the ICCPR and the regional treaties would be meaningless. To successfully argue that the effects of climate change amount to an exercise of extraterritorial jurisdiction, one would have to focus on concrete ways that climate change places those affected under the control of the states causing the harm. Although that argument would be very difficult to make in many contexts, it might be possible with respect to particularly extreme impacts, such as the effect of climate change on small island states … As their territory literally disappears, they are arguably at the mercy of the larger, more powerful countries that have caused the harm.
>
> (2009)

Diagonal human/environmental rights?

The former Special Rapporteur on human rights and the environment, John Knox, has advanced the notion of diagonal rights in relation to transboundary environmental damage. He contends that in the Inuit petition filed by Canadian and US Inuit against the United States (discussed in Chapters 4 and 9), the claim by Canadian Inuit against the United States is a diagonal claim (Knox, 2010). According to Knox, "environmental rights are 'diagonal' if they are held by individuals or groups against the governments of states other than their own" (Knox, 2010: 82). While international human rights law has not adopted a comprehensive approach to extraterritorial obligations, he believes that both human rights law and international environmental law have begun to develop some possible bases for diagonal environmental rights.

In his Mapping Report, he recognized that many grave threats to human rights are due to transboundary environmental harm. He pointed out that the application of human rights obligations to transboundary environmental harm is not always clear, and the difficulty is because different human rights instruments address jurisdiction in different ways.

As noted in this textbook, human rights law operates vertically between states and individuals and is primarily territorial in nature, while international environmental law (IEL) operates primarily horizontally between states. Unlike human rights law, IEL encompasses both transboundary and global environmental issues, and one of the cardinal principles of IEL is *not* to cause extraterritorial damage. Given this difference in focus, Knox argues, grounding diagonal environmental rights in either discourse is problematic: "If the challenge for human rights law is to extend rights from the vertical axis to the diagonal, the challenge for IEL is to derive diagonal rights from horizontal ones" (Knox, 2010: 83). However, not all IEL operates horizontally. Many environmental obligations are applicable internally – those relating to EIA, information, public participation, and remedies are applicable internally. Moreover, where environmental harm is felt beyond territorial limits and impinges on the enjoyment of rights in another state, there is no logical reason to avoid responsibility, simply because the harm was felt outside its territory. As the Inter-American Court pointed in its 2018 Advisory Opinion, if pollution can travel across borders, so can responsibility. This Opinion is discussed in the next section.

Several Special Rapporteurs have expressed the view that states should avoid causing harm extraterritorially and protect human rights extraterritorially and take action to ensure that non-state actors do not nullify or impair the enjoyment of economic, social and cultural rights. This translates into an obligation to avoid contamination of watercourses in other jurisdictions and to regulate non-state actors accordingly (United Nations, 2013).

Cases where extraterritorial damage was alleged

In a few cases, the issue of violation of human rights by transboundary environmental damage was alleged. Both the European Court of Human Rights and the Inter-American Court of Human Rights have discussed the possibility of states being responsible for the effects of their actions on human rights of individuals beyond their territory.

In *Saldano v. Argentina*, the Inter-American Commission indicated that a "party to the American Convention may be responsible under certain circumstances for the acts and omissions of its agents *which produce effects* or are undertaken outside that state's own territory" (Knox, 2010: 87). Similarly in *Ilascu and Others v. Moldova and Russia*, the European Court stated, "in exceptional circumstances the acts of Contracting States performed outside their territory, or *which produce effects there*, may amount to exercise by them of their jurisdiction within the meaning of Article 1 of the Convention" (Knox, 2010: 87). While it may be argued that obligations of a state extend extraterritorially to areas over which it has effective control, in *Bankovic and others v. Belgium and 16 other contracting States*, the European Court rejected the argument that the NATO bombing of Serbia in 1999 amounted to "effective control":

> [t]he applicant's submission is tantamount to arguing that anyone adversely affected by an act imputable to a Contracting State, wherever in the world that act may have been committed or its consequences felt, is thereby brought within the jurisdiction of that State for the purpose of Article 1 of the [European] Convention. The Court is inclined to agree with the Governments' submission that the text of Article 1 does not accommodate such an approach to "jurisdiction".
>
> (2001: para. 75)

Knox argues that "if dropping bombs on a city does not amount to effective control of its occupants, allowing pollution to move across an international border almost certainly would not" (Knox, 2010: 87). However, the Inter-American Court has interpreted the jurisdiction clause more broadly. In *Alejandre v. Cuba*, the Commission held that by shooting down an unarmed plane over international waters, agents of the Cuban government placed the civilian pilots under their authority (1999: para. 25). In its 2018 Advisory Opinion, discussed later, the Court seems to have gone even further.

The *Case Concerning Aerial Herbicide Spraying (Ecuador v. Colombia)* (2009) before the ICJ involved the spraying of herbicide by Colombia to eradicate drug-producing crops along the Colombia/Ecuador border. Ecuador alleged that Colombia had caused transboundary pollution and significant harm in the territory of Ecuador, including damage to indigenous communities in Ecuador. Ecuador alleged that the violations of obligations by Colombia included:

- International law obligations including the failure to prepare a transboundary EIA;
- Failure to provide information to both Ecuador and its people, especially the indigenous communities;
- Ecuador alleged that Colombia had violated several human rights treaties, as well as provisions relating to indigenous peoples.

Ecuador alleged that a range of human rights violations has resulted from Colombia's spraying of toxic chemicals along the border which had caused adverse impacts on residents in the border area, including rights to life, health, food, water, property, information, humane treatment, private life, and a healthy environment. What constituted a healthy environment must be determined by reference to the natural, social, economic, and cultural character of the region in question. Interestingly, Ecuador relied on the *Ogoniland* case decided by the African Commission of Human Rights (discussed in Chapter 4). Of course, the Ogoniland decision has no binding force on a case before the ICJ, but this cross-fertilization of international law is significant.

Ecuador also relied on the observation made by the UN Human Rights Committee in *Delia Saldias de Lopez v. Uruguay* (1981): "It would be *unconscionable* to so interpret the responsibility under article 2 of the Covenant [ICCPR] as to permit a State party to perpetrate violations of the Covenant on the territory of another State, which violations it could not perpetrate on its own territory." Ecuador further contended that Colombian aerial sprayings over Ecuadorian territory have destroyed peoples' "most basic conditions of survival" and have resulted in a violation by Colombia of the right to a healthy environment as set out in the 1988 Additional Protocol to the American Convention on Human Rights (OAS). Colombia's failure to take preventive measures, including mechanisms to facilitate access to information and participation, aggravates its international responsibility.

According to Ecuador the relationship between three distinct areas of international law – human rights, environmental protection, and indigenous rights – is at the heart of this case and their inter-relationship has long been recognized.

With regard to the extraterritorial application of the ICCPR, Ecuador relied on the approach taken by the ICJ in its Advisory Opinion in the *Legal Consequences of the Construction of a Wall in the Occupied Palestinian Territory* (ICJ): "while the jurisdiction of the state is primarily territorial, it may sometimes be exercised outside the national territory" (ICJ, 2004). The ICJ noted that this interpretation is consistent with the practice of the Human Rights Committee which has found the Covenant applicable *when states exercise jurisdiction on foreign territory*. However, transboundary environmental damage cannot be considered as "exercising jurisdiction on foreign territory."

In its Advisory Opinion on the Wall, the Court relied on preparatory materials to the Covenant to support the Human Rights Committee's interpretation of Article 2. It noted that the intention of the drafters was not to let states escape their obligations when they exercise jurisdiction outside their national territory, but to prevent persons residing abroad from claiming vis-à-vis their state of origin.

- The question is whether transboundary effects of activities within a state would amount to *exercising jurisdiction* outside the national territory of states;
- The above formulation seems to refer to deliberate action outside the territory of states, rather than to the transboundary impact of activities carried out within the territory of states;
- Foreseeability of significant transboundary harm could be a useful test here to determine the responsibility of the state of origin under international human rights law.

In its reply to Colombia's counter-memorial in which Colombia refuted Ecuador's claims referring to them as being "vague," Ecuador relied on the jurisprudence of the Inter-American Court of Human Rights that has recognized the connection between a number of distinct human rights, such as health, food, access to clean water, and the right to a decent existence. With regard to those living along the border, the persistent spraying of toxic herbicides resulted in a violation of their enjoyment of a decent existence.

Ecuador further alleged that Colombia failed to see that the right to a healthy environment is closely related to the enjoyment of other fundamental rights. It relied on the *Yakye Axa* case which established the connection between the right to a healthy environment and the right to a decent life. Colombia also argued that the right recognized in Article 24 of the Banjul Charter is not related to a right to a healthy environment simply because it is formulated as a right to a satisfactory environment: "This is semantics that focuses on form not substance: The decision of the African Commission on Human and Peoples Rights in the *Ogoniland Case* shows that the objectives pursued by both formulations are the same" (Ecuador, 2011).

Colombia argued in its Counter-Memorial that it is under no international law obligation to respect and protect the rights of local populations and indigenous peoples living outside its territory or not falling under its jurisdiction. It relied on the admissibility decision of the European Court of Human Rights in *Bankovic v. Belgium*, in which the Court observed that there was no jurisdictional link between the persons who were victims of the act complained of and the respondent States. Accordingly, it was not satisfied that the applicants and their deceased relatives were capable of coming within the jurisdiction of the respondent States on account of the extraterritorial act in question. Contrasting with the present case, Ecuador argued that both Colombia and Ecuador were parties to the American Convention, the ICCPR, and the ILO Convention 169 (1989). Ecuador, by contrast, relied on the observation made by the ICJ in the provisional measures phase of *Georgia v. Russia* (2008) that the question whether a human rights treaty applies extraterritorially is a question of interpretation, to be determined according to the ordinary meaning and in light of the object and purpose of each treaty.

According to Ecuador, the issue of territoriality of human rights obligations in the context of two states within the Inter-American system and sharing the same legal space are:

- First, both states are part of a common legal space at the regional level relating to the protection of human rights and indigenous peoples;
- Second, it follows that both states are subject to an international public order of human rights which does not allow for loopholes in the effective protection of human rights for any part of their populations;
- Third, Colombia is under an obligation not to frustrate the protection of human rights in the regions adjacent to its borders.

Contrasting the decision of the European Court of Human Rights in the *Bankovic case*, Ecuador pointed out in that case, the military acts were performed outside the regional scope of the Convention, as the Federal Republic of Yugoslavia was not a party to the European Convention when the attacks occurred. It is not a rejection of the extraterritorial application of human rights as alleged by Colombia. By contrast, in this case, both Ecuador and Colombia are parties to the American Convention of Human Rights and ILO Convention 169 (1989).

Moreover, the jurisdictional clause in the American Convention of Human Rights provides in Article 1(1) that: "The State Parties to this convention undertake to respect the rights and freedoms recognized herein and to ensure to all persons subject to their jurisdiction the free and full exercise of those rights and freedoms." Ecuador is of the view that the obligation to "respect" is not synonymous with the obligation to "ensure." Thus:

> Even if indigenous peoples within Ecuador are not subject to the jurisdiction of Colombia, Colombia nevertheless has an obligation to respect their rights – an obligation identical to the requirements of Article 14(2) of the 1988 Narcotics Convention – and Ecuador is entitled to invoke such respect.
>
> (ILO, 1989)

Despite the variation in wording in the ICCPR, the UN Human Rights Committee accepted the same argument in *Delia Saldias de Lopez v. Uruguay* (1981), where it stated that it would be unconscionable to interpret the responsibility under Article 2 to permit a state to perpetrate violations on the territory of another state, which violations it would not perpetrate on its own territory.

Ecuador believed that to exclude acts committed in one state but which produce effects in the territory of another contracting party from the protection offered by the American Convention, the ICCPR and ILO 169 "would run counter to the object and purpose of these instruments" (1989). It also referred to the Preamble of the American Convention which emphasizes that the essential rights of man are not derived from nationality but are based upon attributes of human personality. It further argued that the existence of an obligation to ensure the rights enshrined in the Convention by the territorial state does not contradict the existence of an obligation to respect human rights extraterritorially.

In *Coard v. United States* (1999), the Inter-American Commission stated, with regard to extraterritorial application of the American Declaration, that under certain circumstances, the exercise of jurisdiction over acts with an extraterritorial locus will be required by the norms which pertain. While most commonly, the phrase "subject to its jurisdiction" refers to persons within a state's territory, it may refer to conduct with an extraterritorial locus where the person concerned is present in the territory of one state, but subject to the control of another state – usually through the acts of the latter's agents abroad:

> In principle, the inquiry turns not on the presumed victim's nationality or presence within a particular geographic area, but on whether, under the

> specific circumstances, the state observed the rights of a person subject to its authority and control.
>
> (*Coard v. United States*, 1999)

Relying on this principle, Ecuador argued that spraying of toxic herbicides that foreseeably drifts over the territory of another state "will amount to the exercise of authority or control over the victims for the purposes of applying the American Declaration extraterritorially" (*Ecuador v. Colombia*, 2009).

Ecuador also relied on the Advisory Opinion of the Court in *The Effect of Reservations on the Entry into Force of the American Convention on Human Rights* (1982), in which the Court spelled out the objective character of obligations arising out of multilateral human rights treaties, and pointed out that modern human rights treaties are not like traditional treaties that embody reciprocal rights and obligations:

> Their object and purpose is the protection of the basic rights of individual human beings *irrespective of their nationality, both against the State of their nationality and all other contracting States*. In concluding these human rights treaties, the States can be deemed to submit themselves to a legal order within which they, for the common good, assume various obligations, not in relation to other States, but towards all individuals within their jurisdiction.
>
> (emphasis in original)

Thus, this Advisory Opinion makes it clear that because human rights obligations are *erga omnes* obligations, states are under an obligation to respect the human rights of persons irrespective of nationality.

The Court had occasion to further refine its position in a more recent Advisory Opinion. Colombia requested an advisory opinion from the Inter-American Court of Human Rights on the scope of state responsibility for environmental harm under the American Convention. The Court affirmed that human rights depend on the existence of a healthy environment. It ruled that states must take measures to prevent significant environmental harm to individuals inside – *and outside* – their territory. In other words, if pollution can travel across the border, so can legal responsibility. The Court further recognized the existence of an "autonomous" right to a healthy environment under the American Convention.

The Court clarified the extraterritorial scope of the American Convention in environmental matters. Whether the term "jurisdiction" in Article 1(1) of the American Convention – which obligates states parties "to respect the rights and freedoms recognized [in the Convention] and to ensure to all persons subject to their jurisdiction the free and full exercise of those rights and freedoms" – could encompass extraterritorial obligations relating to environmental harm, the term "jurisdiction" can embrace activities within a state that cause cross-border effects, noting that states have a duty to prevent transboundary environmental damage that could impair the rights of persons outside their territory.

In case of transboundary environmental harm, a person will be deemed to be subject to the "jurisdiction" of the state in which the harm originates if there is a "causal relationship" between the polluting activities in the state's territory and the cross-border impact on rights. The exercise of jurisdiction arises because the state has effective control over the activities that caused the damage and is in a position to prevent harm. The Court, however, emphasized that the exercise of extraterritorial jurisdiction is "exceptional" and must be interpreted on a case-by-case basis. The precise contours of states' extraterritorial obligations will be developed in concrete cases, and will significantly depend on the legal context (such as applicable international environmental law obligations).

With regard to the duty to prevent transboundary environmental harm, the Court pointed out that the duties to "respect" and "ensure" the rights to life and personal integrity, which were the focus of Columbia's request, must be interpreted in the light of international environmental law principles. The Court emphasized in particular the importance of (a) the duty to prevent environmental harm; (b) the precautionary principle; (c) the duty to cooperate with potentially affected states; and (d) procedural environmental rights. While these elements are well established in international environmental law, the Advisory Opinion provides guidance on their application in the Inter-American human rights system:

> States have a duty to prevent "significant" harm to the environment of other States or the global commons. To comply with this obligation of prevention, States must:
>
> - regulate, supervise, and monitor activities under their jurisdiction that could cause significant harm to the environment;
> - carry out environmental impact assessments when there is a risk of significant damage to the environment (EIAs);
> - prepare contingency plans to minimize the possibility of major environmental disasters; and
> - mitigate any significant harm to the environment which takes place despite preventive measures.

The duty to ensure the rights to life and personal integrity implies a duty to act with due diligence. Thus, states must act in accordance with the precautionary principle, even in the absence of scientific certainty, and adopt measures to prevent serious or irreversible damage to the environment.

The general duty to cooperate with the potentially affected states in case of transboundary environmental harm (including the related obligations of notification, consultation, and information-sharing) remains applicable under the American Convention. Specifically, states must cooperate, in good faith, to protect against environmental damage and to protect the rights of persons under their jurisdiction.

The Court reaffirmed that procedural environmental rights – access to information, public participation, and justice – are fundamental for the fulfillment of other rights in the American Convention. Implementing procedural rights in a cross-border context (and enabling affected individuals to take part in foreign proceedings) will be challenging. Nonetheless, states are expected to provide non-discriminatory treatment to persons affected by transboundary environmental harm. The Court pointed out that these obligations were established in relation to the rights to life and personal integrity because they were the subject of Colombia's request. However, the Court indicated that the same obligations could exist in relation to other rights that are particularly vulnerable to environmental degradation.

This Advisory Opinion made the position clear with regard to state's human rights obligations vis-à-vis transboundary environmental damage. While common sense dictates that when human rights obligations are combined with environmental obligations, states have an obligation to refrain from violating human rights of people in neighboring states due to transboundary environmental damage, the real question arises with regard to the remedies that the victims have under human rights law. Would the victims in a neighboring state be able to bring a claim under human rights law against the perpetrator state, say for example, before the Inter-American Commission of Human Rights if both the perpetrator state and the victim's state of nationality are parties to the American Convention? The answer is not very clear-cut. Because human rights are universal and because states have an obligation to prevent transboundary environmental damage, the victims in one state should be able to bring a claim against the perpetrator state, regardless of nationality, provided the damage was foreseeable and the harm suffered is significant. Human rights law does not operate in a vacuum and the Advisory Opinion of the Inter-American Court showed how the parameters of human rights obligations can be elucidated by using international environmental law principles. A lesson from history is discussed in the next section, where a group of Nordic countries provided remedies to victims of environmental damage (referred to as nuisances then) in any contracting state, regardless of nationality, and provides a useful model to address extraterritorial environmental damage that impinge on human rights of people in neighboring countries.

Novel approaches to extraterritoriality – a lesson from history

The Nordic Environmental Protection Agreement of 1974 between Denmark, Finland, Norway, and Sweden is an interesting example of a treaty that provides a framework to address damage caused by a "nuisance" in any other signatory country. The Agreement provides that "In considering the permissibility of environmentally harmful activities, the nuisance which such activities entail or may entail in another Contracting State shall be equated with a nuisance in the States where the activities are carried out" (Article 2). The Agreement further provides that:

> Any person who is affected or may be affected by a nuisance caused by environmentally harmful activities in another Contracting State shall have the right to bring before the appropriate Court or Administrative Authority of

> that State the question of the permissibility of such activities including the question of measures to prevent damage, and to appeal against the decision of the Court or the Administrative Authority to the same extent and on the same terms as a legal entity of the State in which the activities are being carried out.

In regions where a regional human rights system exists, it would be quite easy to facilitate this type of access to victims of transboundary environmental damage. The Inter-American Court's Advisory Opinion may possibly have paved the way for expanded standing for such victims, although it is too early to predict whether it will actually happen. The Nordic Environmental Agreement provides us with an example where it has actually been done, albeit within a small group of states with similar legal systems.

Conclusion

According to the former UN Special Rapporteur on Human Rights and the Environment, John Knox, work remains to be done to clarify the content of extraterritorial obligations with regard to human rights. The Framework Principles proposed by him include a provision on transboundary harm: "States should cooperate with each other to establish, maintain and enforce effective international legal frameworks in order to prevent, reduce and remedy transboundary and global environmental harm that interferes with the full enjoyment of human rights" (OHCHR, 2018: Principle 13).

Thus, whether or not states accept extraterritorial human rights obligations, states have an obligation to cooperate internationally with respect to human rights. This is mentioned in many human rights treaties, including the UN Charter itself, as well as international environmental law. Principle 7 of the Rio Declaration includes this: "States shall cooperate in a spirit of global partnership" (1992). In fact, much of international environmental law is based on this principle. Does this give an extraterritorial dimension to human rights? This would radically change the nature of human rights which is based on rights-holders and duty-bearers bound together by the link of nationality.

Furthermore, do states have an obligation to control the activities of their citizens/corporations abroad? Would this amount to a violation of sovereignty of another state where the citizen resides or where the corporation is registered? (Seck, 2012). Can the state of the parent company incur liability for harmful activities of their subsidiaries abroad? This is one of the biggest challenges in relation to both human rights law and international environmental law. Some argue that the obligation of due diligence has an extraterritorial dimension: "A state may be responsible for a violation of an international human rights treaty obligation where 'acts of their authorities, whether performed within or outside national boundaries … produce effects outside their own territory' and for 'the extraterritorial consequences of its intra-territorial decisions'" (McCorquodale and Simons, 2007: 618, footnotes omitted).

The word "extraterritoriality" implies something illegal or an extension of power by a state. Should we simply say that states have a duty to cooperate with one another when protecting human rights and addressing transboundary environmental damage? As pointed out earlier, international human rights law does not operate in a vacuum. Can we combine Principle 21 of the Stockholm Declaration with international human rights law and argue that states have to prevent transboundary environmental damage that will impact on the human rights of people in neighboring states? The Advisory Opinion of the Inter-American Court seems to have done precisely that. Will the recognition of a distinct right to a healthy environment under international law help us to address the issue of transboundary environmental damage that leads to a violation of human rights in neighboring states?

DISCUSSION QUESTIONS

1. What are extraterritorial obligations? Why are they important?
2. Do you think human rights law should recognize extraterritorial obligations? Why or why not?
3. If extraterritorial obligations are recognized, how will the human rights institutions facilitate individual complaints?
4. Will the recognition of extraterritorial obligations help victims of climate change?
5. What are the lessons from history that will help us design a legal regime with regard to extraterritorial obligations?

References

Advisory Opinion The Effect of Reservations on the Entry into Force of the American Convention on Human Rights (1982). Advisory Opinion OC-2/82, September 24, 1982, Inter-Am. Ct. H.R. (Ser. A) No. 2 [Online]. Available at: http://hrlibrary.umn.edu/iachr/b_11_4b.htm [Accessed 1 October 2018].

African Charter on Human and People's Rights (1981). Also referred to as the Banjul Charter [Online]. Available at: www.achpr.org/files/instruments/achpr/banjul_charter.pdf [Accessed 18 October 2018].

Boyle, A. (2015). Human Rights and the Environment: Where Next? In: Boer, B. (ed.), *Environmental Law Dimensions of Human Rights*, *The Collected Courses of the Academy of European Law*. Oxford: Oxford University Press, p.201.

Committee on Economic, Social and Cultural Rights (CESCR) (1999a). General Comment No 12.

Committee on Economic, Social and Cultural Rights (CESCR) (1999b). General Comment No 13.

Committee on Economic, Social and Cultural Rights (CESCR) (2000). General Comment No 14.

Committee on Economic, Social and Cultural Rights (CESCR) (2002). General Comment No 15.

CORTEIDH (2017). The Environment and Human Rights (State Obligations in Relation to the Environment in the Context of the Protection and Guarantee of the Rights to Life and to Personal Integrity – Interpretation and Scope of Articles 4(1) and 5(1) of the American Convention on Human Rights), Advisory Opinion OC-23/18, Inter-Am. Ct. H.R., (ser. A) No. 23 (Nov. 15, 2017) [Online]. Available at: www.corteidh.or.cr/docs/opiniones/seriea_23_esp.pdf (in Spanish). Available at: http://www.corteidh.or.cr/docs/opiniones/resumen_seriea_23_eng.pdf (in English) [Accessed: 1 October 2018].

Dankwa, V., Flinterman, C., and Leckie, S. (1997). Commentary on the Maastricht Guidelines on Violations of Economic, Social and Cultural Rights SIM Special 20, 13–34 [Online]. Available at:www.icescr-pakistan.org/wp-content/uploads/2013/11/20-02.pdf [Accessed: 1 October 2018].

Declaration of the United Nations Conference on the Human Environment (1972) [Online]. Available at: www.un-documents.net/unchedec.htm [Accessed on 18 October 2018].

De Schutter, O., Eide, A., Khalfan, A., Orellana, M., Saloman, M., and Seiderman, I. (2012). Commentary to the Maastricht Principles on Extraterritorial Obligations of States in the Area of Economic, Social and Cultural Rights. *Human Rights Quarterly* 34(4): 1084–169.

Earthjustice (2005). The Petition to the Inter American Commission on Human Rights Seeking Relief from Violation Resulting from Global Warming Caused by Acts and Omissions of the United States, Summary of the Petition (the Inuit petition) [Online]. Available at: http://earthjustice.org/sites/default/files/library/legal_docs/petition-to-the-inter-american-commission-on-human-rights-on-behalf-of-the-inuit-circumpolar-conference.pdf [Accessed: 1 October 2017].

ESCR (1986). Limburg Principles on the Implementation of the International Covenant on Economic, Social and Cultural Rights [Online]. Available at: www.escr-net.org/resources/limburg-principles-implementation-international-covenant-economic-social-and-cultural [Accessed: 1 October 2017].

ETO Consortium (2011). Maastricht Principles on Extraterritorial Obligations of States in relation to Economic, Social and Cultural Rights [Online]. Available at: www.etoconsortium.org/nc/en/main-navigation/library/maastricht-principles/?tx_drblob_pi1%5BdownloadUid%5D=23 [Accessed: 20 October 2018].

ICJ (1996). Legality of the Threat or Use of Nuclear Weapons, Advisory Opinion [Online]. Available at: www.icj-cij.org/files/case-related/95/095-19960708-ADV-01-00-EN.pdf [Accessed: 1 October 2018].

ICJ (2004). Legal Consequences of the Construction of a Wall in the Occupied Palestinian Territory, Advisory Opinion, ICJ Rep [Online]. Available at: www.icj-cij.org/files/case-related/131/131-20040709-ADV-01-00-EN.pdf [Accessed 10 October 2018].

ILO (1989). Indigenous and Tribal Peoples Convention, No. 169 [Online]. Available at: www.ilo.org/dyn/normlex/en/f?p=NORMLEXPUB:12100:0::NO::P12100_ILO_CODE:C169 [Accessed: 1 October 2018].

International Law Commission (2001). Draft Articles on Responsibility of States for Internationally Wrongful Acts adopted by the International Law Commission at its fifty-third session [Online]. Available at: http://legal.un.org/ilc/texts/instruments/english/commentaries/9_6_2001.pdf [Accessed: 1 October 2017].

ITLOS (2011). Responsibilities and Obligations of States Sponsoring Persons and Entities with Respect to Activities in the Area, Advisory Opinion, Seabed Disputes Chamber of the International Tribunal for the Law of the Sea [Online]. Available at: www.itlos.org/fileadmin/itlos/documents/cases/case_no_17/adv_op_010211.pdf [Accessed: 1 October 2018].

Knox, J.H. (2009). Climate Change and Human Rights Law. *Virginia Journal of International Law* 50 (1): 163.

Knox, J.H. (2010). Diagonal Environmental Rights. In: Gibney, M. and Skogly, S. (eds.), *Universal Human Rights and Extraterritorial Obligations*. Philadelphia, PA: University Pennsylvania Press, p.82.

Knox, J.H. (2013). *Mapping Human Rights Obligations Relating to the Enjoyment of a Safe, Clean, Healthy and Sustainable Environment*, Individual Report on the UN General Assembly and the Human Rights Council, including the Universal Periodic Review Process, Report No. 6. A/HRC/25/53. Geneva: Office of the High Commissioner for Human Rights (OHCHR).

McCorquodale, R. and Simons, P. (2007). Responsibility Beyond Borders: State Responsibility for Extraterritorial Violations by Corporations of International Human Rights Law. *Modern Law Review* 70 (4): 598–625.

Meron, T. (1995). Extraterritoriality of Human Rights Treaties. *American Journal of International Law* 89: 78–82.

Nordic Environmental Protection Agreement (1974) [Online]. Available at: http://sedac.ciesin.org/entri/texts/acrc/Nordic.txt.html [Accessed: 1 October 2018].

OAS (1988). Additional Protocol to the American Convention of Human Rights in the Area of Economic, Social and Cultural Rights. [Online]. Available at: www.oas.org/juridico/english/treaties/a-52.html [Accessed 10 October 2018].

OAS (1969). American Convention of Human Rights [Online]. Available at: www.cidh.oas.org/basicos/english/basic3.american%20convention.htm [Accessed 1 October 2018].

OHCHR (2018). Framework Principles on Human Rights and Environment. [Online]. Available at: www.ohchr.org/EN/Issues/Environment/SREnvironment/Pages/FrameworkPrinciplesReport.aspx [Accessed: 10 October 2018].

Sands, P., Peel, J., Fabra A., and MacKenzie, R. (2012). *Principles of International Environmental Law*. 3rd ed. Cambridge: Cambridge University Press.

Seck, S. (2012). Home State Regulation of Environmental Human Rights Harms as Transnational Private Regulatory Governance. *German Law Review* 13: 1363

United Nations (1992). Rio Declaration on Environment and Development. A/CONF.151/26 (Vol. I) [Online]. Available at: www.un.org/documents/ga/conf151/aconf15126-1annex1.htm [Accessed: 10 October 2018].

United Nations (2013). Report of the Special Rapporteur on the Human Right to Safe Drinking Water and Sanitation A/68/264 [Online]. Available at: www.etoconsortium.org/nc/en/main-navigation/library/documents/detail/?tx_drblob_pi1%5BdownloadUid%5D=98 [Accessed: 1 October 2017].

Cases

Alejandre v. Cuba (1999). 42 F. Supp. 2d 1317 (S.D. Fla. 1999).

Bankovic and Others v. Belgium and 16 Other Contracting States (2001). Application no. 52207/99, 12 December 2001.

Coard et al. v. United States (1999). Report N. 109/99 Case 10.951, Inter-American Commission on Human Rights (IACHR), 29 September 1999.

Delia Saldias de Lopez v. Uruguay (1981). CCPR/C/13/D/52/1979, UN Human Rights Committee (HRC), 29 July 1981.

Ecuador *v. Columbia (Aerial Herbicide Spraying)* (2009). No. 2011/31, Memorial (2009). [Online]. Available at: www.icj-cij.org/files/case-related/138/17540.pdf [Accessed: 1 October 2018].

Ecuador v. Colombia (Aerial Herbicide Spraying) (2011). Reply of Ecuador, vol 1, ICJ [Online]. Available at: https://www.icj-cij.org/files/case-related/138/17554.pdf [Accessed: 18 October 2018].

Georgia v. Russian Federation (2008). I.C.J. 140.

Georgia v. Tennessee Copper Co. (1907). 206 U.S. 230.

Ilascu and Others v. Moldova and Russia (2004). 48787/99, Council of Europe: European Court of Human Rights, 8 July 2004.

Pulp Mills on the River Uruguay (Argentina v. Uruguay) (2010). ICJ Reports (2010).

Saldano v. Argentina Petition (1998). IACHR Report 38/99, 11 March 1999, Ann. Rep. IACHR 1998, p 289.

Social and Economic Rights Action Center and the Center for Economic and Social Rights v. Nigeria (Ogoniland Case) (2001). Comm. No. 155/96 (2001).

United States v. Canada (Trail Smelter Case) (1941). Arbitral Trib., 3 U.N. Rep. Int'l Arb. Awards 1905 (1941).

Yakye Axa v. Paraguay (2005). June 17, 2005.

14 Business, human rights, and the environment

Introduction

In the course of the 1990s, the United Nations started to raise awareness about the human rights responsibilities of multinational corporations (MNCs) and other business entities. As MNCs were exerting more and more power in an increasingly globalized world, it was argued, they were also expected to respect or even promote human rights within their sphere of influence. These debates resulted from fears of a "race to the bottom" in environmental and social standards. This means, with trade liberalization and increasing business inter-relations across borders, many MNCs decided to move production sites to countries with low wages and taxes, minimal labor and social protection schemes, and deficient environmental legislation. In this context, the slogan "with great power comes great responsibility" is not only Spider-Man's (alias Peter Parker) motto, but has been used by many researchers when trying to grasp the changing role of business in human rights (Nolan, 2005: 581).

Since industrialization, businesses have been responsible for emitting greenhouse gases through their industrial processes. For a considerable period of time, the implications of these emissions were not very well-known, and thus did not lead to any consequences. Today, however, businesses are also considered to be political actors that are accountable for their decisions. Under corporate responsibility, they are required to mitigate and to adapt to climate change and environmental challenges, and to respect human rights. As a result, they try to influence policy-making, voice their concerns in relation to governmental decisions, and participate as observers in the international climate negotiations. Encouraging business actors to take more responsibility for the consequences of their actions on society and the environment is called corporate social responsibility (CSR). This also requires them to adopt practices initiating change towards more sustainability in the area of economic and social development, working conditions, health and safety, environmental protection, and human rights. The responsibilities of businesses at the intersection of human rights and the environment will be discussed in this chapter. We will first look into the United Nations' approaches to determining business responsibility in relation to human rights, including attempts to create binding norms and voluntary standards. We will then look at the work of the former UN Special Representative on Business and

Human Rights, soft law and self-regulative initiatives as well as the UN Guiding Principles on Business and Human Rights (United Nations, 2011). The chapter will close with a case study of the Philippines' "Carbon Majors" petition, under which petitioners are seeking to hold businesses accountable for contributing to climate change and its severe impacts.

Voluntary efforts or binding norms?

One crucial question when debating the role of business actors in the field of human rights and the environment is whether there should be a legally binding commitment for them to act, or whether it is more effective to encourage voluntary engagement. Along these lines, the UN has worked on two main approaches to establish the responsibility of corporate actors: (a) the UN Global Compact, a completely voluntary scheme, as well as (b) the UN Draft Norms that were meant to become a binding international law instrument.

The UN Global Compact

One of the world's largest corporate responsibility initiatives is the UN Global Compact (United Nations Global Compact, 2018a) introduced by the former UN Secretary-General, Kofi Annan, in 1999 and officially launched at the UN headquarters in 2000. It is based on voluntary commitments and public–private cooperation. By becoming a member of the Global Compact, business actors agree to operate in accordance with ten basic principles that are derived from international law and relate to human rights, labor protection, environmental protection, and anti-corruption. The ten principles comprise:

Human rights

Principle 1: Businesses should support and respect the protection of internationally proclaimed human rights; and
Principle 2: make sure that they are not complicit in human rights abuses.

Labor

Principle 3: Businesses should uphold the freedom of association and the effective recognition of the right to collective bargaining;
Principle 4: the elimination of all forms of forced and compulsory labor;
Principle 5: the effective abolition of child labor; and
Principle 6: the elimination of discrimination in respect of employment and occupation.

Environmental protection

Principle 7: Businesses should support a precautionary approach to environmental challenges;

Principle 8: undertake initiatives to promote greater environmental responsibility; and

Principle 9: encourage the development and diffusion of environmentally friendly technologies.

Anti-corruption

Principle 10: Businesses should work against corruption in all its forms, including extortion and bribery.

(United Nations Global Compact, 2018a)

Transnational corporations voluntarily commit to these principles without being subject to the international law instruments they are based on, like the Universal Declaration of Human Rights, the ILO's Declaration on Fundamental Principles and Rights at Work, the Rio Declaration on Environment and Development, as well as the United Nations Convention against Corruption.

At the same time, companies and businesses undertake to become active in advancing broader societal objectives that are clearly linked with the Sustainable Development Goals (SDGs). This means, the ten principles are not viewed in isolation, but as embedded into the SDGs and into an overall vision of promoting sustainable businesses. For instance, Principle 9 of the Global Compact calls for the development and diffusion of environmentally friendly technologies. An example could be a company using solar panels to generate the electricity they need for production. By using solar energy, this company at the same time also contributes to the achievement of SDG 7 (affordable and clean energy), SDG 12 (responsible consumption and production), as well as SDG 13 (climate action). Hence, the principles of the Global Compact and the SDGs are closely interlinked.

The Global Compact can be understood rather as a platform for knowledge exchange on innovative corporate practices and communicating new ideas, than as a regulatory framework. On its website, 9,000 companies and 4,000 non-business organizations (e.g., NGOs) are registered. Interaction and engagement are encouraged through the requirement of submitting an annual "Communication on Progress" (for business partners), or a bi-annual "Communication on Engagement" (for non-business partners). These should comprise a statement of commitment by the respective Chief Executive, a description of practical actions taken to implement the ten principles of the Global Compact, and an explanation on how the outcome of these actions can be measured. These communications are made publicly available on the website and provide key stakeholders, such as governments, civil society, investors, and consumers with valuable information on best practices in relation to sustainable business practices (United Nations Global Compact, 2018b).

The UN Draft Norms

By the end of the 1990s, a sub-committee of the then-Human Rights Commission started to investigate the relationship between businesses and human rights.

In 1998, the commission began working on a document outlining human rights obligations for transnational corporations (Blackburn, 2010). The outcome document was completed in 2003 and tabled for presentation at the 2004 session of the UN Human Rights Commission titled Draft Norms on the Responsibilities of Transnational Corporations and Other Business Enterprises with Regard to Human Rights (United Nations, 2003). As one of the general obligations, the document contained a very clear statement about the responsibilities of business entities in relation to human rights:

> Within their respective spheres of activity and influence, transnational corporations and other business enterprises have the obligation to promote, secure the fulfilment of, ensure respect of, and protect human rights recognized in international as well as national law.
>
> (United Nations, 2003)

By directly assigning the entire spectrum of human rights obligations to transnational corporations, the norms were seen as a radical step to impose binding obligations on businesses and thus, to make them subjects of international law. This was very strongly opposed by a number of important business organizations, among them the International Chamber of Commerce (ICC) and the International Organization of Employers (IOE), as well as the United States Council for International Business (USCIB). The draft norms also initiated new discussions pertaining to whether responsibility to protect human rights rests solely with governments, or if it is possibly shared with non-governmental organizations or other private actors. On the contrary, many feared the "privatization" of human rights norms, if business actors were accountable for human rights protection (Nolan, 2005). As a consequence, the Human Rights Commission remained rather skeptical about the proposal, and highlighted that the draft version of the norms is in no way legally binding. Thus, the proposal was rejected and never adopted as an international law instrument. With the re-structuring of the Commission as the Human Rights Council, the relevant subcommittee also ceased to exist. However, a new resolution adopted in 2005 paved the way for further work in this area. It created the post of a Special Representative to the Secretary-General on Human Rights and Transnational Corporations and other Business Enterprises. John Ruggie, a Professor of International Affairs at Harvard University, was appointed to the post and served as Special Representative from 2005–2011. His main tasks were to map human rights obligations of business actors and to make recommendations outlining further steps to be taken (Blackburn, 2010: 16).

UN Special Representative on Business and Human Rights

In 2007, John Ruggie presented his report on his mandate to the UN Human Rights Council. Scholars argue that this document presents a framework for corporate social responsibility. Ruggie addresses the misalignment between strong economic forces and societies' capacities to manage adverse consequences of these. He suggests that

> This misalignment creates the permissive environment within which blameworthy acts by corporations may occur without adequate sanctioning or reparation. For the sake of the victims of abuse, and to sustain globalization as a positive force, this must be fixed.
>
> (Human Rights Council, 2007: 3)

Ruggie proposed a realignment of the relationships between different social institutions, in which governments should play a key role but need to cooperate closely with a number of institutions and actors to achieve just and sustainable solutions. He also acknowledged that this realignment will be a long-term process. In this first report, he affirmed that states have a duty to protect human rights within their jurisdiction and this includes protection against human right violations committed by non-state entities, including businesses. These obligations are anchored in the core UN human rights treaties, confirmed by the respective treaty bodies and accepted to be part of international customary law (Human Rights Council, 2007). Ruggie identified the interplay of two important developments as an indicator for stronger enforcement mechanisms in relation to corporate responsibility. These developments are, on the one hand, the increased recognition of individual responsibility via international *ad hoc* tribunals and the International Criminal Court, and on the other hand, corporate liability for international crimes has been extended under domestic law and is progressively being enforced by national courts. However, Ruggie suggested that international human rights instruments do not yet impose any legal obligations on private businesses, but that a range of soft law mechanisms will inspire further elaboration of corporate responsibility (Human Rights Council, 2007: 8–14). These include the ILO Tripartite Declaration of Principles Concerning Multinational Enterprises and the OECD Guidelines for Multinational Enterprises.

Soft law and self-regulative initiatives are considered to be innovative schemes fostering corporate responsibility. As we know, soft law is not legally binding. It entails a lower degree of commitment, rules are often not that precisely formulated, there are fewer contract costs involved, and in case of non-compliance, non-judicial procedures are applied, if any are applied at all, which makes it often easier for actors to agree to soft law arrangements (Abbott and Snidal, 2009). These can include inter-governmental guidelines and hybrid multi-stakeholder arrangements. The latter usually involve state governments, business actors, and civil society, including trade unions. Sharing responsibility between these actors seems to be the most promising way ahead. This "shared responsibility" has its origins in moral philosophy (Young, 2004), and it can be understood as one step forward in the realignment of relationships between social institutions that Ruggie emphasized in his first report with a view to making the process of globalization more sustainable.

Soft law and self-regulative initiatives

OECD Guidelines for Multinational Enterprises

The OECD Guidelines for Multinational Enterprises, for instance, can be understood as recommendations provided by governments to multinational companies. They were adopted in 1976 as a soft law mechanism and continuously revised in 1979, 1982, 1984, 1991, 2000, and 2011. The Guidelines comprise voluntary standards and principles for responsible business conduct that are consistent with the legal framework of the respective countries in which corporations operate. They are included as an annex to the *OECD Declaration on International Investment and Multinational Enterprises*, but are not legally binding for business actors – even though some issues addressed in the Guidelines might be legally regulated under domestic frameworks or international law. If domestic laws are in place, business corporations are legally obliged to comply with them. Countries deciding to agree to the Guidelines legally commit to implement them according to the *Decision of the OECD Council on the OECD Guidelines for Multinational Enterprises*. In this way, they constitute a unique system of shared responsibility between public and private actors and responsible governance. The main aim of governments accepting the Guidelines is to balance economic, social, and environmental progress in relation to business activities with a view to fostering sustainable development (OECD, 2011).

The OECD Guidelines for Multinational Enterprises are quite complex, and comprise 11 chapters, including recommendations on human rights, the environment, industrial relations, combating bribery, consumer interests, science and technology, competition, and taxation. These provisions are regularly updated; the human rights chapter, for instance, was updated in 2011 in order to be consistent with the *UN Guiding Principles on Business and Human Rights* proposed by John Ruggie (United Nations, 2011). The human rights recommendations emphasize that multinational enterprises should respect human rights and avoid contributing to adverse human rights impacts within their sphere of activity, and they should prevent and mitigate any adverse rights impacts that are in any way related to their business activities and their products or services. They should also carry out human rights due diligence, i.e., assessing actual and potential human rights infringements relating to the business activities of multinational enterprises, their products, services, or supply chains, and they should provide for the remediation of adverse human rights impacts caused by their conduct (OECD, 2011: 31). As emphasized in the commentary to these Guidelines, multinational enterprises still need to respect and protect human rights even if they operate in countries that do not comply with, or have not even ratified, international (human rights) law. This means, the "race to the bottom" approach described earlier is not acceptable, according to these Guidelines.

Chapter 6 of the OECD Guidelines for Multinational Enterprises, which focuses on the environment, addresses a number of issues that are relevant from a human rights perspective as well. This includes providing workers with adequate information about the environmental health and safety impacts of the respective business activities and consulting communities affected by these. In case their undertakings have adverse effects on the environment, health, and safety, multinational corporations need to conduct environmental impact assessments. The environmental performance of enterprises should continuously be improved via new technologies, awareness-raising among customers, recycling, and efficient resource use. This includes the reduction of greenhouse gas emissions. These examples highlight that the environment chapter of the Guidelines reflects important principles outlined in international legal instruments, such as the 1992 Rio Declaration on environment and Development or the Aarhus Convention on Access to Information, Public Participation in Decision-making, and Access to Justice in Environmental Matters discussed in chapter 6 (OECD, 2011: 42–6).

In contrast to other self-regulative schemes that will be discussed below, the OECD Guidelines can be understood as an intergovernmental initiative. This means governments deciding to adhere to them will take over implementation responsibility and will encourage their use within their territorial and legislative boundaries. This entails the establishment of National Contact Points that promote the Guidelines and at the same time, act as an exchange and discussion forum for all matters relating to them. Governments also take responsibility for appropriate review and consultation procedures, as well as interpreting the Guidelines and taking appropriate measures in case any issues arise. This creates a unique relationship between governments and multinational enterprises that have to comply with the main principles outlined in the Guidelines. As complying governments take the main implementation responsibility, there are stronger enforcement mechanisms in place in this soft law scheme than in other self-regulative initiatives. However, the Guidelines are only applicable in countries that are member states of the OECD.

International Finance Corporation's (IFC) Performance Standards

Another inter-governmental scheme is the International Finance Corporation's (IFC) Performance Standards. The IFC is part of the World Bank Group. The IFC Environmental and Social Performance Standards build an important part of the Corporation's commitment to sustainable development and are relevant for all projects financed by the World Bank Group. These can involve a variety of different actors, including private businesses and investors, non-governmental organizations or other civil society actors, national governments, and local communities. If any financial contribution or investment is made by the IFC as a member of the World Bank Group, the IFC Performance Standards have to be observed. One example of this is the hydroelectric Gibe III dam in Ethiopia as a development project discussed in Chapter 12. In this project, a number of different actors were involved, including the Government of Ethiopia, private investors like the Chinese Export-Import Bank (EXIM), research institutes, civil society organizations, and affected local population groups, mainly indigenous peoples. The World Bank had initially contributed to planning the hydroelectric dam and other development banks, such as the African Development Bank and the European Investment Bank, considered funding the dam. After carrying out preliminary evaluation studies, the World Bank, the European Investment Bank, and the African Development Bank announced in 2010 that they were no longer considering funding Gibe III due to serious doubts relating to adverse social and environmental impacts (International Rivers, 2011). The World Bank used the IFC Performance Standards to evaluate the impact of the hydroelectric dam (see Table 14.1).

Taking a closer look at these standards, we can again find basic principles of international law reflected in them. Performance Standard 7 on Indigenous Peoples, for instance, requires meaningful consultation and the support of affected communities for all project activities. This is in accordance with 1989 ILO Indigenous and Tribal Peoples Convention as well as the 2007 UN Declaration

Table 14.1 IFC Performance Standards

Performance Standard (1) **Assessment and management of environmental and social risks and impacts** Underscores the importance of identifying environmental and social risks and impacts, and managing environmental and social performance throughout the life of a project.
Performance Standard (2) **Labor and working conditions** Recognizes that the pursuit of economic growth through employment creation and income generation should be balanced with protection of basic rights for workers.
Performance Standard (3) **Resource efficiency and pollution prevention** Recognizes that increased industrial activity and urbanization often generate higher levels of air, water, and land pollution, and that there are efficiency opportunities.
Performance Standard (4) **Community health, safety, and security** Recognizes that projects can bring benefits to communities, but can also increase potential exposure to risks and impacts from incidents, structural failures, and hazardous materials.
Performance Standard (5) **Land acquisition and involuntary resettlement** Applies to physical or economic displacement resulting from land transactions such as expropriation or negotiated settlements.
Performance Standard (6) **Biodiversity conservation and sustainable management of living natural resources** Promotes the protection of biodiversity and the sustainable management and use of natural resources.
Performance Standard (7) **Indigenous peoples** Aims to ensure that the development process fosters full respect for indigenous peoples.
Performance Standard (8) **Cultural heritage** Aims to protect cultural heritage from adverse impacts of project activities and support its preservation.

Source: IFC, 2012: 2 and IFC, 2018.

on the Rights of Indigenous Peoples. The World Bank Group does not finalize an investment agreement without asking the client to conduct an impact assessment of environmental and social risks, according to the Performance Standards and the World Bank's Environmental, Health and Safety Guidelines. They will also publicly disclose the project and consult with the local community affected, as well as regularly monitor the project's impact through field visits and evaluation reports that have to be submitted by the implementer (IFC, 2018).

Self-certification schemes

The Kimberley Process Certification Scheme (KPCS) is an example of a self-regulative initiative established in 2000 and aimed at preventing so-called "conflict diamonds" that were used to finance rebel groups engaged in intra-state violence and situations of conflict from entering the rough diamond market. Conflict diamonds (also called "blood diamonds" or "war diamonds") at that time had fueled civil wars and empowered rebel groups acting against democratically elected governments in several African countries, including Angola, the Ivory Coast, Sierra Leone, Liberia, Guinea, Guinea Bissau, the Democratic Republic of the Congo, and others. As a consequence, a diamond certification process was brought into life by UN General Assembly Resolution 55/56 in December 2000 based on information provided by the Fowler report. Robert Fowler, at that time Canada's ambassador to the UN, headed a commission that collected evidence to establish the link between illicit diamond trade and the União Nacional para a Indepêndencia Total de Angola (UNITA). UNITA was one of the conflict parties during the Angolan Civil War and, despite several UN Security Council Resolutions imposing sanctions in an attempt to end the conflict, kept financing the war by selling conflict diamonds (United Nations, 2000).

Based on the recommendations of the Fowler report, the Kimberley Process Certification Scheme was developed with Southern African diamond-producing states meeting in Kimberley, South Africa in 2000. After the adoption of the landmark resolution by the UN General Assembly, negotiations between governments, civil society and the diamond industry (represented by the World Diamond Council), were initiated and the Kimberley Process Certification Scheme was created in 2002 and entered into force in 2003. Today, the Kimberley Process has 54 member states that represent 81 nations (with the EU counting as one member) (Kimberley Process, 2018). Countries wishing to participate need to ensure that their diamonds do not finance any violent or non-violent group looking to overthrow recognized governments. In addition, every exported diamond needs a Kimberley Process Certificate and neither import nor export to a non-member of the scheme is allowed. Moreover, member states need to harmonize national legislation with certain key principles, ensure transparency and share relevant data (Kimberley Process, 2018).

The World Diamond Council as the international diamond industry partner has developed a System of Warranties embraced by all participants in the

Kimberley Process. It asks all partners to make the following affirmative statement on all invoices:

> The diamonds herein invoiced have been purchased from legitimate sources not involved in funding conflict and in compliance with United Nations resolutions. The seller hereby guarantees that these diamonds are conflict free, based on personal knowledge and/or written guarantees provided by the supplier of these diamonds.
>
> (World Diamond Council, 2018)

There are also other principles of self-regulation all participating member states and trade partners need to comply with. For example, all company employees need to be informed about governmental regulations with respect to trade restrictions on conflict diamonds, or to abstain from buying from unknown or even suspect sources. Monitoring of implementation takes place via field visits and annual reports and by sharing relevant data in working groups and committees involving member states, civil society, and the diamond industry (Kimberley Process, 2018). Nevertheless, there has been some criticism regarding the issuing of fake Kimberley Process certificates and the inability of the scheme to stop the flow of conflict diamonds (BBC, 2011).

A similar example is the Extractive Industries Transparency Initiative (EITI) founded by Tony Blair in 2002 with important validation procedures agreed in 2008. By committing to and monitoring self-regulative standards, EITI ensures that natural resources, such as oil, gas, and minerals, are governed responsibly in order to benefit the citizens of resource-rich countries. The EITI Standards require transparency and accountability. This means participating countries need to clarify how licenses have been allocated and provide information about tax payments and social contributions of companies, and the ways in which these are used by the government. In each country, a multi-stakeholder group comprised of government representatives, industrial partners, and civil society actors makes decisions on how the EITI process should work. Information about governance decisions in this sector needs to be made transparent, and recommendations are issued by the multi-stakeholder group and subsequently used for public awareness and follow-up procedures regarding the recommendations (EITI, 2018). Here again, we can see how different actors, both governmental and non-governmental, cooperate in unique partnerships to establish a regulatory scheme using key principles that are not legally binding.

Where soft law mechanisms and self-regulative schemes remain most challenged is in relation to accountability. The number, diversity, and adoption of these instruments have grown significantly, but they also pose serious questions about accountability and enforcement mechanisms and how these can be established or improved. Partner companies in the supply chain especially cannot always be effectively monitored regarding compliance with key principles and standards. Often, supplier codes are developed to strengthen accountability in the supply chain, but, in many cases, these lack credibility without independent external monitoring. Against this background, the UN Special Representative on business and human rights, John Ruggie, concentrated on further developing a unique soft law scheme that would contribute to clarifying corporate responsibility in relation to human rights.

UN Guiding Principles on Business and Human Rights

After his first three-year mandate from 2005–2008, Special Representative of the UN Secretary General on Business and Human Rights, John Ruggie, presented the "Protect, Respect and Remedy" Framework to the UN Human Rights Council. This framework was an attempt to conceptualize corporate responsibility in relation to human rights. The Human Rights Council embraced the framework and extended Ruggie's mandate until 2011 with the main task of "operationalizing" it (United Nations, 2011; Davis, 2012: 962). On this basis, Ruggie further developed the UN Guiding Principles on Business and Human Rights (UNGPs), which were unanimously endorsed by the Human Rights Council in 2011. Making reference to the framework, the principles are based on three main pillars:

- The state's duty to *protect* human rights
- The corporate responsibility to *respect* human rights
- Access to effective *remedy* for rights abuses in relation to business activities

Thus, the "Protect, Respect and Remedy Framework" tells us more about *what* should be done and the Guiding Principles complement this by outlining *how* this should be done (Davis, 2012: 962). Informally, the UNGPs became known as the "Ruggie Principles."

The UNGPs were welcomed by a number of actors working at the intersection of human rights and business activities, including state governments, civil society organizations, and even corporate actors. Even though they did not create any new standards or legally binding norms, they clearly spelled out the respective obligations and responsibilities of governments and businesses. Thus, Ruggie has created an authoritative focal point, sometimes called a global standard, for enhancing good governance in relation to human rights and business conduct, with a view to contributing to sustainable globalization (United Nations, 2011). One key achievement of the UNGPs was a subsequent convergence of standards in the field with the OECD revising its Guidelines for Multinational Enterprises in 2011 adding a new human rights chapter that reflects the Ruggie Principles. The International Financial Corporation (IFC) has reacted by including corporate responsibility to respect human rights in its Sustainability Framework and Performance Standards. Moreover, the European Commission released a new communication pertaining to corporate social responsibility, calling upon companies conducting business in the European Union to comply with the UN Guiding Principles on Business and Human Rights (Davis, 2012: 963). Thus, the

Principles can be understood as a global benchmark defining corporate responsibility in relation to human rights.

The 31 principles of the UNGP comprise foundational as well as operational principles for each of the three pillars identified there. The detailed Guidelines contain clear recommendations on the implementation of the framework and its principles (see Table 14.2 for foundational principles).

Table 14.2 UN Guiding Principles on Business and Human Rights

The state duty to protect human rights

- States must protect against human rights abuses within their territory and/or jurisdiction by third parties, including business enterprises. This requires taking appropriate steps to prevent, investigate, punish, and redress such abuse through effective policies, legislation, regulations, and adjudication.
- States should set out clearly the expectation that all business enterprises domiciled in their territory and/or jurisdiction respect human rights throughout their operations.

The corporate responsibility to respect human rights

- Business enterprises should respect human rights. This means that they should avoid infringing on the human rights of others and should address adverse human rights impacts with which they are involved.
- The responsibility of business enterprises to respect human rights refers to internationally recognized human rights – understood, at a minimum, as those expressed in the International Bill of Human Rights and the principles concerning fundamental rights set out in the International Labor Organization's Declaration on Fundamental Principles and Rights at Work.
- The responsibility to respect human rights requires that business enterprises:
  (a) Avoid causing or contributing to adverse human rights impacts through their own activities, and address such impacts when they occur;
  (b) Seek to prevent or mitigate adverse human rights impacts that are directly linked to their operations, products, or services by their business relationships, even if they have not contributed to those impacts.
- The responsibility of business enterprises to respect human rights applies to all enterprises regardless of their size, sector, operational context, ownership, and structure. Nevertheless, the scale and complexity of the means through which enterprises meet that responsibility may vary according to these factors and with the severity of the enterprise's adverse human rights impacts.
- In order to meet their responsibility to respect human rights, business enterprises should have in place policies and processes appropriate to their size and circumstances, including:
  (a) A policy commitment to meet their responsibility to respect human rights;
  (b) A human rights due diligence process to identify, prevent, mitigate, and account for how they address their impacts on human rights;
  (c) Processes to enable the remediation of any adverse human rights impacts they cause or to which they contribute.

Access to remedy

- As part of their duty to protect against business-related human rights abuse, States must take appropriate steps to ensure, through judicial, administrative, legislative, or other appropriate means, that when such abuses occur within their territory and/or jurisdiction those affected have access to effective remedies.

Source: United Nations, 2011: 3–27.

Foundational principles of the UN Guiding Principles on Business and Human Rights

The human rights due diligence process mentioned as one foundational principle under the Corporate Responsibility to Respect Human Rights is particularly interesting when investigating environmental challenges and human rights. The Guiding Principles make it clear that a human rights due diligence process includes assessments of potential human rights impacts of any business activities prior to major decisions being made, consultation with affected population groups, and consideration of the situation of the most vulnerable within societies. These processes can be incorporated in social and environmental impact assessments, as long as they have international rights standards as their major reference point (United Nations, 2011: 20). This is particularly relevant and could potentially have a major impact in relation to many issues we have discussed in this textbook, including the prevention of environmental discrimination, like in the case of the Ogoni people in the Niger delta, or the avoidance of human rights abuses in the case of indigenous peoples affected by the Gibe III hydroelectric dam in Ethiopia.

After the UN Guiding Principles on Business and Human Rights were endorsed by the Human Rights Council in 2011, the Council established a "working group on the issue of human rights and transnational corporations and other business enterprises" (Business and Human Rights Resource Centre, 2018). Its mandate comprises the promotion of the Guidelines and their implementation, identification and exchange of good practices in relation to the Guidelines, capacity-development, and advancing legislation and policies pertinent to human rights and business conduct, integrating a gender perspective and the views of particularly vulnerable population groups, as well as developing a dialogue and cooperation with other UN agencies, multinational companies, national human rights institutions, civil society organizations, and affected actors (Business and Human Rights Resource Centre, 2018).

The Guiding Principles are viewed critically by a number of scholars and civil society organizations. Penelope Simons (2012), for instance argues, that the principles reflect an unjust international legal system that reproduces corporate impunity and lack of accountability. She argues that there is a crucial need for binding human rights obligations for private businesses and suggests how these can be developed incrementally (Simons, 2012).

The discussion about a legally binding instrument at the United Nations, however, is not over yet. In 2014, the Human Rights Council adopted a resolution

> to establish an open-ended intergovernmental working group on transnational corporations and other business enterprises with respect to human rights, whose mandate shall be to elaborate an international legally binding instrument to regulate, in international human rights law, the activities of transnational corporations and other business enterprises.
>
> (OHCHR, 2018a)

The OHCHR emphasized that the development of this new treaty should not be understood as conflicting with the Guiding Principles, but rather building on them. However, a legally binding treaty is still considered to be relevant with a view to regulating the activities of transnational corporations and improving access to remedies for everyone suffering from corporate abuse. During the fourth session of the open-ended intergovernmental working group, stakeholders including governments, international organizations, civil society, and national human rights institutions, discussed a draft of that legal instrument, as well as a draft optional protocol annexed to it. Based on the input received at that meeting, the draft instrument will now be further developed (OHCHR, 2018b).

Case Study: The Philippines' "Carbon Majors" Petition

A unique case to consider when studying corporate responsibility in relation to human rights is the Philippines Petition against Carbon Majors. In September 2015, Greenpeace Southeast Asia, the Philippine Rural Reconstruction Movement, and other non-governmental organizations requested an investigation of the responsibility of 50 major fossil fuel companies for human rights violations resulting from the impact of climate change in the Philippines. The population of the Philippines is already suffering from severe adverse effects of climate change on their human rights, particularly in relation to extreme weather events like Typhoon Haiyan, considered the strongest tropical cyclone recorded in human history (Lucia, 2018). The "Carbon Majors" are multinational corporations including Chevron, Exxon, British Petroleum, and Royal Dutch Shell, among others. The Commission on Human Rights of the Philippines accepted the petition, and in 2015, launched the first-ever investigation into the responsibility of these companies for the impact of climate-related consequences on the human rights of its population.

From a legal perspective, there are very interesting aspects when considering the petition. One is whether the Philippine Commission on Human Rights (CHR) has jurisdiction over the petition. The Carbon Majors who have responded to the Commission claim that it does not, because the CHR does not have adjudicatory powers and because they do not conduct their business in the Philippines (Lucia, 2018). The petition, however, clearly states that

> "extraterritoriality is not a bar to the Commission's exercise of authority, considering the transboundary and global nature of climate change and other environmental problems and the associated human rights implications"
>
> (Greenpeace, 2017)

ClientEarth states that, in contrast to small and medium enterprises, Carbon Majors have contributed largely to climate change consequences by emitting greenhouse gases and advancing global warming:

> Carbon Majors, as demonstrated by the evidence … provided by the Petitioners, have, through their operations and products, contributed to

> the emission of such significant quantities of greenhouse gas emissions into the earth's atmosphere so as to impose upon them specific and particular obligations in respect of climate change.
>
> (2016: 45)

In its *amicus curiae* brief, ClientEarth explicitly refers to the UN Guiding Principles on Business and Human Rights as the global standard on the behavior of private business actors in relation to respecting human rights, and demands that the Guidelines should be considered. It stated that, according to the Principles, businesses should avoid causing or contributing to adverse human rights impacts through their activities, and if they have caused or contributed to adverse rights impacts, they should provide for or cooperate in their remediation (ClientEarth, 2016: 29).

In 2017, the Philippines Commission on Human Rights announced that it is assuming jurisdiction, and is investigating allegations made in the petition by conducting hearings at home and abroad in the course of 2018. A resolution of the petition, which is expected in 2019, could potentially be a substantial contribution to climate justice litigation (Lucia, 2018).

DISCUSSION QUESTIONS

1. Do you think multinational businesses should become subjects of international law? Why, or why not?
2. What are the opportunities and the challenges of soft law mechanisms and self-regulative initiatives?
3. Do the Guiding Principles on Business and Human Rights constitute a new approach to corporate responsibility?
4. How do the Guiding Principles on Business and Human Rights differ from other soft law mechanisms and self-regulative schemes?
5. Can the Guiding Principles be applied in the Philippines "Carbon Major" Petition?

References

Abbott, K. and Snidal, D. (2009). Strengthening International Regulation through Transnational New Governance: Overcoming the Orchestration Deficit. *Vanderbilt Journal of Transnational Law* 42(2): 501–78.

BBC (2011). Global Witness Leaves Kimberley Process Diamond Scheme [Online]. Available at: www.bbc.co.uk/news/business-16027011 [Accessed: 10 September 2018].

Blackburn, D. (2010). What is Ruggie Doing? *International Union Rights* 17 (2): 16–17.

Business and Human Rights Resource Centre (2018). About the Working Group [Online]. Available at: www.business-humanrights.org/en/working-group/about-the-working-group [Accessed: 12 September 2018].

ClientEarth (2016). *Republic of the Philippines Commission on Human Rights, In Re: National Inquiry on the Impact of Climate Change on the Human Rights of the Filipino People*. Amicus curiae brief presented by ClientEarth. Brussels, London and Warsaw: ClientEarth.

Davis, R. (2012). The UN Guiding Principles on Business and Human Rights and Conflict-Affected Areas: State Obligations and Business Responsibilities. *International Review of the Red Cross* 94 (887): 961–79.

EITI (2018). The Global Standard for the Good Governance of Oil, Gas and Mineral Resources [Online]. Available at: https://eiti.org/sites/default/files/documents/eiti_factsheet_en_feb2018.pdf [Accessed: 10 September 2018].

Greenpeace (2017). The Climate Change and Human Rights Petition [Online]. Available at: www.greenpeace.org/seasia/ph/press/releases/Worlds-largest-carbon-producers-ordered-to-respond-to-allegations-of-human-rights--abuses-from-climate-change/The-Climate-Change-and-Human-Rights-Petition/ [Accessed: 28 February 2017].

Human Rights Council (2007). *Business and Human Rights: Mapping International Standards of Responsibility and Accountability for Corporate Acts* A/HRC/4/035, 9 February 2007. Geneva: Human Rights Council.

IFC (2012). *IFC Performance Standards on Environmental and Social Sustainability*. Washington, DC: International Financial Corporation, World Bank Group.

IFC (2018). Understanding IFC's Economic and Social Due Diligence Process. [Online]. Available at: www.ifc.org/wps/wcm/connect/b58ead804942ee5da7a5ff4f5ddda76e/IFC+Process.pdf?MOD=AJPERES [Accessed: 9 September 2018].

ILO (1989). C169 – Indigenous and Tribal Peoples Convention, 1989 (No. 169). Available at: www.ilo.org/dyn/normlex/en/f?p=NORMLEXPUB:12100:0::NO::P12100_ILO_CODE:C169 [Accessed: 19 December 2018].

ILO (1998). ILO Declaration on Fundamental Principles and Rights at Work. [Online]. Available at: www.ilo.org/declaration/lang--en/index.htm [Accessed: 19 December 2018].

International Rivers (2011). *Ethiopia's Gibe III Dam: Sowing Hunger and Conflict*. Berkeley, CA: International Rivers.

Kimberley Process (2018). About [Online]. Available at: www.kimberleyprocess.com/en/about [Accessed: 10 September 2018].

Nolan, J. (2005). With Power Comes Responsibility: Human Rights and Corporate Accountability. *University of New South Wales Law Journal* 38 (3): 581–613.

OECD (2011). *OECD Guidelines for Multinational Enterprises*. Paris: OECD Publishing. [Online]. Available at: http://dx.doi.org/10.1787/9789264115415-en [Accessed: 23 October 2018].

OHCHR (2018a). Open-ended intergovernmental working group on transnational corporations and other business enterprises with respect to human rights [Online]. Available at: www.ohchr.org/en/hrbodies/hrc/wgtranscorp/pages/igwgontnc.aspx [Accessed: 23 October 2018].

OHCHR (2018b). Legally Binding Instrument to Regulate, in International Human Rights Law, the Activities of Transnational Corporations and Other Business Enterprises [Online]. Available at: www.ohchr.org/Documents/HRBodies/HRCouncil/WGTransCorp/Session3/DraftLBI.pdf [Accessed: 23 October 2018].

Simons, P. (2012). International Law's Invisible Hand and the Future of Corporate Accountability for Violations of Human Rights. *Journal of Human Rights and the Environment* 3(1): 5–43.

Sta. Lucia, R. (2018). Update on the 'Carbon Majors' Petition: The Role of the Philippine Commission on Human Rights [Online]. Available at: http://blogs.law.columbia.edu/

climatechange/2018/03/08/update-on-the-carbon-majors-petition-the-role-of-the-philippine-commission-on-human-rights/ [Accessed: 12 September 2018].

UNDRIP (2007). United Nations Declaration on the Rights of Indigenous People. [Online]. Available at: www.un.org/esa/socdev/unpfii/documents/DRIPS_en.pdf [Accessed: 19 December 2018].

United Nations (1948). Universal Declaration of Human Rights. [Online]. Available at: www.un.org/en/universal-declaration-human-rights/ [Accessed: 19 December 2018].

United Nations (2000). Final Report of the UN Panel of Experts on Violations of Security Council Sanctions Against UNITA [Online]. Available at: www.globalpolicy.org/component/content/article/202/41606.html [Accessed: 10 September 2018].

United Nations (2003). Draft Norms on the Responsibilities of Transnational Corporations and Other Business Enterprises with Regard to Human Rights, E/CN.4/Sub.2/2003/12 [Online]. Available at: http://hrlibrary.umn.edu/links/NormsApril2003.html [Accessed: 31 July 2018].

United Nations (2004). United Nations Convention against Corruption. [Online]. Available at: www.unodc.org/documents/brussels/UN_Convention_Against_Corruption.pdf [Accessed: 19 December 2018].

United Nations (2011). *Guiding Principles on Business and Human Rights: Implementing the United Nations 'Protect, Respect and Remedy' Framework.* New York and Geneva: Office of the High Commissioner for Human Rights.

United Nations Conference on Environment and Development (1992). The Rio Declaration on Environment and Development. [Online]. Available at: www.unesco.org/education/pdf/RIO_E.PDF [Accessed: 19 December 2018].

United Nations Global Compact (2018a). The Ten Principles of the UN Global Compact [Online]. Available at: www.unglobalcompact.org/what-is-gc/mission/principles [Accessed: 1 August 2018].

United Nations Global Compact (2018b). The Communication on Progress (CoP) in Brief [Online]. Available at: www.unglobalcompact.org/participation/report/cop [Accessed: 1 August 2018].

World Diamond Council (2018). Core Values [Online]. Available at: www.worlddiamondcouncil.com/ [Accessed: 10 September 2018].

Young, I.M. (2004). Responsibility and Global Labor Justice. *Journal of Political Philosophy* 12(4): 365–88.

15 Inter-generational rights, animal rights, and rights of nature and ecosystems

Introduction

When studying human rights, it is important to realize that they are not static, but dynamic. Throughout their history and development, human rights have responded to various political and social challenges, with a view to protecting the dignity of human beings. In the Universal Declaration of Human Rights (UDHR) adopted during the post-Second World War political order, "equal and inalienable rights of all members of the human family" are understood to be the foundation for freedom, justice, and peace in the world (UDHR, 1948). After the atrocities of the Second World War, the idea was to create international human rights "above the state" that all governments have to comply with and that would protect citizens regardless of their nationality. In a world that is severely threatened by environmental and climate challenges, one might argue that we once more need to reconsider our understanding of rights as the basis for conflict resolution, peace, development, poverty reduction, and a sustainable environment. To address contemporary challenges in relation to the environment, scholars have begun to rethink notions of justice and rights (Beckman and Page, 2008). What we can observe in discourse and practice is a changing character of *human* rights: from individual civil and political rights (with their origins in the 1215 Magna Carta) to collective inter-generational rights that are strongly related to ecosystem integrity. Many of the debates that have initiated this rethinking on rights originate in political theory and political philosophy, as scholars from these disciplines ask what ought to be changed in order to achieve more justice in the world. In this chapter, we will discuss three developments that are characteristic of the changing nature of human rights: inter-generational rights, animal rights, and ecosystem rights (or rights of nature).

Inter-generational rights

Inter-generational rights have, for a long time, been viewed rather skeptically. This is due to classic liberal ideas of a social contract between the government and its people. Establishing a social contract, according to classic liberal thinkers, like Jean-Jacques Rosseau, Thomas Hobbes, and John Locke, means that people who live under a government give up their right to self-protection and agree to follow the law. The government, in turn, protects the rights of its subjects, i.e., its citizens. Thus, human rights determine the relationship between the state and its citizens. However, can they also determine the relationship between a state and its future citizens? Who exactly can speak for future generations, how can they be represented, and in which way can we assess their interests? (Dudai, 2009). Protecting the rights of future generations in a changing environment means reshaping current political systems and institutions, which is quite an ambitious endeavor and often too far away from current political realities and agendas (Beckman and Page, 2008). Moreover, inter-generational rights are often understood as collective rights. Collective rights belong to the third generation of rights and are not anchored in the UN human rights edifice (except for the right to self-determination and rights of minorities), due to many states questioning their universal applicability.

The key central argument pertaining to inter-generational rights is that future generations have interests that need special protection and cannot be compromised because they conflict with certain desires of present generations (Vernon, 2009: 9). Some of these interests have the status of rights, including access to clean drinking water, the highest attainable standard of health, and the right to adequate food. The idea of inter-generational justice implies that future generations will still be able to enjoy these rights when they become the current generation, and that we should not, by taking decisions today, compromise their ability to do so in the future.

Many scholars have argued that the temporal and spatial dimensions of climate change make it difficult to impose certain obligations on others as it is not always clear who the duty-bearers are. The spatial dimension relates to climate change being mainly caused in industrialized countries but having adverse effects predominantly in developing countries. The temporal dimension refers to responsibilities for greenhouse gas emissions (and their effects) of the past and present generations to the younger generation or to generations who are not yet born. Some scholars address both the spatial and temporal dimension at the same time, to make normative claims for diminishing climate injustice. One very important line of argument to protect the interests of future generations by imposing duties has been developed by Simon Caney. He contends that we usually ascribe less value to the rights and interests of people in the future. Caney argues for a "zero-discount rate" in the context of rights protection (2008: 540). This means the rights of future people do not have less meaning or significance than the rights of contemporary human beings. If we take a zero-discount rate into account, we cannot simply put today's interests in greenhouse gas-intensive activities first, without considering the consequences for future generations. In his seminal work on climate justice, Henry Shue has emphasized that CO_2-intensive development and lifestyles of people in industrialized countries have almost exhausted the world's "cumulative carbon budget" (Shue, 2014: 324–7). This has led to a situation of climate injustice between developed and developing countries, between more and less affluent societal groups within these countries, and also between current and future generations. Shue also introduces the concept of "compound injustice" (2014: 4) to highlight how an initial situation of injustice, like colonialism and the exploitation of resources that have contributed to development and wealth in industrialized countries, paves the way for further injustices, such as the adverse impacts of climate change that are hampering further processes of development in many countries of the Global South. Thus, Shue argues, one key dilemma of our time is the need to reduce the use of fossil fuels, but at the same time, guarantee the right to development in countries with increasing energy demands.

In light of the exhausted cumulative carbon budget, Shue has revised his argument that developing countries have emission rights to provide for the subsistence of poor population groups. If these economies now start to emit greenhouse gases without restrictions, we could not guarantee basic human rights for future generations anymore. Thus, Shue argues that developing countries need to have energy rights. In order for them to enjoy these rights today and in the future, the use of fossil fuels needs to be substantially reduced and taxed, and renewable energy sources need to be further developed for them to become competitive and accessible as well as affordable for less affluent population groups in developing countries. Tim Hayward (2007) supports a similar point of view: Both emissions and subsistence, can be understood as central questions of justice. However, we need to be cautious and cannot understand emission rights as human rights. This is against the idea of a human right to a sustainable environment because it would mean to exploit natural resources to an extent that is not sustainable. Hayward thus suggests "an equitable distribution of rights to ecological space" as a more comprehensive framework of all natural resources and environmental services in relation to economic and welfare goods (Hayward, 2007: 433).

Several scholars argue that not only environmental rights, but all human rights claims are future-oriented to a certain extent as they try to improve existing standards of behavior: "Their principle aim is to challenge or change existing institutions, practices, or norms, especially legal institutions" (Donnelly, 1989: 14). Thus, adopting and committing to rights always implies an intention to guarantee them to citizens in the future (Hiskes, 2005: 1355). Even though rights of future generations are collective rights, we need to realize that we are not talking about an abstract group of people but about future citizens in our very own societies (Hiskes, 2005: 1363). In this way, these rights seem to be less abstract. They can be understood as emerging rights that are characteristic of a changing relationship between current and future generations, between individuals and communities within society, as well as between human beings and their natural environments (Hiskes, 2005: 1356).

But how can future generations and their interests be represented in democratic societies? Do we need to reshape our current political systems to accommodate considerations of intergenerational rights, or do we need to develop a different understanding of democracy? (Beckman and Page, 2008). One avenue to realizing intergenerational interests is the adoption of constitutional rights, e.g., a fundamental right to an adequate environment (Hayward, 2004). Constitutional rights limit the power of state governments and protect states' citizens today and in the future; they can prevent governmental decisions that are based on short-term economic interests. Today, environmental rights are included in more than 100 national constitutions (Boyd, 2012: 3) and were discussed in Chapter 7.

Some countries have established institutions to protect future generations. Hungary is a good example. Other more radical ideas include the proposal to reserve parliamentary seats for parties or representatives acting as trustees of future generations (Ekeli, 2005), or ensuring that the environmental lobby is granted special access and status in legislative processes (Dobson, 1996). The 2015 Paris Agreement acknowledges inter-generational equity and several cases that were discussed in Chapter 8 refer to rights of future generations.

Animal rights

Similar to debates about inter-generational rights, the idea that animals are rights-holders has been proposed by scholars from political theory, philosophy, and other disciplines. Until today, the discourse and practice on animal rights, including the animal rights movement, has been inspired by normative thinking on interspecies justice, i.e., justice between different species including humans and animals (Regan, 1983; Singer, 1975; Nussbaum, 2006; Kymlicka and Donaldson, 2011). In this regard, Paula Cavalieri proposes that we should remove the word "human" from human rights (Cavalieri, 2001), and expand our understanding of rights to different species.

Two very influential monographs on animal rights were published in the 1970s and early 1980s, *Animal Liberation* by Peter Singer (1975) and *The Case for Animal Rights* by Tom Regan (1983). Singer proposes a more sophisticated account of equality, extending it to all beings, including black and white, masculine and feminine, human and non-human. What he calls "speciesism," in analogy with racism, is to disadvantage the members of one species against those members of another species. Whereas racists pay more attention to the interests of members of their own race, he argues, speciesists disregard the interests of members of other species. Singer suggests that we should not ask whether animals can reason or talk, but the question should be whether animals can suffer. Based on the work of Jeremy Bentham, Singer (who is also a utilitarian) argues that the capacity to suffer gives species the right to equal consideration. To avoid a vast amount of suffering among animals, human beings need to make radical changes to their diet, farming methods, scientific experiments, attitude to hunting, trapping, and wearing furs, but also entertainment, including circuses, zoos, and rodeos (Singer, 1975).

Tom Regan (1983) agrees with Singer that speciesism is unjust and the way we treat animals is simply wrong. However, he denies that it is wrong because of animal suffering. What he conceives as wrong is to view animals as human resources, that is, to eat them, to exploit them for entertainment, sport, or any commercial activity, and to surgically manipulate them and use them for medical research. Regan denies that animal rearing methods should become "more humane"; he supports the complete abolition of commercial animal husbandry (Regan, 1983: 337).

Similar to the advocates of inter-generational rights, Regan argues that contractarianism, i.e., the idea that citizens enter into a contract with the government to be a rights-holder, cannot be the reason for animals to not enjoy their rights. Regan, who also criticizes the utilitarian perspective brought forward by Singer, claims that all beings with inherent value, that are experiencing subjects of life, have rights. The value of animals cannot be reduced to their usefulness for others (Regan, 1983: 343). This means our duty is not to use them, but to recognize their rights. Regan views the animal rights movement as part of the human rights movement; a movement that does not simply improve conditions of animal farming or treatment, but that does not treat animals as resources at all.

A newer account of animal rights in political theory was published in 2011 by Will Kymlicka and Sue Donaldson. In their book *Zoopolis: A Political Theory of Animal Rights*, they argue that we need a more comprehensive approach to animal rights that varies according to the relationship humans have with a particular animal. Such an approach should integrate universal negative rights with differentiated positive rights depending on the character of the human–animal relationship (Kymlicka and Donaldson, 2011: 11). Both scholars argue that liberalism today is already combining universal human rights with more relational, bounded, and group-differentiated rights, such as the rights of indigenous peoples, children's rights, and the rights of persons with disabilities. Based on that, they claim, citizenship theory can be fruitfully used to "combine traditional animal rights theory with a positive and relational account of obligations" (Kymlicka and Donaldson, 2011: 14).

When referring to human–animal relationships, Kymlicka and Donaldson make the following differentiation between (a) animals living in the wild, forming sovereign communities on their own territories; (b) animals that, similar to migrants or denizens, decide to move into areas of human habitation; and (c) domesticated animals that have been bred over generations for interdependence with human beings. Kymlicka and Donaldson argue that Category (c) should be seen as full citizens with citizenship rights (Kymlicka and Donaldson, 2011: 14). Understanding them as citizens with rights is a moral obligation that arises from including them into human societies – and from taking away their independence and ability to survive in the wild. The citizenship framework suggests that humans and domesticated animals are connected through various ways of interaction and interdependence (Kymlicka and Donaldson, 2011: 61). Wild animals, by contrast, can be understood as citizens of their own sovereign communities – and we should respect their autonomy and their territory (just like the principle of non-interference in international affairs). As they are independent of human beings, wild animals rather need to be protected from humans, and they do not need to be included in human citizenship regimes (Kymlicka and Donaldson, 2011: 62). Non-domesticated animals living amongst humans are being compared to denizens, according to Kymlicka and Donaldson. They need to be respected as co-residents of urban spaces, but they are not included in the citizenship scheme of humans and domesticated animals (Kymlicka and Donaldson, 2011: 15). By employing political concepts, such as citizenship, denizenship, sovereignty, territory, migration, and membership, and exploring their use or adaptation in the context of animals, these scholars make a clear attempt to establish animal rights. This comes with obligations to respect, protect, and provide for animal rights.

The animal rights movement

The modern animal rights movement was heavily influenced by the work of Singer and Regan – it is even said that it is the first social reform movement that was inspired by philosophers (Wise, 2016). Later on, many other professionals, such as lawyers, scientists, academics, veterinarians, theologians, and psychologists joined the movement. Since the beginning of the 21st century, there have been a number of lawsuits brought to protect the interests and rights of animals. Legal scholars have continuously attempted to advance basic animal rights, often accompanied by scientific evidence that provides a better understanding of the capacities and behavior of animals (Wise, 2016).

An increasing number of animal rights groups have tried to raise awareness about the abusive conditions in keeping animals, mass animal farming, and medical research laboratories. Rights groups can be active at various levels, in local animal shelters, or organized at the international level, like PETA (People for the Ethical Treatment of Animals). PETA was founded in 1980 by Ingrid Newkirk and Alex Pacheco, who were both directly influenced by Peter Singer's book *Animal Liberation* (1975). Their initial efforts focused on litigation and awareness-raising in relation to animals used in testing in research laboratories. Advocacy efforts subsequently expanded to the cosmetics and pharmaceutical industries which also criticized the automobile industry for using animals in crash tests. PETA was successful in their work and, especially in cosmetics, the pressure was so high that testing products on animals was widely limited when "500 cosmetics companies signed a pledge of assurance that they would not engage in animal experimentation" (Kort, 2017).

PETA also pressured the fur and fashion industry and celebrated a major success when the company leaders of Calvin Klein, Giorgio Armani, and Ralph Lauren decided to refrain from using fur. Other relevant initiatives included the entertainment industry, such as circuses, which consequently reduced the use of animals or promoted animal-free alternatives, like Cirque du Soleil. The fast-food industry was pressured to provide transparency regarding abusive practices of their suppliers who were often in countries that lacked any protective legislation (Kort, 2017). PETA applied Singer's terminology and directly fought against "speciesism." They argue that animals' interest to not experience any unnecessary pain needs to be protected. Like many other social movements, they use creative campaigns and a rights frame in order to create awareness, change attitudes, and reform existing legal and political practices.

Animal rights in national constitutions

At the national level, the animal rights movement was successful in achieving stronger legal protection for animals by lobbying for an inclusion of animal rights into national constitutions. Two prominent examples are Switzerland and Germany. In Switzerland, animal protection has long been an issue of debate in the Swiss parliament. The "dignity of creatures" ("die Würde der Kreatur") was first mentioned in the constitution of the canton Aargau in 1980. It initiated a wider debate about the need to include animal welfare and dignity in the federal constitution (Goetschel, 2000: 12). The discourse on animal protection in Switzerland was strongly linked to debates about the legal boundaries of gene technology. On the basis of a successful animal rights campaign, a constitutional amendment was passed in 1992 that clearly stated that researchers need to respect the "dignity of creatures" (Jaber, 2000). In the course of creating a new constitution in 1998, animal activists even tried to strengthen this amendment, but were unsuccessful. However, in 2000, the wording of the 1992 amendment was included in the revised constitution (Evans, 2010: 239).

In Germany, a decade-long battle between campaigners and conservative politicians ended with paragraph 20a of the German Basic Law, i.e., the constitution, stating that animals have to be respected and protected by the state and judicial decisions (Connolly, 2002). The campaign was started because the Basic Law protected freedom of research and freedom of the professions and thus, courts usually ruled in favor of scholars and researchers, even if they conducted cruel experiments involving animals (Evans, 2010: 235). After animal activists' efforts to include animal rights in the constitution was constantly blocked by the Christian–Democrat majority in parliament during the 1990s, campaigners used the political structures when a Social Democrat–Green government coalition was in power from 1998 until 2002. In 2002, activists created increased public awareness after the Supreme Court granted a Muslim person permission to practice a traditional slaughter ritual that – according to many campaigners – involved unnecessary cruelty (Judd, 2003: 22). Public opinion relating to this decision and the support of the Green Party led to a successful constitutional amendment in 2002. Article 20a of German Basic Law now reads:

> (t)he state protects, in the interest of future generations, the natural basis of life, and the animals, within the framework of constitutional laws and through the making of laws and in accordance with ordinances and through judicial decision.

Here, we can see the strong link between rights of nature, animal rights and intergenerational justice. Even though the German Animal Protection League was hoping that this constitutional amendment would lead to a number of relevant changes in protecting animals in Germany (Connolly, 2002), there are still many problems, mostly relating to animals kept in mass stocks. However, many legislative changes at the federal and state level following the constitutional amendment of 2002 have almost completely eliminated cruelty in relation to research involving animals, keeping animals for fur farming, and certain conditions in animal husbandry.

Case study: The Great Apes Project

An interesting case study for the advancement of animal rights is the Great Ape Project (GAP) established in 1994. It considers itself as an international movement promoting the protection of the basic right to life, freedom and non-torture of great apes, including chimpanzees, gorillas, orangutans and bonobos. These rights have been incorporated into the World Declaration on Great Apes that can be signed by anyone who supports the campaign (GAP, 2018). The idea for the GAP was born in a book of the same title written by Peter Singer and Paula Cavalieri. It is based on the realization that human beings and primates share a large percentage of DNA as well as many characteristics, like communication, intelligence, and affectionate relationships. This means that, similar to humans, great apes should have rights as well.

The international Great Ape Project consists of a number of independent institutions in different countries, including Brazil, Chile, Cote d'Ivoire, Germany, Mexico, Spain, Uruguay, and the United Kingdom. The GAP is explicitly fighting against the exploitation of primates in various settings, such as laboratories, entertainment, and zoos, and they condemn these as forms of slavery. One key objective is to defend great apes' liberty to live in their natural habitats (GAP, 2018).

The project has been successful in initiating legislative and policy changes in a number of countries. In New Zealand, there was an attempt to confer rights on primates in 1999, and even though this was not entirely successful, the government at least granted legal protection from animal experimentation (Glendinning, 2008). Later on, in 2015, New Zealand passed an Animal Welfare Amendment Bill "recognizing animals as sentient beings" (McIntyre, 2015). This includes a ban on using animals for testing in the cosmetics industry and substantially raised standards on animal welfare, such as a governmental requirement to undertake assessments whether "non-sentient or non-living alternatives" are available for research. After New Zealand's first attempt of anchoring the rights of primates in national legislation, Spain followed suit in 2008. Also inspired by the Great Apes Project, the Spanish parliament approved the right to life and freedom, as well as protection from torture, for chimpanzees, gorillas, orangutans, and bonobos (Glendinning, 2008).

Rights of nature and ecosystems

Scientists believe we have entered a new geologic era called the Anthropocene, where human beings are the major driving force in modifying the biosphere and causing global environmental change. It is characterized by human domination and disruption of Earth system processes beyond its regenerating capacity. We are nearing tipping points where the Earth is rapidly moving into a critically unstable state, with Earth systems becoming less predictable, leading to mass extinction of species, and changes in the global climate. That there is a close relationship between human beings and environment (nature) is indisputable – human beings cannot survive without nature, without water, air, and food, all of which are provided by nature.

A number of environmental activists and scholars have proposed to treat animals (or humans) in relation to the ecosystems in which they persist and thrive. They suggest that we should take a holistic view of ecosystems and natural communities, and consider them as rights-holders. Thus, they do not only have to be subjectively managed by human beings, or passively treated as their property, but they have to be recognized as rights-bearing entities that need to be respected and protected. In this context, one key argument is that in the face of global warming and environmental challenges, ecosystems might collapse and species that are an inherent part of it might not be able to survive. Thus, a human right to a healthy environment may not be enough – and it cannot be guaranteed without the protection of ecosystem rights. The ideological assumption behind this is that our unsustainable behavioral and consumption patterns require a rethinking from an anthropological approach that places human beings at the center of all relevant decisions, to an eco-centric approach that focuses on preserving ecosystems as a priority in decision-making. On the basis of these arguments, there have been several attempts to introduce rights of nature and ecosystems at the international level. The World Charter for Nature of 1982 was the first international attempt to recognize the intrinsic value of nature. The Preamble recognizes that (a) mankind is part of nature and life depends on the uninterrupted functioning of natural systems, and (b) civilization is rooted in nature. It further notes that (a) every form of life is unique, warranting respect regardless of its worth to man, and (b) man can alter nature and exhaust natural resources by his action or its consequences. The general principles recognized in the Charter include: nature shall be respected and its essential processes shall not be impaired; genetic viability shall not be compromised; all areas of the earth shall be subject to principles of conservation.

In 2000, the Earth Charter, endorsed by more than 6,000 organizations including civil society, international organizations and national governments, was adopted. In an inclusive and participatory drafting process, four main principles and several sub-principles for "a sustainable global society founded on respect for nature, universal human rights, economic justice, and a culture of peace" (Earth Charter, 2000: 1) were adopted. These principles include: (a) respect and care for the community of life; (b) ecological integrity; (c) social and economic justice; and (d) democracy, non-violence, and peace (Earth Charter, 2000).

Other relevant international instruments include the Convention on Biological Diversity adopted in 1992, which refers to the intrinsic value of biological diversity, and the Paris Agreement of 2015 that refers to the integrity of all ecosystems. In addition to these international developments, a number of countries, including Ecuador, Columbia, India, Nepal, and New Zealand and several communities in the United States have recognized rights of nature.

Ecuador's constitutional rights of nature

The first country that incorporated rights of nature into its constitution is Ecuador. The constitution was rewritten during 2007 and 2008 – and was ratified by a referendum by the people in September 2008. Rights of nature, or *Pacha Mama*, are mentioned in different parts of the constitution. Section II, Article 14 of the constitution recognizes the right to a healthy environment. In addition to recognizing people's "right to live in a healthy and ecologically balanced environment that guarantees sustainability and the good way of living (*sumak kasway*)" (Constitution of the Republic of Ecuador, 2008: Section II, Article 14), it is supplemented by a provision that guarantees environmental protection, ecosystem integrity and biodiversity conversation:

> Environmental conservation, the protection of ecosystems, biodiversity and the integrity of the country's genetic assets, the prevention of environmental damage, and the recovery of degraded natural spaces are declared matters of public interest.
>
> (Constitution of the Republic of Ecuador, 2008: Section II, Article 14)

This supplementary provision demonstrates that human rights are not necessarily prioritized in Ecuador's constitution, but they are in balance with ideas concerning ecosystem conservation and the protection of biodiversity.

The most important chapter in this regard is chapter 7 that explicitly addresses rights of nature. Article 71 states:

> Nature, or Pacha Mama, where life is reproduced and occurs, has the right to integral respect for its existence and for the maintenance and regeneration of its life cycles, structure, functions and evolutionary processes.
>
> (Constitution of the Republic of Ecuador, 2008: chapter 7, Article 71)

Ecuador's people and communities can call upon public authorities to ensure the enforcement of these rights. The constitution even states that the government can provide incentives to anyone who protects nature and preserves ecosystems or its components. The remaining articles under chapter 7 include the right of nature to be restored (Article 72), preventive and restrictive measures to protect species, ecosystems, and natural cycles (Article 73), and the protection of ecosystem services from appropriation (Article 74) (Constitution of the Republic of Ecuador, 2008: Article 72–4).

The first cases involving the rights of nature have been brought to court in Ecuador. One example is a lawsuit filed by Richard Frederick Wheeler and Eleanor Geer Huddle to defend the rights of the Vilcabamba river. The river can be considered the plaintiff in this case and people acted to enforce its constitutional rights to existence and preservation. In this case, the flowing and functioning of the river as an ecosystem was adversely affected by a road construction project. The Provincial Justice Court of Loja ruled in favor of the river Vilcabamba in 2011 – and this was the first time the constitutional rights of nature were upheld by a court decision (CELDF, 2015).

The Universal Declaration on the Rights of Mother Earth

Ecuador's constitutional process has inspired a number of other civil society initiatives to establish rights of nature in different settings. An important initiative was the 2010 World People's Conference on Climate Change and the Rights of Mother Earth held in Cochabamba, Bolivia. The outcome document was a Universal Declaration on the Rights of Mother Earth (CELDF, 2010) that has been introduced at the UN General Assembly for further consideration. The Declaration was modeled according to the Universal Declaration of Human Rights (CELDF, 2015). Next to the Preamble, it contains four articles on the rights of Mother Earth. Article 5.5, for instance, declares that "no distinction shall be made between organic and inorganic beings, species, origin, use to human beings, or any other status" (CELDF, 2010: Article 5.5). Here, we can clearly see a departure from an anthropological approach towards an ecocentric approach, in which the preservation and functioning of nature – and all relevant components of an ecosystem – is at the center of protection efforts.

Article 2 outlines all the inherent rights of Mother Earth and of its components. These comprise: (a) the right to life and to exist; (b) the right to be respected; (c) the right to continue their vital cycles and processes free from human disruptions; (d) the right to maintain its identity and integrity as a distinct, self-regulating, and inter-related being; (e) the right to water as a source of life; (f) the right to clean air; (g) the right to integral health; (h) the right to be free from contamination, pollution, and toxic or radioactive waste; (i) the right to not have its genetic structure modified or disrupted in a manner that threatens its integrity or vital and healthy functioning; and (j) the right to full and prompt restoration of the violation of the rights recognized in this Declaration caused by human activities (CELDF, 2010: Article 2).

These articles are at the core of the Declaration, whereas the remaining ones focus on definitions and enforcement. When analyzing the articles, it again becomes clear that the focus of the Declaration is on protecting the various components of nature from destruction and disturbance caused by human activities.

After the adoption of the Universal Declaration on the Rights of Mother Earth in 2010, the Global Alliance for the Rights of Nature was founded at a meeting in Patate, Ecuador. This is an umbrella organization comprising civil society groups that are promoting the rights of nature and ecosystems. Participating advocacy organizations include the Community Environmental Legal Defense Fund (CELDF), the Pachamama Alliance, Centro para el Desarrollo del Indígena Amazónico (CEDIA), Fundación Ambiente y Sociedad, the Gaia Foundation, and EnAct International. These two events – the adoption of the Declaration and the establishment of the Global Alliance – marked the birth of a worldwide movement for ecosystem rights.

New developments in ecosystem rights

Since 2010, there have been many interesting new developments that strengthened the rights of nature locally, nationally, and globally. Since 2014, four International Rights of Nature Tribunals have taken place. These tribunals reconsider relevant environmental and social justice cases on the basis of jurisprudence that takes rights of nature into account. The tribunal judges include environmental activists, lawyers, civil society experts, and rights advocates. They reframe and re-interpret cases of alleged violations of the Universal Declaration of the Rights of Mother Earth and provide a platform for hearing, discussing, and judging these cases. They are also raising awareness and educating the general public, politicians, and civil society in relation to ecosystem rights. The first International Rights of Nature Tribunal was held in 2014 in Quito, Ecuador. The subsequent ones all took place at the Conferences of the Parties (COP) in parallel to the international climate negotiations in 2014 in Lima, in 2015 in Paris, and in 2017 in Bonn (GARN, 2018).

In 2017, rivers in two different countries were granted legal status. These are the Whanganui river in New Zealand and the Ganga and Yamuna rivers in India. The Whanganui River Treaty Settlement in New Zealand is an important achievement, as it also an agreement between the Māori people and the Crown government. It recognized that the river is a rights-bearing entity that needs care and protection and cannot be the property of the government or any citizens. This is in accordance with the indigenous understanding of components of nature, like mountains and rivers, as ancestors inseparable from the cultural identity of the people who have traditionally lived in harmony with them. In India, the recognition of legal personhood for the Ganga river has been the culmination of a broad coalition cooperating in the National Ganga Rights Movement together with the US-based Community Environmental Legal Defense Fund. Their aim is to even extend river protection and work towards a national Ganga Rights Act entailing additional legal mechanisms preserving the river as an intact ecosystem (Biggs, 2017).

In its Advisory Opinion on the Environment and Human Rights, the Inter-American Court of Human Rights pointed out that the human right to a healthy environment is a right with both individual and collective connotations. Environmental degradation may cause irreparable damage to human beings. Therefore, a healthy environment is a fundamental right for the existence of humankind. The Court further recognized that the right to a healthy environment, as an autonomous right, protects components of the environment such as forests, rivers, seas, and others, as juridical interests themselves, even in absence of certainty or evidence about the risk to individual persons (Banda, 2018, English translation). The Court stated that it is about protecting nature and the environment not only because of its utility for human beings or the effects that its degradation might cause on other rights of the individuals, like health, life, and personal integrity, but for its importance for the rest of living organisms with whom the planet is shared and that are entitled to protection themselves. The Court took notice of a tendency to recognize the legal personhood of rights to nature, not only in judicial decisions, but also in constitutions (Banda, 2018).

In a climate case that a group of children brought against Colombia challenging deforestation in the Amazon, the Supreme Court ruling in favor of the petitioners said that the government must come up with an action plan to stop deforestation in the Amazon and escalate the fight against climate change. The Court further said the forest is an entity subject of rights, essentially conferring rights upon the ecosystem (DeJusticia, 2018).

Conclusion

The examples and developments discussed in this chapter demonstrate how inter-generational rights, rights of animals, and ecosystems are closely interwoven. We are departing from a traditional understanding of human rights that focuses on the liberal assumption of individual civil and political rights, towards a more holistic understanding of rights encompassing communities of human beings in balance and peaceful interaction with nature. This new understanding of rights places an emphasis on protecting the environment, biodiversity, and various ecosystem components, and on taking the interests of future generations into account.

DISCUSSION QUESTIONS

1. How can we represent the interests of future generations in democratic societies? Which innovative forms of governance do we need for that?
2. In which ways can an international human right to a healthy and sustainable environment protect future generations?
3. Do you think animals should have citizenship rights as Kymlicka and Donaldson (2011) have suggested?
4. Why is the movement for rights of nature particularly strong in Ecuador and other Latin American countries?
5. Do we need specific ecosystem rights in the way they were adopted in India and New Zealand? How can we govern these rights?

References

Banda, M.L. (2018). Inter-American Court of Human Rights' Advisory Opinion on the Environment and Human Rights. *ASIL Insights* 22(6) (20 May 2018) [Online]. Available at: www.asil.org/insights/volume/22/issue/6/inter-american-court-human-rights-advisory-opinion-environment-and-human [Accessed: 1 September 2018].

Beckman, L. and Page, E.A. (2008). Perspectives on Justice, Democracy and Global Climate Change. *Environmental Politics* 17(4): 527–35.

Biggs, S. (2017). *When Rivers Hold Legal Rights*. [Online]. Available at: www.earthisland.org/journal/index.php/elist/eListRead/when_rivers_hold_legal_rights/ [Accessed: 30 July 2018].

Boyd, D. (2012). The Constitutional Right to a Healthy Environment. *Environment: Science and Policy for Sustainable Development* 54 (4): 3–15.

Caney, S. (2008). Human Rights, Climate Change, and Discounting. *Environmental Politics* 17(4): 536–55.

Cavalieri, P. (2001). *The Animal Question: Why Non-Human Animals Deserve Human Rights*. Oxford: Oxford University Press.

CELDF (2010). Universal Declaration on the Rights of Mother Earth [Online]. Available at: https://celdf.org/wp-content/uploads/2016/03/universal-declaration-of-the-rights-of-mother-earth-april-22-2010.pdf [Accessed: 30 July 2018].

CELDF (2015). Rights of Nature Going Global [Online]. Available at: https://celdf.org/rights/rights-of-nature/ [Accessed: 30 July 2018].

Connolly, K. (2002). German Animals Given Legal Rights. [Online]. Available at: www.theguardian.com/world/2002/jun/22/germany.animalwelfare [Accessed: 26 July 2018].

Constitution of the Republic of Ecuador (2008). Constitution of 2008 [Online]. Available at: http://pdba.georgetown.edu/Constitutions/Ecuador/english08.html [Accessed: 30 July 2018].

DeJusticia (2018). *Demanda Generaciones Futuras v. Minambiente* (Future Generations v. Ministry of the Environment and Others) (Colombian youth case). Unofficial summary in English by DeJusticia. [Online]. Available at: http://blogs2.law.columbia.edu/climate-change-litigation/wp-content/uploads/sites/16/non-us-case-documents/2018/20180405_11001-22-03-000-2018-00319-00_decision-1.pdf [Accessed: 1 October 2018].

Dobson, A. (1996). Representative Democracy and the Environment. In Lafferty, W. and Meadowcraft, J. (eds.), *Democracy and the Environment*. Cheltenham, UK: Edward Elgar, pp.124–39.

Donnelly, J. (1989). *Universal Rights in Theory and Practice*. Ithaca, NY: Cornell University Press.

Dudai, R. (2009). Climate Change and Human Rights Practice: Observations on and around the Report of the Office of the High Commissioner for Human Rights on the Relationship between Climate Change and Human Rights. *Journal of Human Rights Practice* 1 (2): 294–307.

Ekeli, K.S. (2005). Giving a Voice to Posterity – Deliberative Democracy and Representation of Future People. *Journal of Agricultural and Environmental Ethics* 18 (5): 429–50.

Earth Charter (2000). *The Earth Charter*. [Online]. Available at: http://earthcharter.org/invent/images/uploads/echarter_english.pdf [Accessed: 23 October 2018].

Evans, E. (2010). Constitutional Inclusion of Animal Rights in Germany and Switzerland: How Did Animal Protection Become an Issue of National Importance? *Society and Animals* 18: 231–50.

GAP (2018). Great Primates Await Your Protection [Online]. Available at: www.greatapeproject.uk/worlddeclaration.html [Accessed: 27 July 2018].

GARN (2018). What is an International Rights of Nature Tribunal? [Online]. Available at: http://therightsofnature.org/rights-of-nature-tribunal/ [Accessed: 30 July 2018].

German Bundestag (1949). *Basic Law for the Federal Republic of Germany*, 23 May 1949 (amended June 2008). [Online]. Available at: www.refworld.org/docid/4e64d9a02.html [Accessed: 19 December 2018].

Glendinning, L. (2008). Spanish Parliament Approves 'Human Rights' for Apes. [Online]. Available at: www.theguardian.com/world/2008/jun/26/humanrights.animalwelfare [Accessed: 27 July 2018].

Goetschel, A.F. (2000). Animal Cloning and Animal Welfare Legislation in Switzerland. Report by the Foundation for the Animal in the Law. Presented at a Symposium at the University of Lüneburg, Germany, January 28–30 2000.

Hayward, T. (2004). *Constitutional Environmental Rights*. Oxford: Oxford University Press.

Hayward, T. (2007). Human Rights Versus Emissions Rights: Climate Justice and the Equitable Distribution of Ecological Space. *Ethics and International Affairs* 21 (4): 431–50.

Hiskes, R. (2005). The Right to a Green Future: Human Rights, Environmentalism and Intergenerational Justice. *Human Rights Quarterly* 27(4): 1346–64.

Jaber, D. (2000). Human Dignity and the Dignity of Creatures. *Journal of Agricultural and Environmental Ethics* 13 (1–2): 29–42.

Judd, R. (2003). The Politics of Beef: Animal Advocacy and the Kosher Butchering debates in Germany. *Jewish Social Studies* 10 (1): 117–50.

Kort, E.D. (2017). People for the Ethical Treatment of Animals. [Online]. Available at: www.britannica.com/topic/People-for-the-Ethical-Treatment-of-Animals [Accessed: 27 July 2018].

Kymlicka, W. and Donaldson, W. (2011). *Zoopolis: A Political Theory of Animal Rights*. Oxford: Oxford University Press.

McIntyre, S. (2015). Animals Are Now Legally Recognized as "Sentient" Beings in New Zealand. [Online]. Available at: www.independent.co.uk/news/world/australasia/animals-are-now-legally-recognised-as-sentient-beings-in-new-zealand-10256006.html [Accessed: 27 July 2018].

Nussbaum, M. (2006). *Frontiers of Justice: Disability, Nationality, Species Membership*. Cambridge, MA: Harvard University Press.

Regan, T. (1983). *The Case for Animal Rights*. Berkeley, CA: University of California Press.

Shue, H. (2014). *Climate Justice: Vulnerability and Protection*. Oxford: Oxford University Press.

Singer, P. (1975). *Animal Liberation*. New York, NY: HarperCollins.

UDHR (1948). Universal Declaration of Human Rights. Adopted by the United Nations General Assembly G.A. res. 217A (III), U.N. Doc A/810 at 71. United Nations General Assembly, Paris.

United Nations (1982). The World Charter for Nature. [Online]. Available at: www.un.org/documents/ga/res/37/a37r007.htm [Accessed: 19 December 2018].

Vernon, R. (2009). Intergenerational Rights? *Intergenerational Justice Review* 9(1): 8–12.

Wise, S. (2016). Animal Rights. [Online]. Available at: www.britannica.com/topic/animal-rights [Accessed: 27 July 2018].

16 Human rights and the environment

Square pegs in round holes?

Introduction

The International community has come a long way since the Stockholm Conference on the Human Environment was held in 1972. International environmental law has blossomed into a separate branch of international law with many treaties, principles, and soft law instruments. Institutions have been developed, globally, regionally, and nationally. Similarly, international human rights law has developed into a sophisticated body of law and institutions, globally and regionally, since the adoption of the UN Charter in 1945 and the Universal Declaration of Human Rights soon after. While their origins are very different and the organizing principles are also different, an important development in recent years is the merging of these two areas of international law, and this textbook discussed the evolution of environmental rights under international law.

The close relationship between the enjoyment of human rights and environmental degradation can no longer be denied. There are many examples from across the globe. From industrial plants that spew toxic substances into the atmosphere and pollute waterways to mining activities that have destroyed the environment and peoples' lives, from deforestation to electronic waste, and to more global issues like climate change, we have unfortunately seen many examples of environmental degradation that have infringed the enjoyment of human rights. Some of these issues affect only a few individuals, while others affect the entire international community. Both human rights law and environmental law have had to address this reality.

From framework principles on human rights and environment to the global pact on the environment, and from the UN Environment Programme's environmental rights initiative to the remarkable constitutional entrenchment of environmental rights, many believe that sufficient momentum has been created for the recognition of a human right to a healthy environment under international law. As the former UN Special Rapporteur on Human Rights and Environment, John Knox, observed: "The interdependence of human rights and the environment is an idea whose time is here" (2018).

UN Environment notes with regard to its environmental rights initiative:

> Environmental rights have grown more rapidly than any other human right and are enshrined in over 100 constitutions. Tragically, whilst the right to

> a healthy environment is increasingly recognized, this right is increasingly violated. Irresponsible development projects and agri-businesses destroy our prospects of a safe, clean and healthy environment, as those fighting for these rights are killed at an average rate of four per week.
>
> (UN Environment, 2018)

It further notes that "human rights are intertwined with the environment; environmental harm interferes with the enjoyment of human rights, and the exercise of human rights helps to protect the environment and promote sustainable development" (UN Environment, 2018).

As Knox and Pejan noted:

> Two of the great achievements of international law have been to define the human rights integral to a life of dignity, freedom, and equality, and to develop rules and institutions that protect the global environment. Because these two areas of the law developed separately and at different times, the relationship between them was at first unclear. In the last two decades, however, it has become more and more evident that human rights and environmental protection have a fundamental interdependence: A healthy environment is necessary for the full enjoyment of human rights and, conversely, the exercise of rights (including rights to information, participation and remedy) is critical to environmental protection.
>
> (2018: 1)

Many civil society organizations are trying to raise awareness about this interdependence. They support local community groups who have to be relocated in the context of green economy projects, such as hydroelectric dams or agricultural planning, in litigation at the national level or at regional human rights courts. They advocate for human rights in international climate agreements and for considering climate concerns in human rights policies. In addition, they collaborate with transnational advocacy networks where environmental groups and human rights activists collaborate for a common cause: to promote environmental justice.

This interdependence was the subject of this textbook. Because there is a close relationship between human rights, environment, and development, as many of the violations occur in the name of development, and because of the disparity in the global community both vis-à-vis states as well as within states, we discussed three frameworks that informed our approach: human rights, environmental justice, and sustainable development. One related issue that we did not discuss in depth in this volume is the link with good governance, although the procedural rights we discussed in Chapter 6 are directly relevant here.

Environmental rights, human rights, and good governance

José Ugaz, Chair of Transparency International noted: "In too many countries, people are deprived of their most basic needs and go to bed hungry every night because of corruption, while the powerful and corrupt enjoy lavish lifestyles with impunity" (Transparency International, 2018). This highlights the importance of good governance.

There is no single and exhaustive definition of "good governance." However, there is a significant degree of consensus that good governance relates to political and institutional processes and outcomes that are deemed necessary to achieve the goals of sustainable development. Good governance is the process whereby public institutions conduct public affairs, manage public resources, and guarantee the realization of human rights in a manner essentially free of abuse and corruption, and with due regard for the rule of law. Processes of good governance include decision-making at local, national, and global levels, and can include state as well as non-state actors. Thus, governance can be understood as "systems of rule at all levels of human activity – from the family to the international organization – in which the pursuit of goals through the exercise of control has transnational repercussions" (Rosenau, 1995: 13).

UN Environment recognizes the link between human rights, environmental protection, and good governance:

> In our globalized world, environmental threats require effective responses that promote peace, justice, development and the fulfilment of environmental and human rights. UN Environment promotes sound environmental governance – the rules, policies and institutions that shape how humans interact with the environment. By building robust governance systems, we can safeguard the environment and human rights and work toward all 17 UN Sustainable Development Goals.
>
> (2018)

The true test of "good" governance is the degree to which it delivers on the promise of human rights: civil, cultural, economic, political, and social rights. The key question is: are the institutions of governance effectively guaranteeing the right to health, adequate housing, sufficient food, quality education, fair justice, and personal security?

The concept of good governance has been clarified by the work of the former Commission on Human Rights. The Commission identified the key attributes of good governance as: transparency; responsibility; accountability; participation; responsiveness (to the needs of the people) (Human Rights Commission, 2000: resolution 2000/64).

Thus, good governance encompasses: full respect for human rights and the rule of law; effective participation; multi-actor partnerships, political pluralism, and transparent and accountable processes and institutions; an efficient and effective public sector; legitimacy; access to knowledge, information, and education; political empowerment of people; equity; sustainability; and attitudes and values that foster responsibility, solidarity, and tolerance.

According to the United Nations Development Programme (UNDP), the strongest test of good governance is the degree to which state institutions effectively guarantee the right to basic needs such as adequate housing, sufficient food, quality education, justice, social protection, and a secure and clean environment. The interconnection between good governance, human rights, and sustainable development has been recognized in a number of international documents. For example, the UN Declaration on the Right to Development proclaims that every human person and all peoples "are entitled to participate in, contribute to, and enjoy economic, social, cultural and political development" (1986: Article 1). The ICCPR recognizes (as does the ICESCR) the right to self-determination: "All peoples have the right of self-determination. By virtue of that right they freely determine their political status and freely pursue their economic, social and cultural development" (International Covenant on Civil and Political Rights, 1966: Article 1). In addition, Article 25 of the ICCPR affirms everybody's right to take part in the conduct of public affairs, directly or through freely chosen representatives. In Chapter 6, we discussed Principle 10 of the Rio Declaration and how participatory rights are incorporated in other treaties (UN, 1992).

The Council of Europe has adopted 12 principles of good governance, including transparency, accountability, sustainability, fair conduct of elections and participation, respect for the rule of law, sound fiscal management, and respect for human rights, cultural diversity, and social cohesion (Council of Europe, 2008).

The UN Convention against Corruption adopted in 2003 is an important step forward in the fight against corruption (UN Convention Against Corruption, 2003). It recognizes that the prevention and eradication of corruption is the responsibility of all States and that they must cooperate with one another, and that principles of proper management of public affairs and public property, fairness, responsibility, and equality before the law, and the need to safeguard integrity and to foster a culture of rejection of corruption are necessary to adopt in the fight against corruption. The purposes of the Convention are:

(a) To promote and strengthen measures to prevent and combat corruption more efficiently and effectively;
(b) To promote, facilitate and support international cooperation and technical assistance in the prevention of and fight against corruption, including in asset recovery;
(c) To promote integrity, accountability and proper management of public affairs and public property.

(Article 1)

The latest recognition of the link are the Sustainable Development Goals which include a goal on good governance. SDG 16 refers to the need to promote peaceful societies for sustainable development, provide access to justice for all, and build effective, accountable, and inclusive institutions at all levels. Targets include ensuring responsive, inclusive, participatory, and representative decision-making at all levels; ensuring public access to information and protecting fundamental freedoms, in accordance with national legislation and international agreements; broadening and strengthening the participation of developing countries in the institutions of global governance; and promoting and enforcing non-discriminatory laws and policies for sustainable development.

According to Transparency International – the global watchdog on corruption – transparency ensures that public officials, civil servants, managers, board members, and business people act visibly and understandably, and report on their activities. It means that the general public can hold them to account. It is the best way of guarding against corruption and helps increase trust in the people and institutions on which our futures depend. It points out that corruption impacts societies in a multitude of ways. In the worst cases, it costs lives. Short of this, it costs people their freedom, health, or money. The cost of corruption can be divided into four main categories: political, economic, social, and environmental. On the political front, corruption is a major obstacle to democracy and the rule of law. In a democratic system, offices and institutions lose their legitimacy when they are misused for private advantage. This is harmful in established democracies, but even more so in newly emerging ones. It is extremely challenging to develop accountable political leadership in a corrupt climate. Economically, corruption depletes national wealth. Corrupt politicians invest scarce public resources in projects that will line their pockets rather than benefit communities, and prioritize high-profile projects such as dams, power plants, pipelines, and refineries over less spectacular but more urgent infrastructure projects such as schools, hospitals, and roads. Corruption corrodes the social fabric of society. It undermines people's trust in the political system, in its institutions, and its leadership. A distrustful or apathetic public can then become yet another hurdle to challenging corruption. Environmental degradation is another consequence of corrupt systems. The lack of, or non-enforcement of, environmental regulations and legislation means that precious natural resources are carelessly exploited, and entire ecological systems are ravaged. From mining to logging to carbon offsets, companies across the globe continue to pay bribes in return for unrestricted destruction.

In many cases, human rights violations, environmental degradation, and corruption (or lack of transparency) are intertwined. Corruption is often discussed separately, but good governance is necessary for both human rights and environmental protection. An open, transparent government where the rule of law is upheld and human rights are protected is necessary for sustainable development. The SDGs have finally recognized the link.

Where next?

We seem to be at an exciting moment in history when the international community finally adopts a human right to a healthy environment, 70 years after the adoption of the Universal Declaration of Human Rights. Whether or not this happens within the next few months or next few years, states' obligations are sufficiently clear with regard to environmental rights. Moreover, constitutional and judicial developments point to the emergence of a customary international law principle, or at least a general principle of law, on a right to a healthy environment. As many regional human rights systems have recognized this right, it is time that international law did too.

Looking back at the history of human rights, we can see how rights have changed over the course of time to encompass new rights. They are not static, but dynamic, and they often constitute a response to the challenges of a particular time. In the aftermath of the atrocities of the Second World War, civil and political rights were important to limit the power of the state and to protect citizens and their dignity. The first "generation" of human rights can be understood as an attempt to safeguard individuals – with the help of the international community – even if the state government failed to do so. The most pressing global challenges we currently face, however, are climate change, environmental degradation, and severe weather events. In this context, we need new forms of cooperation, including a variety of state and non-state actors at different governance levels. We are required to make sustainable decisions based on standards that serve as a guidepost for tackling the above-mentioned challenges. These standards are likely to differ from the ones adopted during the post-Second World War order, as they have to be adapted in light of the complex issues we face today.

In order to guarantee the enjoyment of basic human rights to future generations, we need to ensure that delicate ecosystems remain intact as they build the foundation for a clean and healthy environment, in which humans, animals, and other species can thrive. As environmental injustices do not only occur between developed and developing countries, but also within societies, we have to adequately protect particularly vulnerable groups, like children, women, indigenous peoples, the elderly, the displaced, and persons with disabilities. In order to achieve this, we cannot simply view them as passive victims, but need to hear their voices and consider them as active participants in just transition processes and in adequate adaptation and mitigation policies. Moreover, we need to make sure that multinational corporations, as powerful and influential actors in today's world, act in a responsible way in protecting human rights as well as adhering to environmental and social standards. Thus, our vision of sustainable development requires a rebalancing of the economic, social, and environmental dimensions. Given that economic development is prioritized in the dominant neoliberal paradigm, social and environmental considerations have to be strengthened as equal partners with a view to achieving sustainability.

Against this background, the rights discourse has also undergone profound changes. It now includes inter-generational and collective concerns, as well as animal rights, rights of nature, and ecosystem rights. A human right to a healthy environment would embrace many of these aspects, as it would be a collective right obliging today's decision-makers to make choices that preserve the environment, not just for the current generation, but also for future generations.

There is a real need to adopt a right to a healthy environment at the international level, and not simply for its symbolic value. Since all rights are universal, indivisible, interdependent, and inter-related, there is a gap in the protection of rights because a healthy and sustainable environment is necessary for the enjoyment of the full gamut of rights recognized under international human rights law.

Whether or not the international community adopts a stand-alone right to a healthy environment, one thing is clear: victims of environmental degradation will continue to resort to human rights machinery to seek redress. They will be represented by civil society organizations supporting litigation and advocating for stronger recognition and institutionalization of human rights in environmental regimes. Transnational advocacy will become even stronger in the future as human rights groups increasingly collaborate with environmental activists, indigenous peoples' groups, youth groups, trade unions, and gender advocates to foster institutional interaction between the policy fields of human rights, the environment, and development. These alliances use local experiences with climate change and environmental degradation and their respective policies to lobby for change at the national and international levels of decision-making. Regional human rights institutions will continue to be the main foci of action in relation to environmental rights, and as more and more adverse consequences of climate change begin to emerge, victims will not shy away from seeking redress from any mechanism that is available, and this will include human rights institutions. As we discussed in this textbook, human rights law and environmental law have different origins and foundations and do not often fit together, despite synergies and inter-linkages, leading some to argue that using a human rights approach to environmental issues is akin to fitting square pegs in round holes. However, there is no denying that the human rights framework has provided relief to many victims of environmental degradation.

While there is definitely a momentum towards accepting a right to a healthy environment, in the past, there has been a general reluctance to recognize such a right:

> Ultimately, the well-being of our children and grandchildren will depend on what action we take today to address climate change. In an era where even inanimate objects such as rivers and Mother Nature are accorded rights and legal standing, it is baffling as to why there is so much reluctance to accord *human beings* with a right to a healthy environment.
>
> (Atapattu, 2018: 268, footnotes omitted, emphasis in original)

DISCUSSION QUESTIONS

1. What are the links between human rights, environmental protection, and good governance?
2. Do you think the international community should accept a human right to a healthy environment? Why and why not? If you think such a right should be adopted, how should it be drafted?
3. How should the international community adopt such a right? A UN General Assembly resolution, a binding treaty, or other?
4. If the international community fails to adopt such a right, what should the advocates of such a right do next?
5. If the international community adopts such a right, how can various groups use such a right?

References

Atapattu, S. (2018). The Right to a Healthy Environment and Climate Change: Mismatch or Harmony. In: Knox, J. and Pejan, R. (eds.), *The Human Right to a Healthy Environment*. Cambridge: Cambridge University Press, pp.252–68.

Council of Europe (2008). Principles of Good Governance [Online]. Available at: www.coe.int/en/web/good-governance/12-principles-and-eloge [Accessed: 24 October 2018].

Human Rights Commission (2000). The role of good governance in the promotion of human rights, resolution 2000/64 [Online]. Available at: http://ap.ohchr.org/documents/alldocs.aspx?doc_id=4764 [Accessed: 10 October 2018].

International Covenant on Civil and Political Rights (1966) [Online]. Available at: www.ohchr.org/Documents/ProfessionalInterest/ccpr.pdf [Accessed 1 October 2018].

International Covenant on Economic, Social and Cultural Rights (1966) [Online]. Available at: www.ohchr.org/Documents/ProfessionalInterest/cescr.pdf [Accessed: 1 October 2018].

Knox, J.H. (2018). Human Rights Council, Report of the Special Rapporteur on the issue of human rights obligations relating to the enjoyment of a safe, clean, healthy and sustainable environment, Framework Principles on Human Rights and the Environment, A/HRC/37/59 [Online]. Available at: https://documents-dds-ny.un.org/doc/UNDOC/GEN/G18/017/42/PDF/G1801742.pdf?OpenElement [Accessed: 10 October 2018].

Knox, J.H. and Pejan, R. (eds.) (2018). *The Human Right to a Healthy Environment*. Cambridge: Cambridge University Press.

Rosenau, J.N. (1995). Governance in the Twenty-First Century. *Global Governance* 1 (1): 13–43.

Transparency International (2018). Transparency International [Online]. Available at: www.transparency.org/ [Accessed: 24 October 2018].

UN (1992). Rio Declaration on Environment and Development. A/CONF.151/26 (Vol. I) [Online]. Available at: www.un.org/documents/ga/conf151/aconf15126-1annex1.htm [Accessed: 10 October 2018].

UN Convention Against Corruption (2003). United Nations Convention Against Corruption [Online]. Available at: www.unodc.org/unodc/en/treaties/CAC/ [Accessed: 24 October 2018].

UN Declaration on the Right to Development 41/128 (1986) [Online]. Available at: www.ohchr.org/Documents/Issues/Development/DeclarationRightDevelopment_en.pdf [Accessed: 10 October 2018].

UN Environment (2018). Environmental Rights Initiative. [Online] Available at: www.unenvironment.org/explore-topics/environmental-rights-and-governance/what-we-do/advancing-environmental-rights [Accessed: 24 October 2018].

UN General Assembly (2015). Transforming Our World: The 2030 Agenda for Sustainable Development, A/RES/70/1. [Online]. Available at: www.un.org/ga/search/viewdoc.asp?symbol=A/RES/70/1&Lang=E [Accessed: 24 October 2018].

United Nations Development Programme (2011). Why Good Governance Makes for Better Development [Online]. Available at: www.undp.org/content/undp/en/home/ourperspective/ourperspectivearticles/2011/05/20/why-good-governance-makes-for-better-development.html [Accessed: 10 October 2018].

Index

Page numbers in **bold** represent tables.

17 Principles of Environmental Justice 14–15
2030 Agenda for Sustainable Development 21

Aarhus Convention 46, 73, 129–32, 135–7, 139–41, 147
access rights 36–37, 132, 134, 136, 138
Activists Beyond Borders (Keck & Sikkink) 234
Adani's Mundra coal power project 268
Additional Protocol to the American Convention on Human Rights 41, 67, 297
Adelman, S. 281
Adong bin Kuwau and others v. Kerajaan Negeri Johor and Anor 182
Adverse Impacts of Climate Change on the Full Enjoyment of Human Rights (Geneva, 2012) 211–12
Africa 193–4, 250, 316–17
African Charter on Human and Peoples' Rights (1981) 67, 79, 100, 122–3, 186, 273–4
African Commission of Human Rights 100, 121, 123–4
African Court of Human Rights 100
Agenda 21 19
Agyeman, J. 23
Aiken, A. 185, 254
Alston, P. 76
Amazon rainforest 124–5, 189–91, 337
American Commission of Human Rights 121
American Convention on Human Rights 67, 92, 121–2, 186, 299–300
American Declaration of the Rights and Duties of Man (1948) 11, 91, 94, 96–7
Amnesty International 70, 241–2
Angola 158
Animal Liberation (Singer) 328
animal rights 328–32
Anthropocene era 23, 332–3
Anton, D. 146
AP Pollution Control Board II v. MV Nayudu 176
Aparicio Benito v. Spain 90
Arab Charter on Human Rights (2004) 67, 86
Arctic 93–4, 96–7
Argentina 125, 159, 188
Argentina: Saldano v. 296
Armenia 163–4
ASEAN Commission on the Promotion and Protection of the Rights of Women and Children 86
ASEAN Human Rights Declaration 86
ASEAN Inter-Governmental Commission on Human Rights 86
Asia 12, 250
Asia Pacific Forum 87
Asian Human Rights Commission/Charter 86–7
Asian Tsunami (2004) 212
Atapattu, S. 147, 346
Athabaskan Petition 96–7, 220
Australia 12, 124, 170, 195
authoritarian regimes 69–70, 160–1
autonomous rights 98

Bakun Hydro-Electric Dam project (BHEP) 181
Bali Principles of Climate Justice (2002) 14
Bangladesh 163, 176–7, 223
Bankovic and others v. Belgium and 16 other contracting States 296, 298

Barro Blanco project 72–3, 268–70
Basel Convention on International Trade in Hazardous Waste 46, 138
Belarus 159
Belgium 159, 165, 193
Belgium and 16 other contracting States, Bankovic and others v. 296
Belize 125
Belize: Maya Communities of the Toledo District v. 92
Benin 165
Bhutan 162–63, 165
biodiversity 50–1
Bokhari v. Federation of Pakistan 178
Bolivia 17, 120, 157–9
Bonn Convention on Migratory Species of Wild Animals 180
Boyd, D. 155–7, 160, 169–70
Brammer: Wolfson v. 184
Brazil 118, 257
Brundtland Report 18–19
Buergenthal, T. 88
Bulankulama v. Minister of Industrial Development and others 179
Bulgaria: Stoine Hristov v. 90
Bullard, R. 13, 23
"but for" theory 210–11

Cambodia 166–67
Cameroon 124
Canada 96–97, 111, 120, 156, 161, 188, 220, 287–9
Canada: EHP et al. v. 111; *Lubicon Lake Band v.* 117
Cancun Agreements (2010) 15, 79, 211, 218–19, 238, 270, 277
Caney, S. 70, 326
capitalism 24, 78
Carbon Majors petition 196, 321–2
carbon-offset projects 243–4
Case Concerning Aerial Herbicide Spraying (Ecuador v. Colombia) 296–7
Case Concerning the Gabcikovo Nagymaros Project 35
Case of Kichwa Indigenous Peoples of Sarayaku v. Ecuador, Reparations 122
cause and effect 75–6, 80
Centre for Environmental Justice v. Ministry of Agriculture, Environment, Irrigation and Mahaweli Development et al. 180
Cerrell Report 13
charter-based human rights 38–9
Chechnya 160
Chernobyl disaster 64, 161
children 38, 45, 55, 64, 73, 90, 158–9, 212, 219–20, 250, 252–5
Chile 147, 158, 191
Chile: Claude Reyes et al v. 145–6
China 12
civil society 232, 236–7, 341
Claude Reyes et al v. Chile 145–6
Clean Air Council, SB through his Guardian and BB through his Guardians v. USA, Donald Trump as President and others 185–7
Clean Development Mechanism 15, 68, 72–3, 217, 239, 243–4, 267–71
climate change 64–6, 95, 225; adaptation 217–18, 249, 275–7; anthropogenic 75, 253; and constitutions 165; and forest fires 90, 267; and human rights law 49–50; impacts of 208–9, 215, 217; and international injustice 11–12; litigation 183–4, 221–2; and migrations 258–9; obligations around 50; risks to human rights 67, 208, 213–15; threats to island states 67, 208, 210, 221, 259; and water 166
climate injustice 236, 243–4, 253–4, 326–7
climate justice 223–4
climate justice movement 236–7
Climate Justice Now 243
climate refugees 67, 195, 208, 210, 220, 258
Clinton, B. 10
Coard v. United States 299–300
Coffin, T. 184–5, 255
collective rights 8, 67, 69–70, 76, 79, 100, 121, 214, 326–7, 345
Colombia 17, 97–8, 124, 165, 189–91, 296–301
Colombia: EP et al. v. 116–17
colonialism 9, 18, 26, 66
Committee on Economic, Social and Cultural Rights 53, 117, 124, 135, 142, 149, 289–90
Committee on Enforced Disappearances 142
Committee on Migrant Workers 142
Committee on the Elimination of Racial Discrimination 142
Committee on the Rights of Persons with Disabilities 142
Committee on the Rights of the Child 142
common law principles 38, 156
Communications against Zaire (2003) 121

Communities for a Better Environment v. City of Richmond 194
Community of La Oroya v. Peru 93
compliance gaps 75
Comunidad de Chanaral v. Codelco Division el Salvador 191
constitutions 168–9, 327–8; and animal rights 331–2; on climate change 165; and public trust 165; recognizing environmental rights 15–16, 38, 47, 53, 57, 67, 69, 77, 155, **156**, 157–9, 161–2, 168, 175, 181; sustainable development 164–5; *see also individual countries*
Convention Against Torture (CAT) 7, 38, 142
Convention for the Protection of Migrant Workers and Members of their Families 7
Convention on Enforced Disappearances (CED) 38
Convention on Migrant Workers (CMW) 38, 142
Convention on the Elimination of All Forms of Discrimination against Women (CEDAW) 7, 38, 67, 71, 75, 142, 251
Convention on the Elimination of Racial Discrimination 7, 38
Convention on the Protection of Persons with Disabilities 7, 38, 67, 261–2
Convention on the Rights of the Child (CRC) 7, 38, 67, 71, 75, 220, 252–3
Copenhagen Accord (2009) 207, 211
Copenhagen Declaration on Social Development 19, 40
Copenhagen Summit for Social Development (1995) 19
corporate social responsibility 308
corrective justice 10
Council of Europe 85, 88, 90, 343; *see also* European Court of Human Rights
Council of Europe Declaration 146–7
Court of Human and Peoples' Rights 85
Craig, R.K. 217
cross-border pollution 12, 54; *see also* transboundary environmental damage
Cuba 121
Cullet, P. 208
cultural rights 67
customary law 122
Czech Republic 160

Daly, E. 155, 157, 161, 165, 167, 169
de-carbonization 24
de-growth 24
Declaration of Nueva Leon (2004) 146
Declaration on the Right to Development 68, 343
Delia Saldias de Lopez v. Uruguay 297, 299
Demanda Generaciones Futuras v. Minambiente 189–90
Democratic Republic of the Congo 161–2
developing countries 12, 43, 66, 75
disabled people 261–2
discrimination 50–1, 231–32
displacement 91, 210, 220; *see also* climate refugees
distributive justice 10
Doha amendment 216
domestic/sexual violence 250
Dominican Republic 158, 166
Donaldson, S. 329–30
double-discrimination 13, 256
Dr Mohiuddin Farooque v. Bangladesh, represented by the Secretary, Ministry of Irrigation, Water Resources and Flood Control 176–7
Draft International Covenant on the Human Right to the Environment (2017) 53–5

Earth Charter 333
Earthlife Africa Johannesburg v. Minister of Environmental Affairs et al. 194
East Timor 163
Eastern Europe 160–1; *see also specific countries*
Ebbesson, J. 131–2
eco-centric thinking 18
economic growth 24
ecosystem protections 333
ecosystem rights 336
Ecuador 124–5, 258, 296–300, 334–5
Ecuador: Kichwa People of Sarayaku v. 92, 122, 144; *Sarayaku v.* 258
education 21, **22**, 54, 72, 87, 251–3
EHP et al. v. Canada 111
El Salvador 158
elderly people 262
Elver, H. 215
Energy Justice Network 14
English Bill of Rights (1689) 6
environmental advocacy networks 234
environmental challenges 66–7
environmental constitutionalism 164, 167
environmental degradation: impact on enjoyment of human rights 3, 10, 35, 47–8, 64, 340; North/South divide 18, 159; and right to culture 116–17; and right to life 110–11

environmental democracy 36–7
environmental discrimination 233
environmental human rights defenders 244–5
environmental impact assessments 36, 129–30, 181, 194, 293–4
environmental injustice 11–12, 345
environmental justice 8–12, 99, 225
environmental justice movement 10, 14, 37, 231–3
environmental migrants 258–60; *see also* climate refugees
environmental protection 24, 36, 43, 65, 80, 163–4
environmental protection groups 148–9
environmental racism 11, 13, 37, 99, 218, 231–3
environmental rights 4, 38, 76, 109; articulation of 36–8, 56, 64, 175; and constitutions 16–17, 38, 47, 53, 57, 67, 77, 155, **156**, 157–9, 161–4, 167, 169, 175, 181; lack of recognition 42, 65, 154, 167–68; recognition of 3, 36, 38, 47, 67
EP et al. v. Colombia 116–17
Espoo Convention on Environmental Impact Assessment in a Transboundary Context (1991) 138
Ethiopia 162, 165, 256, 271–5, 314–15
European Committee of Social Rights 119
European Convention for the Protection of Human Rights and Fundamental Freedoms 87–8, 90, 113, 115, 145
European Court of Human Rights 85, 87–9, 116, 144–5, 191–2, 295–6; *see also* Council of Europe
European Social Charter 119
European Union 85
Evans, B. 23
Executive Order on Environmental Justice (1994) 11
Extractive Industries Transparency Initiative (EITI) 317
extraterritorial damage *see* transboundary environmental damage
extraterritorial obligations (ETOs) 290–1, 298, 304
extreme weather 212, 216, 261, 280
Exxon Mobil and others: Village of Kivalina and City of Kivalina v. 188

Federación Nacional de Algodoneros y Corporation Autonoma Regional del Cesar: Victor Ramon Castrillon Vega v. 189
Federation of Pakistan: Bokhari v. 178
Finland: Ilmari Lansman et al. v. 117; *Jouni Lasnman et al v.* 117
First National People of Color Environmental Leadership Summit (1991) 14
Flint (Michigan) 37
Florea v. Romania 90
food security 215
Foster, S. 10
Fowlder, R. 316
Framework Principles on Human Rights and the Environment 49, 51–2, 71
France 111–12, 159
France: Vaihere Bordes and John Temeharo v. 111–12
Free, Prior and Informed Consent (FPIC) 92, 143, 144, 150, 238–9, 257, 269
freedom of information 133–4, 136–7, 161; *see also* information
French Declaration on the Rights of Man and Citizen (1789) 5
Friedman, L. 26
Frontiers of Justice (Nussbaum) 17
future generations 77–8, 159, 184–5, 225, 252, 327–8; *see also* children; inter-generational injustice; inter-generational justice; inter-generational rights

Gbemre v. Shell Petroleum Development Company Nigeria Limited and Others 194–95
Geneva Pledge for Climate Action 242
Genocide Convention 7
Georgia v. Tennessee Copper Co. 288
Germany 331–2
Ghai, Y. 86
Ghana 165
Gibe III dam 272–4, 314–15
Global Alliance for the Rights of Nature 335
Global Compact for Safe, Orderly and Regular Migration 260
Global Compact on Refugees 261
Global Environment Facility 278
global environmental constitutionalism 154, 167, 170
Global Legal Action Network (GLAN) 90
Global North 8, 17, 22, 24, 27, 134, 159, 223; *see also* North/South divide
Global Pact on the Environment 55–6
Global South 22, 134, 159, 223, 327; *see also* North/South divide

Global Warming of 1.5°C 206–7
globalization 9, 44
Godavari Marble Industries and Others: Suray Prasad Sharma Dhungel v. 179
good governance 342
good practices: UNEP 48–9
Great Ape Project (GAP) 332
Greece: Marangopoulos Foundation for Human Rights (MFHR) v. 119
Green Climate Fund 278
Green Economy 67–8, 267
greenhouse gas reductions 77, 207, 216, 277
"greening" human rights laws 37–8, 47
Greenpeace Nordic Association v. Norway Ministry of Petroleum and Energy 193
Greenpeace Southeast Asia petition 196
Guerra and others v. Italy 145
Guinea 162
Gunaratne v. Homagama Pradeshiya Sabha 179
Guyana 125, 256

Hatton v. United Kingdom 89–91
Hayward, T. 16, 168, 327
hazardous waste disposal sites 13–14, 232–3
Herz, R. 114
High Commissioner for Human Rights 39–40, 74
Honduras 124, 278–79
Huddle, E.G. 334–5
Human Development Report 217
human rights 5–6, 37–9, 65–70, 75, 210–11, 226, 240–1, 267, 275, 289
Human Rights and Climate Change Working Group (HRCCWG) 237–41, 243
Human Rights Committee 53, 110–12, 214, 290
Human Rights Council 14, 49–50, 52, 76, 110, 186, 209, 211, 221, 223, 238, 312
human rights defense groups 148–9, 244–5
human rights framework 28, 50, 63–6, 70–1, 74–5, 78, 80, 217, 289
human rights laws 4–5, 8, 17, 35–6, 38–9, 47–50, 64, 75–6, 226, 287, 298, 300, 302
human rights movement 78
human rights violations 64, 68, 72, 76–7, 109, 123, 149, 210, 212, 217, 233, 256, 273
Human Rights Watch 70, 241–2, 273–5
Hungary 78, 328
Hungary: Tarsasag a Szabadsagjogokert v. 145
Hunter, D. 218
Hurricane Katrina 37, 212, 216
hydro-electric dams 77, 272–4

Ibeanu, O. 45
Iceland: Thorgeir Thorgeirson v. 145
Ilascu and Others v. Moldova and Russia 296
Ilmari Lansman et al. v. Finland 117
ILO Indigenous and Tribal Peoples Convention 257, 270, 273
Independent Experts 46–9, 148; *see also* Knox, J.
India 12, 17, 134, 147–8, 176, 268, 336
indigenous peoples 130, 182, 220, 244, 255–6; Arctic peoples 25, 118; and customary law 122; discrimination against 50–1; and FPIC 92, 143, 150; and hydro-electric dams 77, 256–7; IFC standards **315**; land ownership 121–2, 273, 336; and nature rights movement 25; participatory rights 143; and REDD+ projects 77, 217, 256, 270; reliance on traditional knowledge 37, 256; rights of 43; *see also* Athabaskan Petition; Inuit petition
individual right to petition 142–3
Indonesia 12, 53
industrialization 9, 15, 21
information 133–6, 144–8, 161, 216
Intended Nationally Determined Contributions (INDCs) 277
Inter-American Commission of Human Rights and Court of Human Rights 11, 85, 91–8, 118, 121–2, 126, 133, 142, 144–6, 149, 220, 294–6, 300–1, 303–4, 336
Inter-American Democratic Charter 146
Inter-Constituency Alliance to Promote Human Rights in the Paris Agreement 240–1
inter-generational injustice 15–16, 254; *see also* future generations
inter-generational justice 18, 20, 77–8, 326; *see also* future generations
inter-generational rights 325–6; *see also* future generations
Intergovernmental Panel on Climate Change (IPCC) 186, 206, 210, 217, 276–7
Internal Displacement Monitoring Centre (IDMC) 259
international assistance 75
International Convention for the Protection of All Persons from Enforced Disappearance 7

International Court of Justice 7, 293–4, 297
International Covenant on Civil and Political Rights (ICCPR) 7–8, 46, 67, 109–11, 115–17, 130–1, 142, 214, 289
International Covenant on Economic, Social and Cultural Rights (ICESCR) 7, 21, 67–8, 71–2, 110, 117, 119–20, 214, 270, 279, 289
international environmental law 295
International Finance Corporation (IFC) 314, **315**, 318
international human rights law 3, 8, 44
international injustice 12–13, 24, 254
International Law Commission 293
international law principles 41, 143, 167, 206, 222, 288–9
International Organization for Migration (IOM) 258–59
International Rights of Nature Tribunals 336
International Rivers 274
intra-societal injustice 12–13, 16, 18, 254
Inuit petition 93–6, 118, 121, 214, 220–1, 234
Ioane Teitiota v. The Chief Executive of the Ministry of Business, Innovation and Employment 195
Ireland 156–7
Irfan v. Lahore Development Authority 178
island states 67, 208, 210, 221, 259
Israel 125
Italy: *Guerra and others v.* 145

Johannesburg Summit (2002) 120
Jouni Lasnman et al v. Finland 117
judicial enforcement 64, 74–5, 99, 140–1, 175
Juliana et al. v. USA et al. 183–4, 221, 254

Keck, M.E. 234–5
Kenya 274–75
Khalfan, A. 242
Kichwa People of Sarayaku v. Ecuador 92, 122, 144
Kimberly Process Certification Scheme (KPCS) 316–17
Kivalina village (Alaska) 37, 187–8, 221, 223
Knox, J. 46–52, 56, 69, 71, 131, 138, 192–3, 214, 223–4, 241, 294–6, 303, 340–1
Kopnina, H. 24
Ksentini, F.Z. 41–2, 47
Kuehn, R. 9–10
Kymlicka, W. 329–30
Kyoto Protocol 205, 208, 216, 243, 267–8
Kyrgyzstan 162

Laos 167
Latin America 148–9, 160–1, 166, 188–93, 242, 245; *see also specific countries*
law and society approaches 26
LCB v. the United Kingdom 113–14
Leander v. Sweden 144–5
Legal Consequences of the Construction of a Wall in the Occupied Palestinian Territory (ICJ) 297
Lliuya v. RWE AG 193
Lopez Ostra v. Spain 89, 115–16
loss/damage 280
Lubicon Lake Band v. Canada 117

Maastricht Principles on Extraterritorial State Obligations (2011) 71–2, 290, 292–3
McArthur River Mine case 195
Macron, E. 55
Madagascar 162
Magill, B. 90
Magna Carta (1215) 5
Magraw, D. 95
Mahuika v. New Zealand 117
Malawi 164
Malaysia 181–2, 257–8
Malé Declaration on the Human Dimension of Global Climate Change 208
Marangopoulos Foundation for Human Rights (MFHR) v. Greece 119
market-based solutions 243–4
Mary and Carrie Dann v. US 92
Mary Robinson Foundation 224
May, J.R. 155, 157, 161, 165, 167, 169
Maya Communities of the Toledo District v. Belize 92
Mayagna (Sumo) Awas Tingni Community v. Nicaragua 92, 121–2
M.C. Mehta v. Union of India and Others 147
Mendoza Beatriz Silva and others v. National Government of Argentina and others in regard to damages suffered 188–9
Metropolitan Manila Development Authority v. Concerned Residents of Manila Bay 181
Mexico 125, 197
Millennium Development Goals 19–21, 120

minorities 48, 50–1, 71, 116–17, 212, 218, 231–2
Minors Oposa Case 168–9, 180–1
mitigation measures 23, 147, 212, 217, 267
Model National Law on Human Rights Defenders 245
Moldova and Russia: Ilascu and Others v. 296
Mongolia 53
Montana Environmental Information Center v. Department of Environmental Quality 182–3
Mossville Environmental Action Now v. United States 11, 37, 98–100
Mother Earth 17, 335
Mozambique 162, 164–5
multinational corporations 43, 72, 308, 310–11, 345
Mundy v. Central Environmental Authority and others 179

naming/shaming 75
Narmada Bachao Andolan v. Union of India 176
National Adaptation Plans 277
National Audubon Society Inc. v. Davis 184
National Human Rights Institutions (NHRIs) 87, 196
Nationally Determined Contributions (NDCs) 165, 216, 277
Nepal 178–9
Nepal Drinking Water Corporation: Prakash Mani Sharma and others v. 179
Netherlands: Urgenda case 90
New Zealand 17, 161, 195, 260, 332, 336
New Zealand: Mahuika v. 117
Newkirk, I. 330
Nicaragua: Mayagna (Sumo) Awas Tingni Community v. 92
Nigeria 167, 194–5, 233–4
Nigeria: The Social and Economic Rights Action Center and the Center for Economic and Social Rights v. 100–1, 121, 123
noise pollution 89
non-traditional rights 154
Nordic Environmental Protection Agreement (1974) 302–3
North/South divide 17–18, 25–7, 159, 207, 211, 223, 243
Norway 159, 193, 242
Norway Ministry of Petroleum and Energy: Greenpeace Nordic Association v. 193
Nussbaum, M.: *Frontiers of Justice* 17

Observer and Guardian v. United Kingdom 145
OECD Guidelines for Multinational Enterprises 313–14
Ogoniland 123, 233–4, 298
Oneryildiz v. Turkey 113, 145
Organization of American States (OAS) 92, 94
Our Common Future report 18

P-10 treaty 37
Pacheco, A. 330
Pakistan 177
Pallemaerts, M. 46
Panama 72–3, 269–70
Papua New Guinea 165
Paraguay 124
Paris Agreement on Climate Change (2015) 36, 55, 64, 69, 73, 79, 166, 190, 205, 207, 216, 220, 225–6, 231, 237, 239–41, 258, 279–80, 328; *see also* Sustainable Development Mechanism (SDM)
participatory rights 42, 65, 138, 143, 212
Paul and Audrey Edwards v. the United Kingdom 113–14
People's Union for Civil Liberties & Another v. UOI & Others 147
Peru 53, 92–3, 193, 270
Peru: Community of La Oroya v. 93
Petro Flores v. Corporacion Del Cobre, Codelco 191
pollution 89, 115, 121, 123–4
population relocation 77, 215, 233, 256–7, 273
Portugal 162
Pound, R. 5
Powell v. United Kingdom 89, 91
Powers, M. 218
Prakash Mani Sharma and others v. Nepal Drinking Water Corporation 179
Prakash Mani Sharma v. His Majesty's Government Cabinet Secretariat and Others 179
Presser, I. 257
procedural justice 10, 15, 36
procedural obligations 48–9, 51, 53
procedural rights 37, 43, 47, 73, 89, 119, 129, 144–5, 147, 150, 160–1, 216, 238, 301
property rights 92, 130
Protect, Respect and Remedy Framework 318
Protocol of San Salvador 98
public access 141, 181

public interest litigation (PIL) 178
public misinformation 46
public participation 138–9, 181
public trust 165
Pulp Mills case 55, 130, 293–4

Rajamani, L. 225
Ramsar Convention on Wetlands 180
Reducing Emissions from Deforestation and Forest Degradation (REDD+) programs 68, 73, 77, 216, 217, 238–9, 256, 267, 270–1
Regan, T. 328–9
Regional Agreement on Access to Information, Public Participation and Justice in Environmental Matters in Latin America and the Caribbean ("ECLAC Agreement") 132–3, 137–42, 148
regional systems of human rights 85–6
right to a healthy environment 37, 56, 79, 87–8, 91, 98, 100–1, 122–3, 133, 154, 167–8, 170, 333
right to development 78
right to food/water 119–20, 166, 214–15
right to health 117–19
right to housing 215
right to information 144–45; *see also* information
right to life 68, 110–11, 114, 182, 190–1, 214
right to privacy/family life 115, 145
right to self-determination 67, 344–5
right to use and enjoy property 121–2
rights 5–6, 133, 154–5, 159, 190, 345
rights of nature 80, 332–3
Rio Declaration on Environment and Development 19–20, 37, 42, 46, 129, 133, 136, 146, 179, 205, 237, 303
Rio+10 conference (2002) 19
Rio+20 conference (2012) 20, 211
rivers: legal status of 336
Romania: Florea v. 90
Roosevelt, Eleanor 6
Rotterdam Convention on Prior Informed Consent Procedure for Certain Hazardous Chemicals and Pesticides in International Trade 46, 135, 138
Ruggie, J. 44, 311–12, 318
Russian Federation 124, 160

Saldano v. Argentina 296
Salt Miners Labor Union v. Industries and Mineral Development 178
Saramaka People v. Suriname 92, 122
Saramaka People v. Suriname (Saramaka), Preliminary Objections, Merits, Reparations 122
Sarayaku v. Ecuador 258
Saro-Wiwa, K. 234
self-certification schemes 316–17
Sendai Framework for Disaster Risk Reduction, 2015–2030 212–13
Senih Ozay v. Ministry of the Environment and Eurogold Madencilik 192
Serbia 160
Shehla Zia and others v. WAPDA 177–8
Shell Petroleum Development Company Nigeria Limited and Others: Gbemre v. 194–5
Shelton, D. 10, 88, 118–19, 146
Shishmaref village (Alaska) 37, 221
Shue, H. 11, 326–7
Sikkink, K. 234–5
Simons, P. 44, 320
Singer, P. 328–9, 332
smoking 90
social development 19–20
social justice 10, 225
social movements 232
soft law 6, 44, 312–13
Solomon Islands 53
South Africa 166, 193–4
South Asia 165, 167, 176–82
Spain: Aparicio Benito v. 90
Spain: Lopez Ostra v. 89, 115–16
special purpose entities (SPEs) 134
Special Representative to the Secretary-General on Human Rights and Transnational Corporations and other Business Enterprises 311
Sri Lanka 134, 157, 161, 163, 165, 179–80
Sri Lanka: Susila Malani Dahanayake and others v. 112–13
State of Bihar: Subhash Kumar v. 176
Statement of Forest Principles 19
states 71, 111, 113, 116, 135, 142, 144, 148–9, 212, 217, 226, 288–90, 292, 295, 301–2
Stern Review 252
Steudtner, J. 270
Stockholm Conference on the Human Environment (1972) 19, 63, 340
Stockholm Convention on Persistent Organic Pollutants 46, 135, 138
Stockholm Declaration on the Human Environment (1972) 35–6, 41–2, 47, 179, 186, 209, 289, 293, 304

Stoine Hristov v. Bulgaria 90
Subhash Kumar v. State of Bihar 176
substantive obligations 51
substantive rights 47, 110, 119, 125–6, 157–8, 160, 214–15; *see also specific rights*
Suray Prasad Sharma Dhungel v. Godavari Marble Industries and Others 179
Suriname: Saramaka People v. 92, 122
Susila Malani Dahanayake and others v. Sri Lanka 112–13
sustainability 21–4
sustainable development 10, 18–24, 40, 77–8, 164–5, 176, 179
sustainable development goals (SDGs) 4, 21–2, **22**, 78, 120, 279, 310, 343–4
Sustainable Development Mechanism (SDM) 15, 73
Swaziland 165
Sweden 125, 161–2
Sweden: Leander v. 144–5
Switzerland 331

Tan Tek Seng v. Suruhanjaya Perkhidmatan Pendididan 182
Tarsasag a Szabadsagjogokert v. Hungary 145
Taskin v. Turkey 89
Tennissee Copper Co: Georgia v. 288
The Philippines 168–9, 180, 196, 245, 321–2
The Social and Economic Rights Action Center and the Center for Economic and Social Rights v. Nigeria (Ogoniland case) 100–1, 121, 123
Thomson v. Minister for Climate Change Issues 195
Thorgeir Thorgeirson v. Iceland 145
toxic products 45, 231–3
traditional knowledge of indigenous peoples 25, 94, 96, 239, 256, 277
Trail Smelter case: US v. Canada 287–89
trans-national corporations 43, 72, 308, 310–11, 345
transboundary environmental damage 11, 18, 28–9, 54, 65–6, 287–8, 294–5, 297, 300–1, 303; *see also* Carbon Majors Petition; Colombia; cross-border pollution; Ecuador
transnational advocacy networks 234–5, 237–41, 243, 346
Transparency International 344
treaty-based human rights 38–9
Turkey 192
Turkey: Oneryildiz v. 113, 145
Turkey: Taskin v. 89

Uganda 165, 167
Ugaz, J. 341–2
UK Climate Change Act (2008) 193
Ukraine 159, 161
UN Commission on Human Rights 40–1, 76; *see also* Human Rights Council
UN Conference on Environment and Development (1992) 19, 207
UN Convention against Corruption 146, 343
UN Convention on Biological Diversity 19, 25, 138, 333
UN Convention Relating to the Status of Refugees (1951) 260
UN Convention to Combat Desertification (1994) 138
UN Declaration on Human Rights Defenders 245
UN Declaration on the Rights of Indigenous Peoples (UNDRIP) 7, 43, 92, 143, 150, 220, 257, 270, 273
UN Development Programme (UNDP) 269, 342–3
UN Draft Norms on the Responsibilities of Transnational Corporations and Other Business Enterprises with Regard to Human Rights 310–11
UN Environment 341–2
UN Environment Programme (UNEP) 20, 48–9, 192, 206, 222, 274, 340
UN Framework Convention on Climate Change (UNFCCC, 1992) 12, 15, 19, 94, 135, 138, 205, 218
UN Global Compact 309–10
UN Guiding Principles on Business and Human Rights 44, 308–9, 318, **319**, 320–1
UN High Commissioner for Human Rights 39–40, 74
UN Permanent Forum on Indigenous Issues 258
UN Special Rapporteurs 36, 39–41, 44–6, 49, 110, 149, 215–16, 244, 278; *see also* Ibeanu, O.; Knox, J.; Ksenti, F.Z.; Pallemaerts, M.;
UN World Charter for Nature 155
Union of India and Others: M.C. *Mehta v.* 147
Union of India: Narmada Bacao Andolan v. 176

Union of India, Vellore Citizens Welfare Forum v. 176
United Church of Christ Commission on Racial Justice (UCC) 232
United Kingdom 156
United Kingdom: Hatton v. 89–91; *LCB v.* 113–14; *Observer and Guardian v.* 145; *Paul and Audrey Edwards v.* 113–14; *Powell v.* 89, 91
United Nations 6–7, **22**, 39, 40, 71, 120
United States: Altgeld Gardens 233; and the American Declaration 99–100; and constitutional environmental protections 156, 159; environmental impact assessments 130; environmental justice movement 37, 231–2; on environmental migrants 260; Executive Order on Environmental Justice (1994) 11, 37; fears of 69–70; greenhouse gas emissions 94–5, 184–5, 222–3; hazardous waste disposal in 13–14, 231–3; Interagency Working Group on Environmental Justice 11; Inuit Petition 93–6, 118, 121, 214, 220–1, 234; lacking commitment by 75, 182; National Climate Assessment 186; Our Children's Trust case 90; Paris Agreement 166; right to water 120; State Constitutions 159–60, 182, 187; Trump Administration 166, 185, 260; Warren County case 231–3
United States: Coard v. 299–300; *Mossville Environmental Action Now v.* 11, 37, 98–100
Universal Declaration of Human Rights (1948) 6, 14, 46–7, 69, 71, 86, 109, 278, 325, 335, 340, 344
Universal Declaration on the Rights of Mother Earth 335–6
Universal Periodic Review (UPR) 39, 53
UOI & Others: People's Union for Civil Liberties v. 147
Urgenda case (Netherlands) 90, 184, 192–3, 196, 221
Uruguay: Delia Saldias de Lopez v. 297, 299
US Bill of Rights 5
US Constitution 5, 156, 159
US: Mary and Carrie Dann v. 92
USA et al.: Juliana et al. v. 183–4

Vaihere Bordes and John Temeharo v. France 111–12
Velasquez Rodriguez case 142
Vellore Citizens Welfare Forum v. Union of India 176
Victor Ramon Castrillon Vega v. Federación Nacional de Algodoneros y Corporation Autonoma Regional del Cesar 189–90
Vienna Declaration on Human Rights (1993) 6, 8
Village of Kivalina and City of Kivalina v. Exxon Mobil and others 188
VZW Klimatzaak v. Kingdom of Belgium 193

Wagner, M. 95
WAPDA: Shehla Zia and others v. 177–78
Ward, T. 92
Warren County case 231–3
Warsaw International Mechanism on Loss and Damage 280
Washington, H. 22–3
water: lack of 250–1
water pollution: deaths due to 64
water privatization 166
Watt-Cloutier, S. 95
Weeramantry, C. 35, 179
Wheeler, R.F. 334–5
Wold, C. 218
Wolfson v. Brammer 184
women 10, 14, 19, 21, 28, 38, 45, 48–9, 56, 64, 71, 73, 125, 196, 209, 212–13, 215, 218–19, 249–52
Wood, M. 184
World Bank Group 314–16
World Bank Inspection Panel 91
World Charter for Nature 333
World Commission on Environment and Development (WCED, 1987) 18, 40
World Health Organization 65
World Summit on Sustainable Development 19

Yemen 125

zero growth 24
Ziegler, J. 215
Zoopolis: A Political Theory of Animal Rights (Kymlicka & Donaldson) 329–30